A FUNCTIONAL BIOLOGY OF ECHINODERMS

Echinvs Diadema Lihnei.
From I.T. Klein (1778) Natvralis dispositio Echinodermatvm. Gleditschiana. Leipsig

Prière de l'étoile de mer

Seigneur,
L'abîme est fermé sur moi.
Ne suis-je pas
quelque Lucifer
tombé du ciel
et délaissé
aux tourments des vagues?
Voyez,
j'ai l'air d'une étoile de sang.
J'essaie de me souvenir
de ma royauté lointaine
mais en vain.
Rampant sur le sable,
j'écarte mes branches
et je rêve, je rêve, je rêve ...
Seigneur,
un ange
ne m'arrachera-t-il pas
du fond des mers
pour m'incruster de nouveau
dans Votre ciel?
Ah! qu'un jour,
ainsi soit-il!

Carmen Bernos de Basztold,
Choral de Bêtes, Editions du Cloitre,
Paris. Reprinted with permission.

Prayer of the starfish

Lord,
The depths have closed over me.
Am I not
like Lucifer
fallen from heaven
and abandoned
to the torments of the waves?
See,
I look like a blood star.
I try to remember
my distant royalty
but in vain.
Crawling on the sand,
I open my arms
and I dream, I dream, I dream ...
Lord,
could not an angel
pull me up
from the bottom of the sea
to place me again
in Your heaven?
Ah! one day,
so be it!

Translated by John Lawrence

A Functional Biology of Echinoderms

John Lawrence
Department of Biology
University of South Florida

The Johns Hopkins University Press
Baltimore

Printed in Great Britain

First published 1987 by
The Johns Hopkins University Press
701 West 40th Street
Baltimore, Maryland 21211

Library of Congress Cataloging-in-Publication Data

Lawrence, John M.
A functional biology of echinoderms.
Bibliography: p.
1. Echinodermata. 2. Echinodermata — Physiology.
I. Title. [DNLM: 1. Echinodermata — physiology.
QL 385.2 L421f]
QL381.L39 1987 593.9′041 87-2843
ISBN 0-8018-3547-X

Contents

Acknowledgements

It is a pleasure to thank those friends and colleagues who so generously gave of their time and knowledge to comment on the various portions of the text: A.C. Campbell (Queen Mary College), M. Downey (Smithsonian Institution), T. Ebert (San Diego State University), P. Fankboner (Simon Fraser University), J. Ghiold (Louisiana State University), M. Jangoux (Free University of Brussels), M. Jensen (University of Copenhagen), J. McClintock (University of California, Santa Cruz), D. Meyer (University of Cincinnati), P. Mladenov (Mount Allison University), M. Shick (University of Maine), M. Telford (University of Toronto), C. Walker (University of New Hampshire), and particularly D. Blake (University of Illinois), G. Hendler (Los Angeles County Museum), D. Nichols (University of Exeter), D. Pawson (Smithsonian Institution) and A.B. Smith (British Museum, National History).

I thank the authors and the following copyright holders for permission to use the figures illustrating this book: A.A. Balkema, Publishers; Aberdeen University Press; Akademie-Verlag; Alan R. Liss, Inc.; *American Midland Naturalist*; American Society of Limnology and Oceanography; Bernice Bishop Museum; *Biological Bulletin*; British Museum (Natural History); *Bulletin of Marine Science*; Cambridge University Press; Century Hutchinson Ltd; the Company of Biologists Ltd; Dansk Naturhistorisk Forening; E. Schweizerbart'sche Verlagsbuchhandlung; Elsevier Science Publishers B.V.; the Geological Society of America and the University of Kansas Press; Gordon & Breach, Science Publishers, Inc.; Gustaf Fischer Verlag; Institute of Oceanographic Sciences; Inter-Research; Koninklijke Nederlandse Akademie; Masson editeur; the Oceanographical Society of Japan; New Zealand Oceanographic Institute; Nytt Magasin for Zoology; Ophelia Publications; the Palaeontological Association; Pergamon Books Ltd; the Royal Society of London; Scandinavian Science Press Ltd; Seto Marine Biological Laboratory; Smithsonian Institution Oceanographic Sorting Center for the National Science Foundation; *South African Journal of Sciences*; Springer-Verlag; *Steenstrupia*; *Tissue and Cell*; the Zoological Society of London.

Acknowledgments

Preface

Biological systems perpetuate themselves through time, both as individuals and as populations. These are essentially the two components of fitness: survival and reproduction. The study of functional biology concerns characteristics of biological systems that increase fitness. It is possible to consider these characteristics in terms of growth rates, mortality, nutrient acquisition and use, fecundity and other similar phenomena without considering the structural and morphological bases for them. However, biological systems do differ in their structural and morphological characteristics, and these characteristics provide the constraints and potentials that affect their life histories. Functional morphology is thus an integral part of functional biology. This is the basis for considering the functional biology of echinoderms as a group. Otherwise the functional biology of all living systems could be considered together without separation by taxonomic group.

My aim in this book has been to provide this concept of the echinoderms and their subdivisions, to describe their functional morphology, and to discuss when possible the ways in which their functional biology promotes their fitness.

John M. Lawrence

1

Characteristics of the Phylum and Classes

... it is rule in nature to suit always in its organic formations the mechanism to the wants. J.S. Miller, 1821. *A natural history of the Crinoidea, or lily-shaped animals.*

1.1. PHYLUM

Bather (1900) stated that the phylum Echinodermata is one of the best characterised and most distinctive in the Animal Kingdom, with the living forms being readily distinguished by a water vascular system derived from the coelom, a mesodermally derived, sub-epidermal skeleton with a characteristic microstructure, and pentamerous radial symmetry. The first characteristic is one that has provided the context in which the evolution of the phylum must be interpreted. It is not only found in all extant members but was also present in the earliest echinoderms. It has proven to be very malleable in many ways, yet has provided constraints to change in others. The skeletal system is present in essentially all echinoderms, seems unique in its characteristics, and has provided the basis for much of the evolution that has occurred in the phylum. However, although calcareous deposits seem essential for most groups, some holothuroids have successfully dispensed with the system. Pentamerism is associated with the radial water canals. It was not a characteristic of all extinct echinoderms and is not symmetrical in the adults of some extant groups.

All members of the phylum Echinodermata possess an elaboration of a coelomic system, the water vascular system. Hyman (1955) and Ubaghs (1967) considered the water vascular system to be the most characteristic morphological feature of echinoderms, and Nichols (1969) stated that it is clearly the outstanding feature of the phylum. All extant classes have the water vascular system developed for ambulacral and/or tentacular systems. However, Sprinkle (1983) has argued that most or all blastozoan (brachiole-bearing) echinoderms and possibly some of the other extinct classes had a greatly reduced water vascular system which lacked tube feet.

Cuénot (1948), Hyman (1955), Nichols (1966) and Binyon (1972) described the structure and function of the water vascular system of echinoderms. The water vascular system in extant echinoderms consists of a circumoral ring, the water canals which extend radially from it, and the tube feet which are extensions from the water canals. The tube feet are the basic functional units of the echinoderm system, and show much specialisation in structure and function. The basis for much of the variation in body form of extant echinoderms involves the length and position of the water canals, for this determines the number and position of the tube feet. There seems to be no inherent limitation to the length of a water canal, and the length is probably controlled by other structural constraints. The total length of the ambulacra is increased by branching of the water canals in some crinoids, ophiuroids, echinoids and holothuroids and by increasing the number of radial water canals in some asteroids and ophiuroids.

Ubaghs (1967) suggested that the appearance of a protective endoskeleton in echinoderms had less to do with a special cause such as the adoption of a benthic existence than with a more general factor causing the production of skeletal structures in many groups of invertebrates in the late Precambrian or the beginning of the Palaeozoic era. Whatever its origin, essentially all members of the phylum Echinodermata are distinguished by the skeletal elements unique to the group. Thus the problematical carpoids of the Lower Cambrian are classified as echinoderms (Ubaghs 1967) because they have calcite plates with a stereom structure. Living cells and organic material form a stroma that is found throughout the stereom. An important characteristic of the system is its plasticity. Individual elements can be produced which vary in size, shape, density and strength. They can be in the form of large single crystals, such as the plates (Figures 3.104, 3.108) or spines (Figure 3.113) of an echinoid; polycrystals, as in the tooth elements (Figure 2.35) of echinoids; or microscopic particles, as in the ossicles of the body wall (Figure 3.84) of holothuroids or other soft tissues such as the gut, gonads and tube feet (Figure 3.123). The elements can be modified by differential deposition and resorption which allows allometric changes to occur with growth. The skeleton may also provide a low-energy-cost means of growth (Emson 1985).

Unlike the exoskeleton of molluscs, for which there is a limited number of basic body forms (Raup 1966), the endoskeleton does not limit the body form of echinoderms. The shape of the molluscan shell can be changed only by differential deposition, whereas the skeleton of echinoderms can be modified by both differential deposition and resorption. Unlike the exoskeleton of crustaceans, the endoskeleton of echinoderms allows growth without moulting. Hence, growth can be continuous and without the requirements and problems associated with moulting. The ability to resorb the skeletal elements allows an echinoderm to decrease its body size when food intake does not meet requirements. The individual remains functional as it regresses allometrically in size.

The endoskeleton of echinoderms cannot be understood adequately without considering the extracellular collagen-based material associated with it. The physical state (viscosity) of this material is controlled, and it can be either rigid or pliable (Motokawa 1984; Wilkie 1984). This provides the basis for controlling the rigidity and flexibility of the body wall and its components. It also provides a mechanism for maintaining the strength of the body wall while reducing the role of the calcareous elements. Most holothuroids have exploited this potential by reducing or in some cases completely eliminating skeletal elements from their body walls.

The third characteristic generally applied to the phylum Echinodermata is a pentamerous radial construction. This is a designation that results from the usual number of ambulacra present in extant groups. Beklemishev (1969), Ubaghs (1967) and Nichols (1969) have given the most complete analysis of symmetry and its development in echinoderms. Ubaghs stated that symmetry in extant echinoderms is very complex and that pentamerously radial symmetry of both overall body form and internal organs is usually neither perfect nor complete. One must distinguish between *pentamerous symmetry,* a correspondence of parts around an axis, and *pentamery,* which may not be symmetrical. Whether symmetrical or not, pentamery in echinoderms is basically a phenomenon associated with the radial water canals and those body structures that are directly (arms and rays) or indirectly (gonads, pyloric caeca) associated with them. The pentamerous (five parts) condition arises embryologically with the development of five water canals, the pentaradiate (five radial water canals) condition. This occurs after metamorphosis prior to the deposition of the skeletal elements.

The evolutionary development of radial symmetry of the adult echinoderm is usually attributed to the assumption of a sessile, attached way of life by a bilaterally symmetrical ancestor. R.B. Clark (1971) pointed out that a sessile way of life is an evolutionary dead end. He concluded that much of the subsequent evolutionary history of echinoderms seems to have involved modification of a body architecture which had already been profoundly modified for a sessile way of life to meet the requirements of other modes. Thus all of the extant classes are primarily free-living, with all but the Crinoidea exclusively so throughout their known fossil record. Despite this, the pentaradiate body form has been maintained, even to the point of essentially pentamerous symmetry in most extant echinoderms.

Beklemishev (1969) pointed out that bilateral symmetry is more suitable for free-living animals, and concluded that the slow movement of echinoderms does not offer great advantages for a bilaterally symmetrical body form. The basis for the failure to develop more rapid movement has not been considered, but may be related to limitations in basic body structure.

Rate of locomotion is not the only factor that can affect symmetry. Mode of locomotion and feeding are involved. The burrowing irregular echinoids, which have a rigid body wall, are bilaterally symmetrical, but the burrowing holothuroids, which have a flexible wall, are pentamerously symmetrical. Epibenthic holothuroids are bilaterally symmetrical whether they are vagile or not. Sessile buried holothuroids are bilaterally symmetrical.

If the explanation for the development of radial symmetry is the assumption by echinoderms of a sessile way of life, the significance of the pentamerism has not been resolved (Nichols 1969). The extinct carpoids (Figure 1.1) had no radial symmetry and were fundamentally asymmetrical. A variety of radial symmetries have occurred in Crinozoa, as in the extinct cystoids which had one to five ambulacral systems (Figure 1.1). The Lower Cambrian helicoplacoids were triradiate (Figure 1.1). Only one structural plan apparently persisted, that of three grooves diverging from the mouth at angles of 120°, one in the anal plane and running to the side opposite the anus, and each of the others dividing into two (Beklemishev 1969; Paul and Smith 1984). This is a restatement of Bather's (1900) concept that the pentaradiate condition arose from the triradiate condition, a concept that was rejected by Hyman (1955) on embryological grounds.

Beklemishev (1969) concluded that pentamerous grooving arose in primitive echinoderms because of the proximity of the anus to the mouth. In these echinoderms which lacked feeding appendages, pentamerism best solved the problem of maximal coverage of the feeding surface by afferent food grooves without the chief grooves being too close to the anus (Figure 1.2). Among the extant echinoderms, only the clypeasteroid irregular echinoids use the body surface proper to feed in this way, and the feeding grooves on the lower body surface show no distinct separation from the anus. Similarly, although the tegmen of crinoids is not used in collecting food, the paths of the ambulacra on the tegmen show patterns similar to those found in the edrioasteroids. However, Beklemishev's explanation concerns the origin of pentamery, not its persistence. As Paul and Smith (1984) pointed out, the development of distinct upper and lower surfaces together with the appearance of five ambulacra in a 2–1–2 pattern were evolutionary innovations that separate and distinguish the pentaradiate echinoderms from their stem-group predecessors.

Thus pentametry does not necessarily involve skeletal elements other than the degree to which they occur to support or protect the ambulacra. Despite this, and in contrast to Beklemishev's explanation, pentamerous symmetry of the body itself can be interpreted as the arrangement of the skeleton that produces the most stable rigid body possible because it has as few sutures as possible which are as short as possible (Nichols 1969; Stephenson 1980). Among the extant echinoderms, only the crinoids and regular echinoids have a rigid pentamerously symmetrical body form to which this explanation seems strictly applicable. The theca of all crinoids is

Figure 1.1: (a) Carpoid (*Heckericystis*) (from Gill and Caster 1960, in Caster 1967). (b) Cystoid (*Pleurocystites*) (from Bather 1900). (c) Helicoplacoid (*Helicoplacus*) (from Durham and Caster 1963, in Durham and Caster 1966)

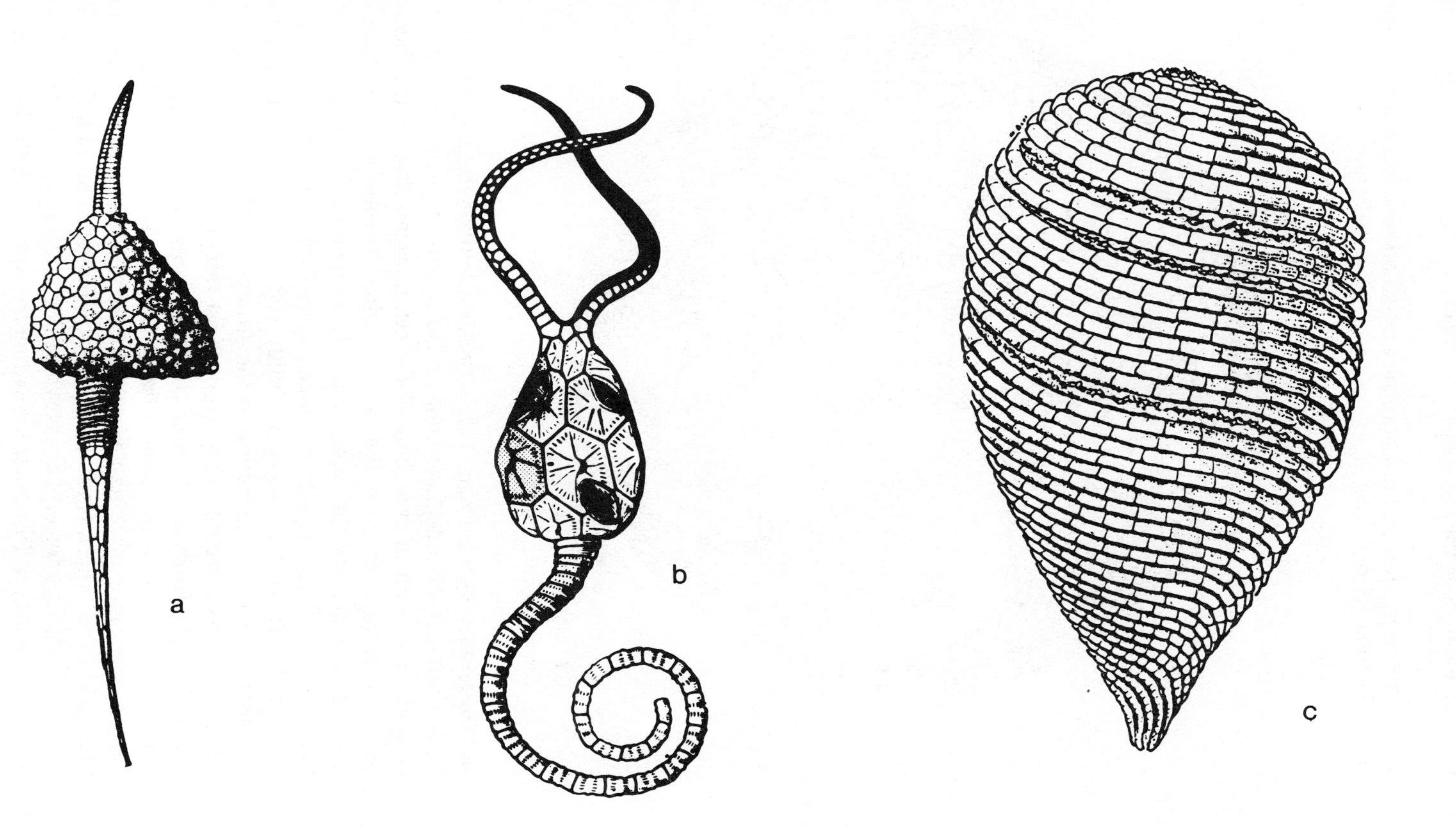

Figure 1.2: Pentamerous food grooves of an edrioasteroid (*Lepidodiscus*) (from Regnéll 1966)

pentamerously symmetrical. This structural pentamerous symmetry of the body does not exclude functional radial symmetry which occurs in most crinoids as a result of branching of the primibrachs distal to the theca. This implies that the strength of the theca required to withstand the stress of the arms is best met by the pentamerously symmetrical arrangement of the thecal plates. That regular echinoids primitively had flexible body walls (Jensen 1981; Kier 1974; A.B. Smith 1984) indicates that pentamerous symmetry was not contributing to the strength of the test although it provided the basis for evolving a strong rigid test. Pentamerous symmetry in asteroids, ophiuroids and flexible regular echinoids may be associated with structural stability of the mouth frame instead of the entire body. In any event, persistence of pentamerous symmetry in those extant echinoderms that lack a rigid theca (most asteroids, ophiuroids and holothuroids) indicates its suitability for organisms whose feeding requires no or only a slow rate of locomotion. Pentamerous symmetry was not itself the basis for the failure to achieve a rapid rate of locomotion. The development of bilateral symmetry in epibenthic holothuroids and burrowing echinoids was not related to rapid locomotion and retained the pentamerous condition of the water vascular system. Either the pentamerous condition of the system is rigidly programmed into the developmental process or it is sufficiently plastic to meet the needs of alternate body forms. The sporadic occurrence of tetramerous individuals in nature (Hotchkiss 1979), the production of trimerous individuals when developing stages of asteroids are exposed to high salinities (Watts, Scheibling, Marsh and McClintock 1983) and the great variety of forms in which the water vascular system exists in echinoderms indicate that developmental plasticity has sufficient potential as an explanation for the development of alternate body forms.

Matsumoto (1929) used general body shape as one criterion distinguishing the Echinozoa, Asterozoa and Crinozoa. H.B. Fell (1963) used body form and growth gradients (Figure 1.3) as two criteria for distinguishing subphyla. Thus the Crinozoa are prevailingly globoid forms with partial meridional symmetry from which ambulacral feeding appendages extend up and out. The Asterozoa are flattened forms with radially symmetrical divergent axes of the body expressed by relatively broad to narrow extensions of the body spread laterally outwards. The Echinozoa are globoid, cylindrical or discoid forms with well marked pentaradial symmetry without outward extensions.

The classification based on body form replaced the division of the phylum into two subphyla: the Pelmatozoa construed to include groups that are attached in a fixed manner to the substratum for all or at least part of their postlarval life and have the oral and anal openings of the gut on the upper surface, and the Eleutherozoa construed to include almost exclusively free-living groups in which the mouth is on the undersurface or anterior. Fell and Moore (1966) stated that the mode of life of the various echinoderm groups was not acceptable as a governing criterion for classification at the subphylum level, even though it might affect morphology in an important way. Thus asterozoans and nearly all echinozoans (Eleutherozoa) are free-living whereas only a minority of crinozoans (Pelmatozoa) are not attached.

Haugh and Bell (1980) concluded that both the pelmatozoan–eleutherozoan criterion and the body-form criteria result in systematic instability. They viewed the body shape as a systematically meaningless character with 'vague' ecological connotation, and the growth gradient with its implied relation to symmetry as a meaningless term. Paul and Smith (1984) and A.B. Smith (1984) combined embryological and anatomical characteristics to derive the relations among the extant classes (Figure 1.4) which identify the crinoids as a primitive sister group to the others, and the asteroids as the primitive sister group to ophiuroids plus the echinoids and holothuroids.

Figure 1.3: Echinoderm body form and growth gradients (from H.B. Fell 1963)

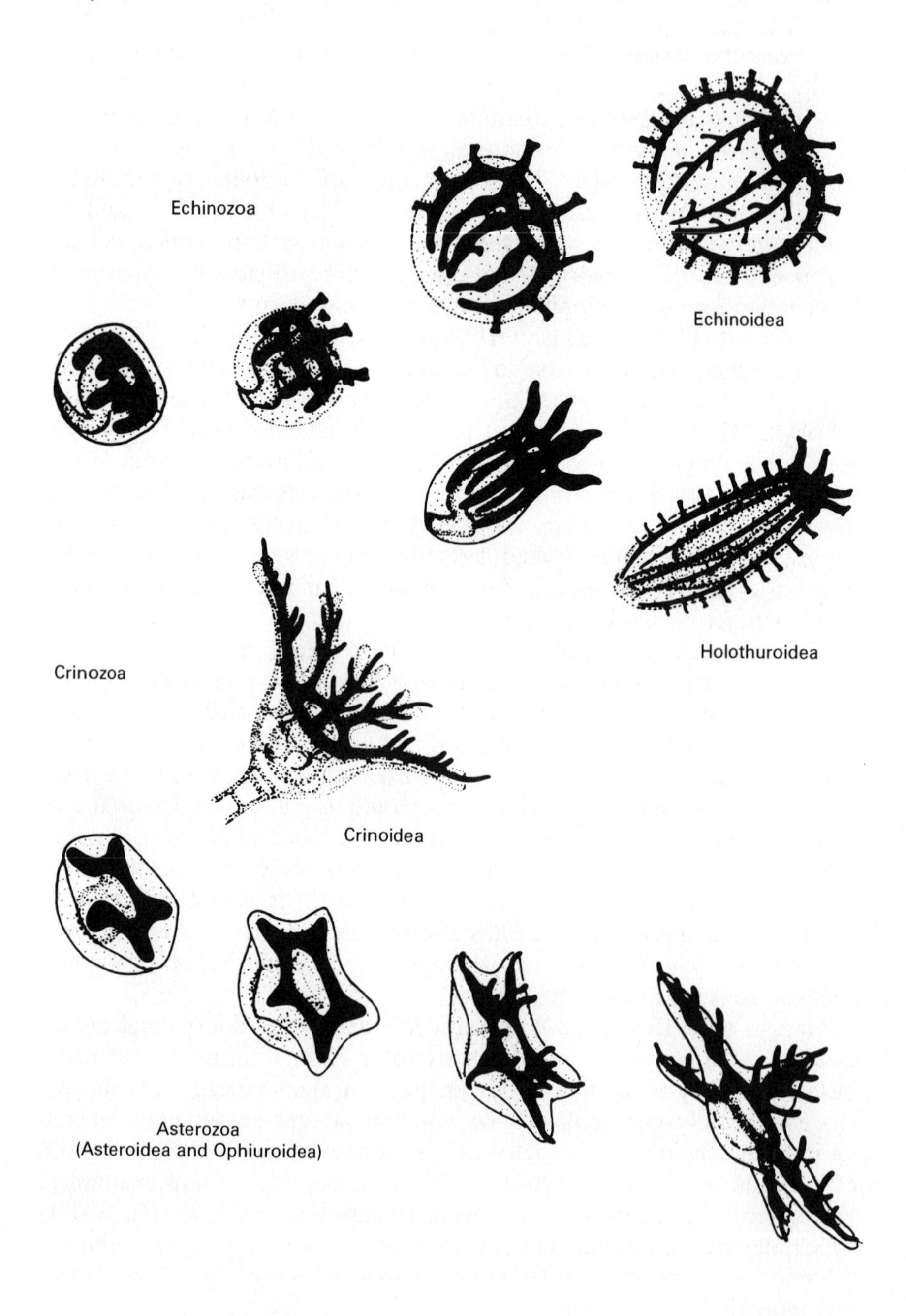

Figure 1.4: Relation between echinoderm classes (from Paul and Smith 1984)

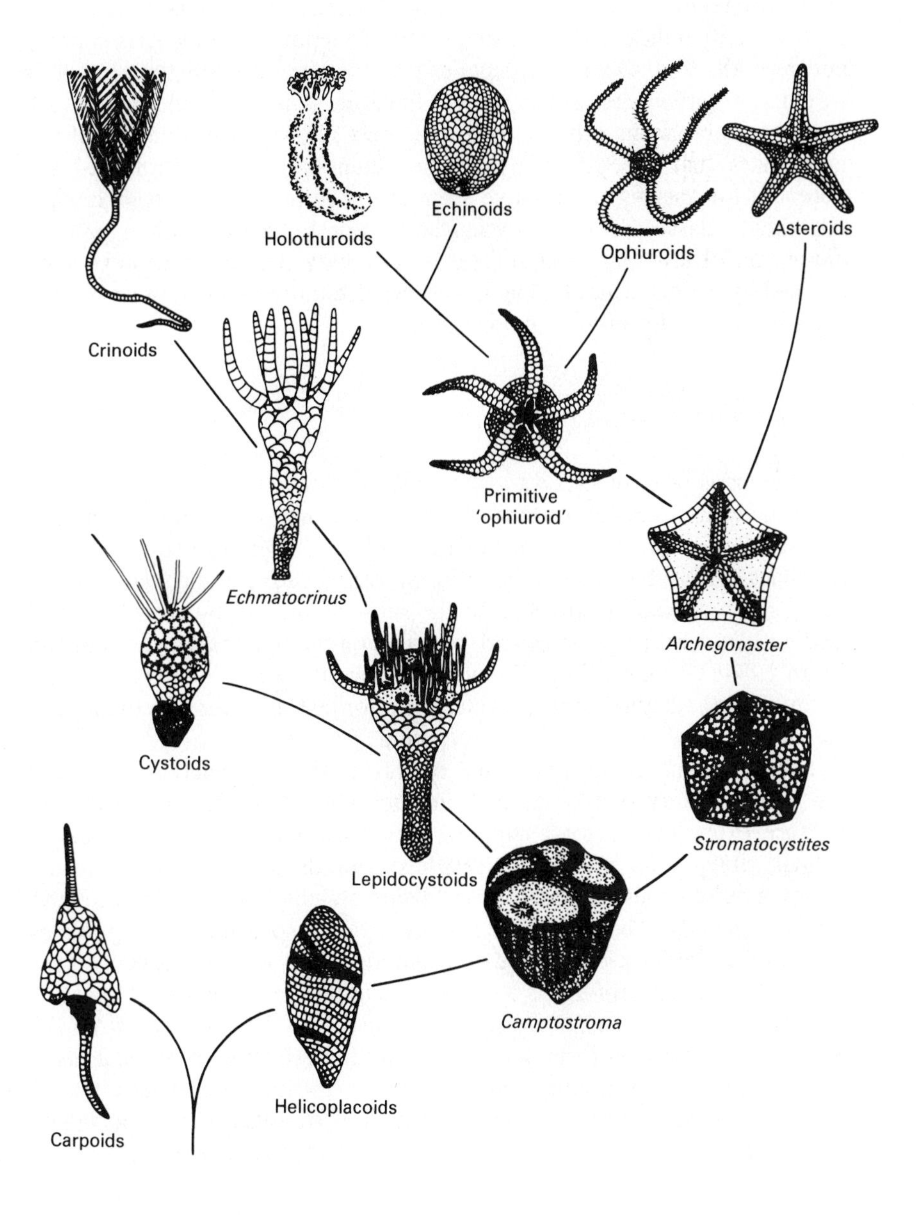

Body size and the relative size of body components are important in considering functional morphology, as so many functions are based on allometric relations of structures. The body size of the adult is important as it presumably indicates the size that is most functional in long-term survival and reproduction. Obviously, smaller juvenile sizes are functional or they would not survive, let alone grow. However, it is the adult size, which seems asymptotic in echinoderms, that best allows an evaluation of the body sizes that are possible with the echinoderm body form and the potential for varying the relative sizes of the different body components. All extant classes of echinoderms have representatives with a greatest dimension of less than 10 mm. The largest body size seems to have been reached by extinct crinoids. The largest representatives of extant classes are the asteroids, followed by the holothuroids.

1.2. CRINOZOA (=PELMATOZOA)

The Crinozoa was construed to include the cystoids, blastoids and crinoids to replace the term Pelmatozoa which had become broadened to include non-stalked forms without arms or brachioles (Ubaghs 1978). Crinozoans are characterised by a stalk at least in the early postmetamorphic stages by which the individual is attached to the substratum, and by erect, calcified feeding structures (brachioles and/or arms) on the upper surface (Paul and Smith 1984; Ubaghs 1978). The orientation of the ambulacra is clearly the distinguishing feature and provides the context in which pelmatozoan functional morphology must be interpreted.

The Crinoidea are characterised by having the ambulacra, mouth and anus on the body surface away from the substratum. The upward- and outward-extending arms are supported by calcareous elements in a longitudinal series. This body form restricts crinoids to suspension feeding, having a more or less rigid body, and being attached or capable of limited locomotion only. The complete reliance upon suspension feeding makes the crinoids the least diversified of echinoderms in feeding habits (Meyer 1980). Associated with this is a lack of specialisation of the tube feet.

The major changes in body forms have involved the transition from the stalked to the stalkless form, an increase in the development of the arms, a decrease in body size, and a change from a vertical to a shallow theca. Wyville Thomson (1878) pointed out the transition from a highly developed supporting structure (the stalk and its appendages) to the crown (the cup and arms) form in crinoids. Among the comatulids, the supporting structure is reduced to a minimum with the loss of the stalk, and the arms are often extremely developed (P.H. Carpenter 1884). These changes seem to be associated with an increase in the degree of free-living, active life. The assumption of a free-living way of life allows spatial and temporal selection of feeding activities and predator avoidance (Meyer and Macurda 1977), but it has not involved a fundamental change in the basic suspension feeding which is characteristic of the group (Meyer 1980). Suspension feeding does not require rapid locomotion, and becoming free-living has not involved the development of rapid locomotion. The body form of crinoids would not lend itself to the development of a high capacity for active locomotion. The stalk and the cirri have not become locomotory organs, and the use of the arms for locomotion is limited in capacity.

The basic form of the arms does not vary, but the length of the arms and their number do vary and thus affect the overall length of the ambulacra (P.H. Carpenter 1884; Gislén 1924). An increase in the number of arms by branching has occurred within the constraints of thecal and arm structures in two important ways. First, the crinoids remain pentaradiate with five radial water canals. The five radii are always expressed by the prominent series of proximal arm segments. The radial water canals branch distally to the theca and thus form additional arms in comatulids and modern, stalked crinoids (P.H. Carpenter 1884, 1888). This implies that the adaptiveness of increasing the number of arms for feeding is better met by branching instead of increasing the number of arms attached directly to the theca. Some extinct groups, e.g. the catillocrinids, did have numerous slender arms attached to a strongly asymmetrical cup produced by the enlargement of two radials (Ubaghs 1978).

The second important feature is that branching is primarily proximal to the theca so that most of the arm length is unbranched. In some extinct melocrinids, the two inner arms of each ray grew closely parallel and branched on their outer side to the end of each arm. This requires a rigidity of the crown, and this is characteristic of the camerates but uncommon in other crinoids (Cowen 1981). Such branching in extant crinoids seems limited to the stalked forms. Proximal branching increases feeding efficiency as it allows better control of arm orientation. With multiplication of arms, superficial pentamerous symmetry is lost and many species show a basic bilateral symmetry due to the specialisation of some of the arms and of the food grooves on the tegmen.

The slenderness of the arms and the provision of internal brachials as skeletal support seem to have eliminated the possibility of increasing the coelomic volume within the arm. Consequently, neither extensions of the viscera nor the gonads are found in the arm. Although a voluminous disc was common among the primitive Articulata (A.H. Clark 1941; Lane and Webster 1980), a major characteristic of extant crinoids is a minimal perivisceral coelomic volume in the theca. The volume of the theca has been increased in some crinoids by incorporation of the proximal part of the arms into the calyx. The small thecal volume in extant crinoids is associated with the lack of intestinal development and the absence of gonads there. Crinoids have a large body surface to volume ratio.

The size of some extinct crinoids was very great (e.g. *Extracrinus subangularis* had a stalk of *c.* 21.5 m in length, P.H. Carpenter 1884), but extant forms are much smaller. The evolutionary trend towards a decrease in body size can be correlated with the assumption of the vagile, free-living state which would be difficult for a large crinozoan. Among the comatulids, the smallest disc diameter is *c.* 3 mm and the largest is *c.* 50 mm. The length of the arms can vary considerably, ranging from 50 to 550 mm. The ratio of arm length to disc diameter is not constant. *Actinometra nobilis* can have a disc diameter of 50 mm and an arm length of 300 mm (ratio of 6.0), whereas *Antedon flexilis* can have a disc diameter of 10 mm and an arm length of 550 mm (ratio of 55.0).

1.3. ASTEROZOA AND ECHINOZOA

The asterozoans (asteroids and ophiuroids) and the echinozoans (echinoids and holothuroids) comprise the eleutherozoans. The major change associated with their development was the change in orientation so that the ambulacra and the mouth are on the body surface in contact with the substratum. The significance of this development was not only in the potential for the development of more efficient locomotion, but also in the potential for subsequent adaptation to feeding on the benthos. An indication of the major shift associated with this change in lifestyle is apparent in the change in the water vascular system involved in the new functions. No longer do the tube feet arise directly from the radial water canals as in the crinoids, but from lateral branches. The extant asterozoans have ambulacra restricted to the surface of the body in contact with the substratum, and almost always have arms. In contrast, the extant echinozoans have ambulacra that extend from the circumoral water ring to the other pole of the body and never have arms.

1.3.1. Asteroidea

Most variation in the body form of asteroids involves the degree of development of the length and breadth of the arms. The arms may be narrow and sharply set off from the disc, may be broad and merge with the disc, or may even be lacking. That pentagonal or armless forms are in the great minority implies that free arms are important, most probably in their primary roles in locomotion and feeding. The frequent distinction between asteroids and ophiuroids by the degree to which the arms are set off from the disc is only a consequence of the mode of skeletal support of the arm. The arms of asteroids are supported by elements in the body wall which merge with those of the disc, whereas the arms of ophiuroids are supported by internal vertebrae which are distinct from the disc. The asteroid arm has little capacity for lateral flexion because of the skeletal support in the body wall. The profile of the asteroid body shows variation also. It may be flattened, raised, cushion-shaped, or even spherical. The body wall may have few calcareous elements and obtain its rigidity from connective tissue, or may be plated to form a test. Asteroids usually have a relatively large perivisceral coelomic volume and a reduced body surface area to volume ratio compared with crinoids and ophiuroids. The large perivisceral coelomic volume within the arms of asteroids allows development of the pyloric caeca and gonads there in a way not possible in crinoids and ophiuroids.

Five arms are usual, but six are not uncommon and some species have large numbers. Unlike the crinoids, the additional arms arise directly from the disc as the result of formation of additional radial water canals and not from branching. The potential for increasing the number of radial water canals results from the lack of a requirement for support of the rays by a rigid theca as in crinoids. The lack of an internal supporting skeleton probably eliminates the potential for branching distal to the disc. However, branched arms would not seem to be functional for the modes of locomotion and feeding found in asteroids.

The increase in number of arms eliminates pentamerous symmetry, but maintains radial symmetry. The breadth of the arms limits the maximal number to fewer than that possible for crinoids. The changes that result from an increase in the number of arms involve the increase in total length of the ambulacra and in the volume of the perivisceral coelom. The changes in total length of the ambulacra have consequences for locomotion, attachment and feeding; and the changes in perivisceral coelomic volume have consequences for the total volumes of the pyloric caeca and gonads. That most asteroids have only five arms indicates that the adaptiveness of an increase in the total number or length of the ambulacra by increasing arm number is not great or is offset by disadvantages. These may include the increase in disc diameter that results from an increase in arm number. Multiple arms would seem to have no disadvantages associated with them.

Asteroids depend upon their tube feet for locomotion, whether they are suckered or not. The presence of suckers on the tube feet allows them to locomote on inclined surfaces and to exist in high-energy environments. These suckered tube feet can also play a major role in feeding.

Asteroids show a great range in body size. Fisher (1928) recalled that Forbes recorded a *Marthasterias glacialis* with a diameter of 840 mm. *Midgardia xandaros* can have a radius of 680 mm but its arms are extremely thin (Downey 1972). The minimal arm length for a species is *c.* 10 mm. The relation of the interradius length (r) to the arm length (R), indicating the degree to which the arm is free from the disc, is quite variable. The arms are not free in *Ceramaster japonicus* which has an R:r ratio of 68 mm:50 mm (1.1:1.0). This contrasts with the other extreme, *Midgardia xandaros*, which has an R:r ratio of 680 mm:13 mm (52:1).

1.3.2. Ophiuroidea

Ophiuroids possess slender arms capable of flexion. This is a consequence of the internal vertebrae which support the arm. Consequently, although similar in body form to asteroids, the ophiuroids are a much different functional system which is conspicuous in the different modes of locomotion and feeding. Locomotion in ophiuroids involves the entire arm, with minimal involvement of the tube feet. No ophiuroid lacks arms, in contrast to the asteroids.

Like the crinoids, there is little variation in the basic body form, with the major variation being in the relative proportion of body parts and the development of branching of the arms in one group. The major coelomic cavity in ophiuroids is in the disc because of the arm structure. Ophiuroids have a large body surface area to volume ratio like that of crinoids, and larger than that of asteroids.

Five arms are usual, but six or seven arms occur in some ophiurid ophiuroids. Larger numbers would probably interfere with the ophiurid mode of locomotion and with burying. The additional arms arise directly from the disc as in asteroids. Additional arms occur in some euryalid ophiuroids, twenty major arms being found in modern basketstars such as *Astrophyton* and *Astroboa* (Meyer and Lane 1976). As in the crinoids, these euryalids remain pentaradiate with the additional arms supported by five primary trunks. However, in contrast to the crinoids, branching is not restricted to the proximal part of the arm, and repeated branching into finer and finer units results.

The ophiuroid body plan leads to the same constraints on the development of the viscera as in crinoids. As in the crinoids, the volume of the disc may be increased by the development of the disc into the interbrachial areas between the arms. In contrast to the crinoids in which the gonads are found on the pinnules and arms, those of ophiuroids are in the disc. In contrast to most asteroids, no extensive pyloric caeca develop.

Ophiuroids range in size from a disc diameter of *c.* 2 to 65 mm. Arm length varies, and, as with crinoids, the relative length of the arm to the disc diameter is not constant. *Ophiacantha granulosa* has a disc diameter of 9 mm and arms 42 mm in length (ratio of 4.6), whereas *Asteroschema tenue* has a disc diameter of 6 mm and arms 200 mm in length (ratio of 33.3).

1.3.3. Echinoidea

Echinoids are characterised by having a body wall composed of plates bearing movable spines, five radial ambulacra extending from the peristome on the under-body surface to the apex of the upper surface, and no arms. The test of primitive echinoids was flexible, and that of some extant groups is flexible to varying degrees (Kier 1974). The spines have shown a great potential for variation in number, size, shape and structure, and provide a major adaptive feature of the group. The tube feet vary little and are not highly developed in the cidaroids. Regional specialisations of the tube feet for attachment and locomotion, feeding and respiration exist in the other groups, particularly in the irregular echinoids.

The echinoids show more basic variation in body form than the crinoids, asteroids and ophiuroids. The echinoids were traditionally divided into two groups, Regularia and Irregularia, on the basis of the difference in general body form (Hyman 1955). The regular echinoids are globular, usually circular but occasionally oval and sometimes slightly pentagonal at the ambitus. Although pentamerous, the irregular echinoids show bilateral symmetry with a mouth that may be displaced anteriorly and an anus that is displaced posteriorly. This condition is related to burrowing with a rigid test. The ambitus of the clypeasteroids may be approximately circular but that of the spatangoids is usually elongated.

The globular form of all regular echinoids and even of some irregular ones results in a large coelomic cavity and a small body surface area to volume ratio. The volume of the coelomic cavity shows considerable variability as a result of the variation in the height and shape of the test. This affects the potential volume of the gut and gonads.

Echinoids vary considerably in the size of the test. The extinct irregular echinoid *Victoriaster gigas* reached dimensions of 220 mm by 205 mm by 110 mm and may have been the largest non-flexible echinoid to have ever lived (McNamara and Philip 1984). The flexible asthenosomid *Hygrosoma hoplacantha* attains a horizontal diameter (HD) of 312 mm (A. Agassiz 1881). The smallest echinoid recorded is the echinacean *Salenia ebroicensis,* which is only 1 mm in length (Mortensen 1935). The internal volume does not show a direct relation to the horizontal diameter in regular echinoids because the relative vertical diameter (VD) varies. The HD:VD ratio is 121 mm:59 mm (2.1:10) in *Echinothrix desori,* and 147 mm:37 mm (4.0:1.0) in *Astropyga radiata.* The relative dimensions of the irregular echinoids also show considerable variation.

1.3.4. Holothuroidea

The holothuroids are characterised by having feeding tentacles extending from the circumoral water ring. The feeding tentacles of holothuroids do not contain a supporting skeleton although spicules may be present. The tentacles are thus completely flexible, but their length and the degree to which the feeding apparatus can be developed are limited. The lack of a skeleton in the tentacle eliminates the need for a body skeleton for support, and only a calcareous ring is present for support in most groups. The calcareous ring is much reduced in some elasipods and absent in apodids.

The primitive holothuroids had plated body walls. The evolutionary trend has been towards a reduction in the development of the skeletal elements except in a few groups (Pawson 1966). This has involved a development of the extracellular tissue which provides flexibility to the body wall and, except in the apodids in which the body wall is extremely thin, a capacity for rigidity.

The body form and characteristics of the body wall are so distinct from those of other echinoderms that holothuroids were long separated from them and classified with sipunculids and worms. The basic body form has been amenable to many variants (Pawson 1982b). It may be cylindrical, U-shaped or flattened or may have extensions appropriate for burrowing, sessile, epifaunal or swimming ways of life, respectively. The mouth and anus may be at opposite ends of the body axis or may be displaced to the upper or lower body surface.

Considerable modification exists in the water vascular system. In actively burrowing holothuroids, the radial water canals are pentasymmetrically arranged. In epibenthic holothuroids, vagile or sessile, bilateral symmetry results from a relocation of the radial water canals so that three (the trivium) are in contact with the substratum while the other two (the bivium) are on the exposed body surface. This occurs whether they are sessile or vagile, or suspension or deposit-feeding forms. The molpadiids have radial water canals but usually lack tube feet. The apodids lack even the radial water canals.

The holothuroids have a large coelomic volume and a body surface area to volume ratio like those of echinoids. Despite having a large coelomic volume, there is no development of digestive diverticula. Holothuroids range in size from *c.* 10 to 2000 mm in length. The body width, and consequently the body volume, is not directly proportional to the body length. The ratio of body length to body width can range from 1.6 in *Scotoplanes papillosa* to 42.5 in *Chrondrocloea macra.*

2

Acquisition of Nutrients

2.1. FEEDING

2.1.1. Energetics of feeding

According to the principles of maximisation and economisation, traits that maximise production in as economical a way as possible should be favoured because they ultimately maximise reproduction (Calow 1984a). Traits involved in feeding should be governed by these principles because they determine the amount of energy and material available to the organism. A major part of the interaction between an organism and the environment involves the acquisition of resources (Calow and Townsend 1981). Maxima are not always reached because of constraints in the organism and because of trade-offs (Calow 1984a). Because of these considerations, it is necessary to consider feeding biology, including functional morphology and behaviour, thoroughly.

Among the higher taxa of echinoderms, differences in feeding seem directed towards the exploitation of different habitats or food sources. Thus the crinoids retain the original passive suspension-feeding way of life. Asteroids feed on benthic surfaces and benthic prey. Ophiuroids feed both on benthic prey and on suspended material. Originally, echinoids fed only by biting or scraping but subsequently in some lineages by ingestion of particulate substrata. Holothuroids first fed on particulate substrata but suspension feeding arose in some lineages. These differences in feeding all involve differences in functional morphology and behaviour, and provide the organism with distinct differences in the characteristics and availability of food that affect the acquisition of material and energy.

2.1.2. Crinoidea

Basically crinoids are passive suspension feeders whose feeding mechanisms rely on environmentally produced movement of water (Meyer 1982). Capture of food is by the tube feet, whose structure and function are similar in the adults of all living groups (Meyer 1982). As a result, crinoids show little variation in the sizes of particles captured, and variation in diet is the result of the place and time of feeding. With modification of the tube feet constrained, crinoids are limited to modifications in the arms and pinnules. Because the basic form of the arms and pinnules does not vary, their modification involves length and number, which in turn affect the total length of the ambulacra (P.H. Carpenter 1884; Gislén 1924) and, along with behavioural adaptations, affect the filtering efficiency.

The stalked crinoids invested material and energy in the stalk to varying degrees. The length of the stalk determines the extent to which the crown is elevated above the substratum. Crowns of individuals with a short stalk would be affected by turbulence near the substratum surface, and those with long stalks would be affected by laminar flow (Roux 1987). Differences in the lengths of the stalks of species-rich assemblages of mid-Palaeozoic stalked crinoids have been interpreted as 'stratification' of the water column by species (Ausich 1980), and would require differential allocation of material to the stalk for feeding purposes.

The decrease in allocation of resources to the stalk noted by Wyville Thomson (1878) in the evolution of crinoids is striking. Not only would the loss of the stalk eliminate the need for energy and material allocation to the structure, but it also provides greater potential for feeding effectiveness because of the capacity for locomotion. Comatulids can 'forage' by attaining locations where food is available. Meyer (1982) noted that this ability to move to elevated, exposed positions makes them 'functionally' stalked crinoids.

Although the arms are involved in locomotion and reproduction, their degree of development is primarily associated with effectiveness of feeding. Because the length and number of arms varies among species, the allocation of energy to the food-gathering system must differ considerably. The relation between allocation of energy to the arms and the efficiency of food gathering is not known.

Each pinnule involved in food capture has groups of three tube feet (triads) which alternate on either side of the ambulacral groove and vary in length and function. The functioning of these triads has been described for *Antedon bifida* (Lahaye and Jangoux 1985a; Nichols 1960) and for *Florometra serratissima* (Byrne and Fontaine 1981, 1983). While awaiting contact with food particles, the long primary tube feet project laterally; the medium-length secondary ones curl up over the vertically held lappets; and the short tertiary ones project vertically over the ambulacral groove (Figure 2.1).

Capture of food particles is passive. Particles impact the mucus produced by the primary tube feet and are either transferred directly to the

Figure 2.1: Inner view of the feeding unit in *Antedon bifida*. l: Lappet, p: primary tube foot, pa: papilla, s: secondary tube foot, t: tertiary tube foot (from Lahaye and Jangoux 1985a)

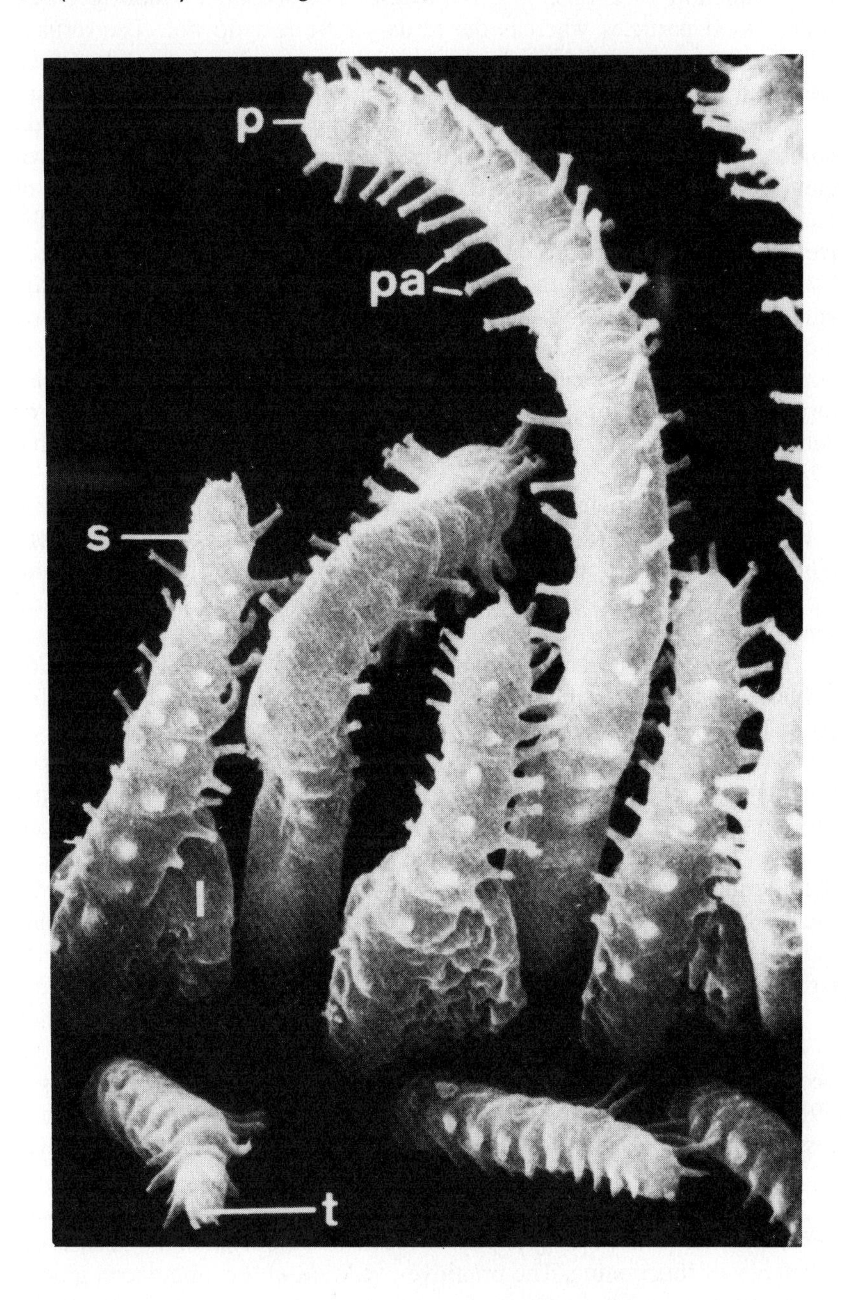

ambulacral groove or are removed by wiping on the tertiary tube feet (Figure 2.2). The tertiary tube feet have two opposed rows of papillae which function as a rake in this process. The secondary tube feet also collect food particles whereas the tertiary tube feet do not. The tertiary tube feet not only remove particles from the primary and secondary tube feet, but also move against the ciliary current. This disrupts the ciliary movement of the particles down the ambulacral groove and produces food boluses. The secondary tube feet are attached to the lappets and the movement of the two are co-ordinated. The tertiary tube feet have a cylinder of muscles which allows them to move in all directions; the secondary tube feet have retractor muscles on their upper and lower sides so that they can bend outwards towards the primary tube feet and inwards towards the ambulacral groove; the primary tube feet have muscles on the upper surface and bend only inwards towards the ambulacral groove. The pentacrinoid of *Florometra serratissima* uses the tube feet to transfer particles directly to the mouth by flicking the tips of the tube feet into the mouth or actually inserting the tube feet into the vestibule (Mladenov and Chia 1983).

Although the structure and function of the feeding tube feet show little variation, the number and orientation of the triads do vary. Several possibilities exist: the distance between triads on a pinnule; the number and length of pinnules and arms; and the posture of the tube feet, pinnules and arms.

The number of triads varies along the pinnules. Meyer and Lane (1976) suggested that the distance between the tube feet along the distal feeding pinnules set the lower size limit of filterable particles, but this had not been demonstrated. Meyer (1979) found that the number of tube feet per millimetre ranged from 4.59 to 9.49 for 22 species of comatulids, and that the number was directly correlated with tube-foot length which ranged from 0.45 to 0.90 mm. The spacing and length of the tube feet would affect the filtering efficiency as well. The total number of triads is also affected by the number and length of pinnules and arms. The isocrinids and comatulids tend to have more complete pinnulation than most millericrinids, bourgueticrinids and cyrtocrinids (Breimer 1978).

Particles found in the guts of crinoids may be inorganic, indicating a lack of qualitative selection in feeding (Liddell 1982). The gut contents of co-occurring crinoids are similar in the Red Sea (Rutman and Fischelson 1969) and the Caribbean (Liddell 1982). However, selection for size may result from the spacing and length of the tube feet of the triads (Meyer 1979).

Specialisation of the pinnules and arms are other important evolutionary developments (Gislén 1924). This division of labour should increase the efficiency of functioning. The primitive condition of the ambulacral groove extending the entire length of the arms and pinnules (still found in the

Figure 2.2: Pinnule of *Antedon bifida* showing the sequence of transfer of particles from the primary tube foot. A: Quiescent stage, B: primary tube foot of middle triad wiping the ciliary tract against the ciliary current, C: primary tube foot wiping the tertiary tube foot. Large arrows indicate the direction of wiping; small arrows indicate the direction of the ciliary tract. p: Primary tube foot, s: secondary tube foot, t: tertiary tube foot, tc: transverse ciliary tract (from Lahaye and Jangoux 1985a)

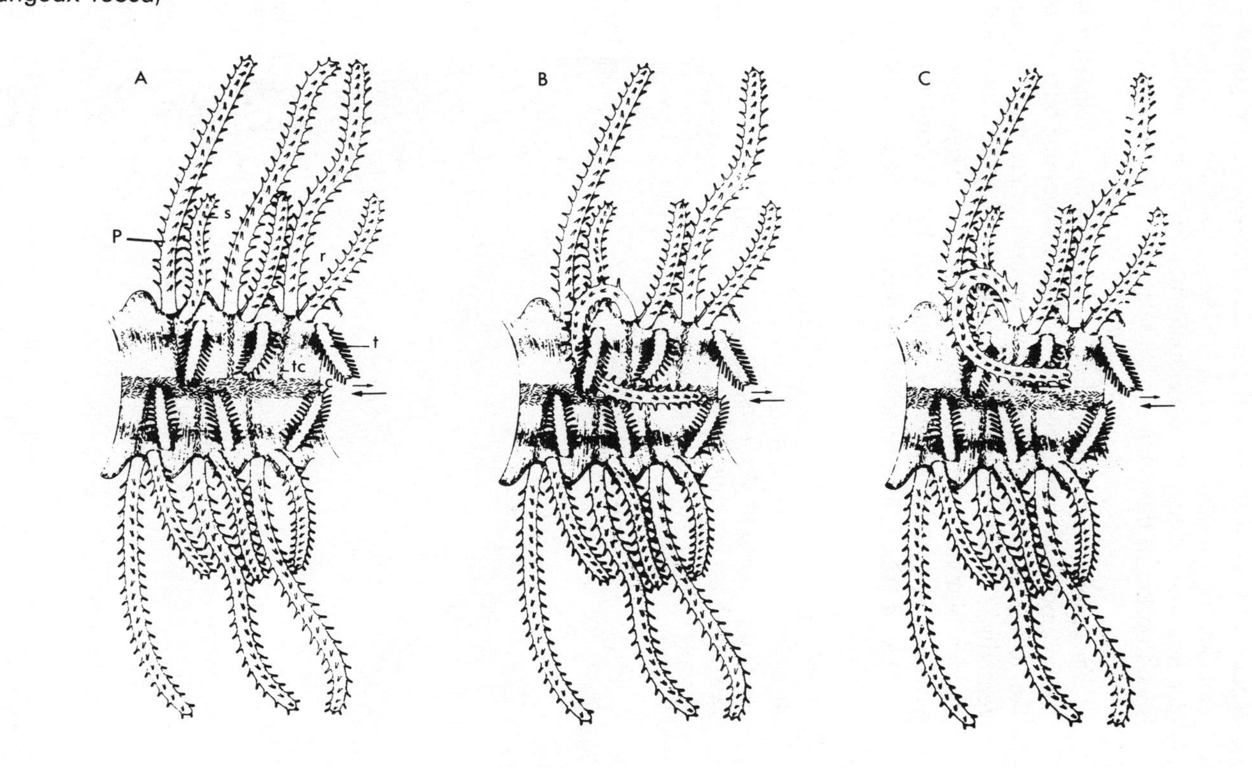

stalked crinoids) has been modified in two ways. In the comatulids, the pinnules are differentiated into proximal oral pinnules (for protection and tactile use), adjacent genital pinnules containing the gonads, and distal pinnules for feeding (Figure 2.3). These distal pinnules are longer, more slender, and with well developed tube feet and ambulacra. Extreme specialisation in this form can be seen with *Tropiometra* (P.H. Carpenter 1884).

The second evolutionary direction taken by the comasterids was specialisation among the arms, with the individual becoming bilateral as a result of the mouth moving to the radius or interradius and the anus moving centrally. The anterior arms (next to the mouth) are longer and have long, slender pinnules like the distal pinnules of *Tropiometra.* The shorter posterior arms have pinnules specialised for reproduction. A large percentage of arms can be removed from functioning in feeding: up to four of

Figure 2.3: Genital and distal feeding pinnules of *Democrinus rawsoni* (from Breimer 1978)

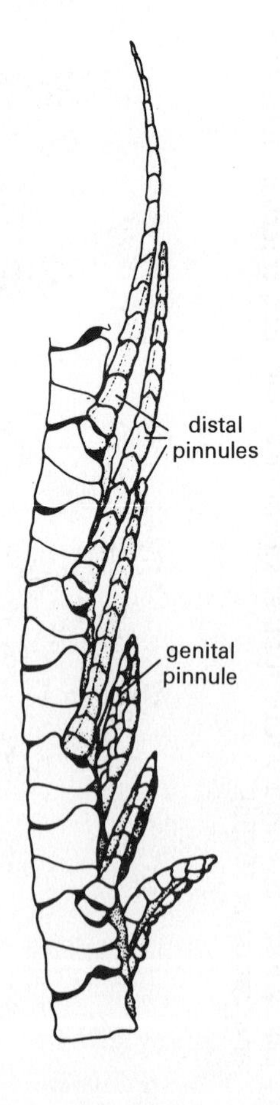

the ten arms of *Comanthus parvicirrus* and six of the ten arms of *Comatula micraster.* The total length of the ambulacra (and thus the number of triads) is determined mainly by the number and length of their distal pinnules and by the number of free arms. Ten-armed crinoids have longer distal pinnules than multibrachiate species, and the relative length of the pinnules decreases in direct proportion to the increase in the number of arms (A.H. Clark 1921).

Although the comatulid *Eudiocrinus* has five arms, the primitive five-armed condition usually occurs in recent crinoids only in such stalked forms as *Rhizocrinus* (Breimer 1978). The comatulid *Comanthina schlegeli* has up to 200 arms and a crown diameter of 50 cm (A.H. Clark 1931).

The number of crinoid arms does not increase by intercalation as occurs in ophiuroids and asteroids, probably because the arms need a sufficient diameter to support themselves and/or need an increase in structural support by the theca. The branches of the arms of extant crinoids occur close to the disc regardless of the number of branches so that the greater part of the crown is formed of free arms. The ability to maintain the arms in the appropriate posture for feeding is better when the arms originate close to a common centre. By contrast, the extinct camerate family Melocrinitidae had an arm pattern with rows of parallel and pinnulate ramules (Figure 2.4). This pattern probably forms a more effective filtration system. However, the system requires a rigid crown not found in other crinoids in which flexibility of the arms has been emphasised (Cowen 1981). Apparently the

Figure 2.4: Filtration networks of *Glyptocrinus ornatus* and a Devonian species of *Melocrinites* (from Browmer 1976)

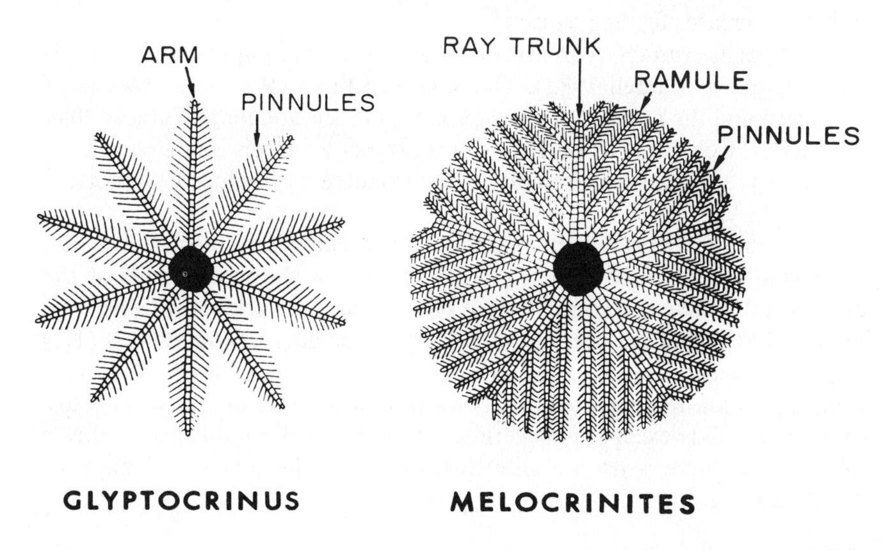

flexibility of the arms provides a versatility in feeding and a consequent potential for exploiting a wider variety of habitats and current regimes.

An increase in the number of free arms is a dominant trend even in several lineages of the extinct Camerata and has been similarly interpreted to indicate evolution for a greater number of pinnules and consequently for improved feeding ability (Ubaghs 1978). The stellate body form of crinoids allows an increase in the number of arms by branching. This increases the ratio of area of ambulacra (food-gathering ability) to body volume (Strathmann 1975).

The branching mode of increasing ambulacral length is efficient when total length rather than breadth of the ambulacrum is important and an increased length from the tip of the ambulacra to the mouth is a disadvantage, as in food collecting (Strathmann 1975). Branching of the arms also affects their function as a filtering system. A.H. Clark (1921) noted that most comatulid species with more than ten arms are tropical and subtropical and that species with usually no more than ten arms are deep sea or cold water, and suggested that the number was probably related to food supply. The strategic response would be an increase in arm number with a decrease in food availability. This would require an increased allocation of material to the arms in areas where food availability is low.

A similar variability in the arms occurs in the stalked Pentacrinidae (Bourseau and Roux 1985). The number of arms possessed by *Metacrinus* species increases from 300 to 600 m depth, but is least in species at greater depths. In pentacrinid species as a whole, the length of the arms decreases with depth to 800 m, and the number of arms decreases after 600 m. This relation between arm length and number has been interpreted in terms of energy requirements for the arms and feeding capacity: a decrease in arm length decreases energy needs for their production, and an increase in arm number increases feeding capacity.

Interspecific variability in arms also occurs among populations of tropical comatulids (Liddell 1982). The length of the arms of three species of *Nemaster* and the number of arms of one species is greater at Curaçao than at Jamaica, Colombia, or Barbados. It is not known if this is related to food availability and thus an example of environmentally induced phenotypic plasticity.

The relation between arm number and feeding effectiveness has not been studied, although Gislén (1924) calculated the total length of the ambulacral groove and its relation to body volume for crinoids (Table 2.1). He concluded that the average value is much smaller in comasterids (1.7) than in the rest of the comatulids (5.0). Consequently, the ambulacral food groove in comasterids is abnormally short in proportion to the animal's size in all comatulids except comasterids. The functional significance of this is unknown as there is no overall evaluation of the feeding effectiveness of the different groups.

Table 2.1: The total length of ambulacral food grooves in crinoids. The total number of arms and the number of arms with ambulacra (in parentheses) are given for *Comatula pectinata* and *Comanthus parvicirrus*. (Modified from Gislén 1924)

Species	Number of arms	Length of arms (mm)	Length of amb[a] groove (m)	Wet body weight (g)	Body volume (cc)	m amb[a] / cc
Comissia parvula	10	35	1.95	0.89	0.74	2.64
Comatula pectinata	10(6)	90	4.37	4.83	3.46	1.26
Comantheria grandicalyx	48	125	102.68	89.6	73.6	1.39
Comaster delicata	46	115	58.92	39.1	30.0	1.96
Comanthus parvicirrus	37(20)	50	7.16	6.65	5.40	1.33
Catoptometra magnifica minor	27	95	34.50	6.70	5.42	6.37
Eudiocrinus loveni	43	2.47	0.65	0.47	1.38	5.26
Amphimetra crenulata	30	75	43.05	11.55	8.28	5.20
Lamprometra protectus	43	50	23.44	7.15	5.75	4.08
Cyllometra pulchella	14	40	6.02	0.72	0.60	10.02
Tropiometra afra macrodiscus	10	255	47.69	39.8	34.4	1.39
Asterometra anthus	10	70	10.60	3.12	2.34	4.52
Pectinometra flavopurpurea	20	80	17.84	4.60	3.55	5.02
Perissometra aranea	11	100	7.04	1.93	1.46	4.83
Antedon petasus	10	110	16.28	2.63	2.29	7.10
Heliometra eschrichti	10	200	54.5	50.0	45.0	1.21
Rhizocrinus lofotensis	6		0.14–0.22			
Comissia ignota	10	20	1.30			
Compsometra parviflora	10	25	1.37			
Pentametrocrinus diomedeae	5	55	1.85			
Thaumatometra comaster	10	45	2.98			
Oligometra chinensis	10	55	7.15			
Stephanometra spicata	26	130	38.48			
Liparometra grandis	26	120	44.72			
Himerometra magnipinna	49	90	61.23			
Metacrinus rotundus	56	210	71.68			

[a]amb: Ambulacra.

Orientation as well as the number affects the ability of triads to intercept particles. The long tube feet, the pinnules and the arms should all be at right angles to the direction of water movement for maximum effectiveness (Meyer 1973). The orientation of the pinnules can also be changed. The pinnular articular facet of the brachial element has a short, oblique cross-ridge with a conspicuous fossa on either side. The ligaments of the external fossa and the muscles of the internal fossa are antagonistic (Meyer 1971). Muscle contraction pulls the pinnule into a closed or vertical position over the arm; muscle relaxation pulls the pinnules out at an angle to the arm into the feeding posture.

The pinnules of *Tropiometra carinata* are held in a planar, two-row arrangement perpendicular to the current (Figure 2.5) (Meyer 1973). In *Nemaster grandis*, the pinnules of each arm form four rows when individuals are in conditions with little or no unidirectional water movement, but form two rows in a plane normal to unidirectional water movement (Figure 2.5) (Meyer 1973). By contrast, *Nemaster rubiginosa* apparently cannot bend its pinnules to form the planar, two-row arrangement, and the pinnules are always held in the radial, four-row arrangement (Figure 2.5). In the radial, two-row arrangement, the pinnules along one side of the ambulacral groove alternate in direction. This makes the spacing between adjacent pinnules in each row greater than in the planar arrangement in which a single row of pinnules lies on each side of the ambulacral groove (Meyer 1979).

Considerable variation in the posture of the arms is possible (Figure 2.6) (Meyer 1982; Roux 1987). A deep-sea comatulid can have the oral side upwards and the arms held horizontally (Reyss and Soyer 1965). Similarly, stalked isocrinids have the oral side up in negligible current flow, but the arms are recurved (Macurda and Meyer 1974). Most comatulids in low-velocity, multidirectional flow hold the arms arched over the disc in a meridional posture and the pinnules are non-planar (Meyer 1973). In unidirectional flow, the arms are aligned into a filtration fan which can vary in form but results in a posture normal to the current flow (Meyer 1973).

The functioning of these structural characteristics of the tube feet, pinnules and arms of crinoids in feeding have been interpreted in terms of the aerosol filtration theory (Meyer 1982; Roux 1987). Particles are captured by mechanisms of gravitative settling and motile-particle deposition with low velocity, and by mechanisms of direct interception and inertial impaction with high-velocity flow. Meyer interpreted the decreased length and spacing of tube feet as adaptations to high-velocity flow because it would provide a denser filter. Radial positioning of the pinnules and the meridional orientation of the arms would be behavioural responses to low-velocity flow. The capture of food by crinoids is also enhanced by 'baffling' mechanisms which increase the settling out of particles by overlapping of pinnules and arms of an individual or of two or more neighbouring conspecific individuals.

The food grooves of the pinnules and arms are invariably down current (Magnus 1963). The pinnules swivel passively in response to the current (Meyer and Macurda 1976), but the arms must twist up to 180° to achieve this orientation and form filtration-fan postures (Meyer 1973). This orientation may facilitate the transfer of food from the long tube feet to the

Figure 2.5: (A) *Tropiometra carinata carinata* with pinnules in planar arrangement (from Meyer 1973). (B) *Himerometra* sp. with pinnules in planar arrangement with extended tube feet (from Meyer 1979). (C) *Nemaster rubiginosa* with pinnules in the four-row radial posture (from Meyer 1973). (D) *Comaster gracilis* with pinnules in the four-row radial posture with extended tube feet. (From Meyer 1979)

Figure 2.6: Principal types of filtration fans of comatulid crinoids, a: Arcuate fan. Left, view parallel to current; right, view perpendicular to current (large arrow); small, circular arrow indicates twisting of arm on right. b: Radial fan of individual attached to a wire coral. Left, view parallel to current; right, view perpendicular to current. c: Parabolic fan. Left, view down current towards concavity of fan. Above right, view during slack current. Below right, view parallel to current. d: Arm postures of multibrachiate individual lacking cirri. e: Arm fan of individual with calyx attached beneath ledge. Planar pinnules held perpendicular to current. (From Meyer 1982)

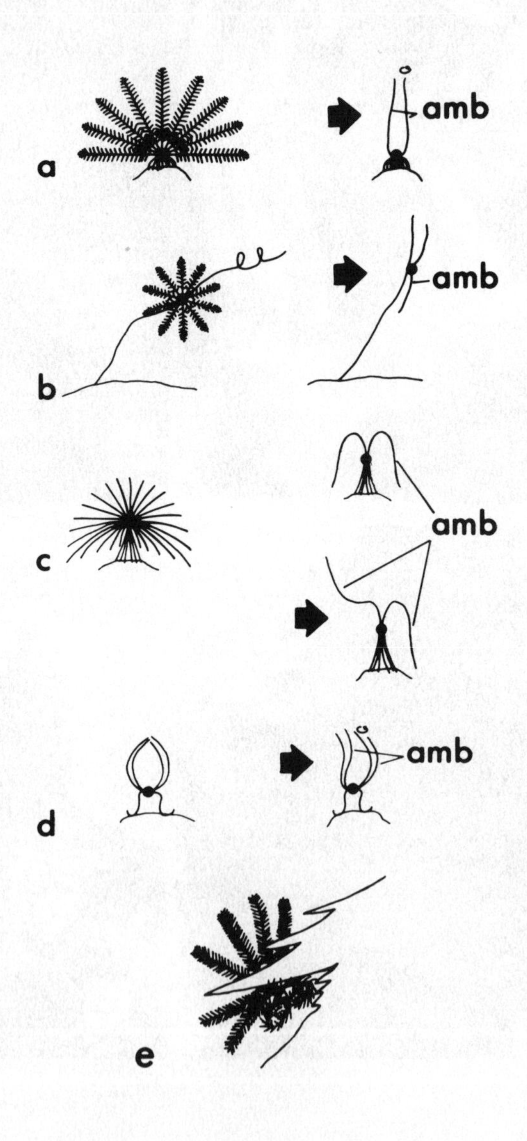

shorter ones and thence to the food groove because the mucous chain is not exposed to direct impact of the current (Meyer 1982).

An important aspect of feeding in currents is the ability to maintain the arm in the extended position. Species in such situations have a more robust arm form for strength and a greater number of oblique muscular articulations for flexibility (Meyer 1973). The collagenous nature of the ligaments in such forms is thought to provide a locking mechanism (Meyer 1971). This mechanism would decrease the energy requirements for maintaining the feeding posture.

The arm filtration fan of rheophilic, stalked crinoids should produce a lift from the water pressure of the current if the fan is oriented obliquely into the current (Figure 2.7) (Breimer and Webster 1975). Roux (1987) pointed out that the crinoid crown produces down-current turbulences (a wake) that develop a drag force which is countered by the arms and pinnules taking an orientation to increase lift. Active movement of the pinnule not only aids feeding but also regulates lift.

Curvature of the arm affects lift and requires many articular surfaces with muscles. When the muscles relax, the aboral ligament bundles contract and give the arm its curvature. The stalk is also affected by current (Roux 1987). With no current the ligaments contract and the stalk becomes rigid. With a current ligament tensile stress is developed and the stalk becomes flexible and capable of orientation. With a boundary layer with turbulent flow, a flexible stem with synarthrial joints is necessary. An ontogenetic change is associated with stalk length. The short stalk of a young individual would be near the bottom and in the boundary layer, whereas most of the longer stalk of the adult would be in the laminar flow. The young of some pentacrinids have stalk joints that allow differential movement, whereas the adults have stalk joints that restrict movement.

Particles transferred to the food grooves of the pinnules move in a mucous string down the food grooves of the arms to the tegmen. The food grooves from the two arms on each ray extend on to the tegmen, join and go to the mouth in the endocyclic crinoids for which the mouth is centrally located (Figure 2.8). The food grooves are asymmetrical on the tegmen of the exocyclic comasterids in which the mouth is displaced towards the margin and the anus is centrally located.

The gut of crinoids has two distinct forms. The primitive, stalked, endocyclic crinoids and the endocyclic, antedonid comatulids have a gut with a single coil which is greatly plicated and shows no superficial differentiation into regions (Figure 2.8) (P.H. Carpenter 1884). By contrast, the exocyclic, comasterid comatulids have a longer gut with four coils which are little plicated. The correlation between a high degree of plication and a single coil, and little plication with an increase in length, could be a mechanism to increase the surface area of the gut (P.H. Carpenter 1884), and may also be associated with an increase in the amount of plant food in the diet of

Figure 2.7: Theoretical postures of stalked crinoids to obtain lift from currents. A (a): During slack current with the stalk held vertical for direct support. (b): With slight current, arms held in a filtration-fan posture, crown tilted to provide lift, stalk held vertical. (c) and (d): With strong currents, arms held in a filtration-fan posture, stalk receiving lateral pressure that eventually cannot be resisted. B: Passive orientation with no lift derived from current. Arms held as a planar brachial fan (a), a conical fan held down current (b), or up current (c), or a curved fan (d). Crown down stream, oriented more or less perpendicularly to the current and receiving no lift because upward and downward pressures are equal (small arrows). C: Active orientation with lift derived from current by changing angle between plane of brachial fan and current and position of lower part of fan (a, b, c), or by recurving arms into current (d). As a result of the lift, the stalk can be less strong, more flexible, longer and either stand obliquely (b and c) or with the concave side facing the current (d). (From Breimer and Webster 1975)

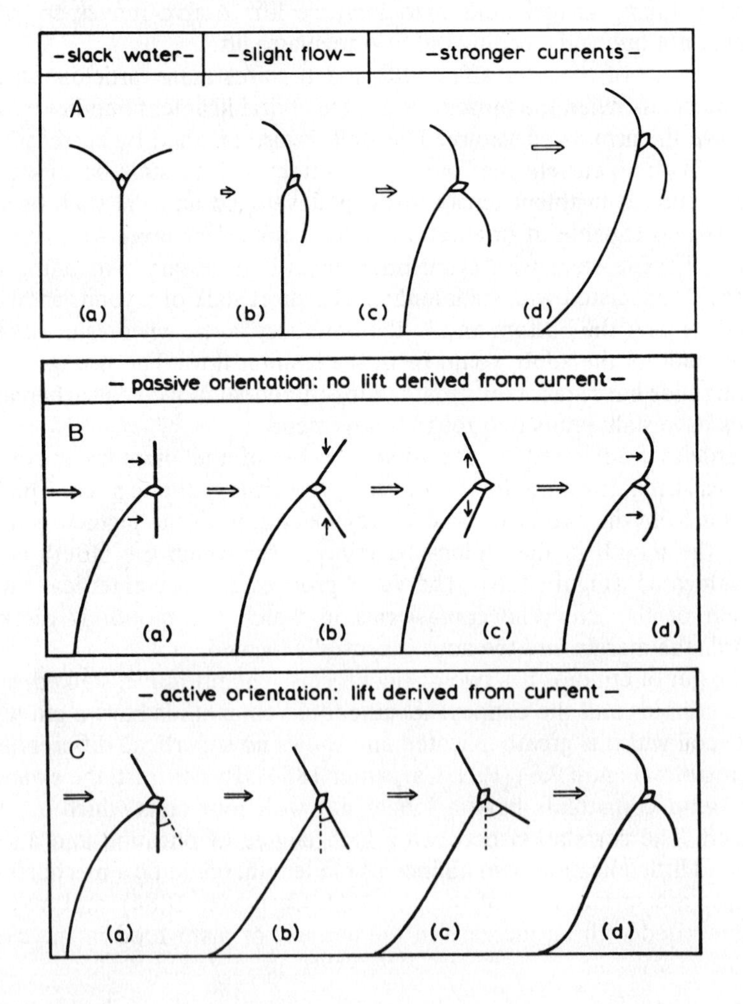

Figure 2.8: The form of the gut and the position of the feeding grooves of (A) endocyclic, and (B) exocyclic crinoids. m: Mouth, a: anus. (From P.H. Carpenter 1884)

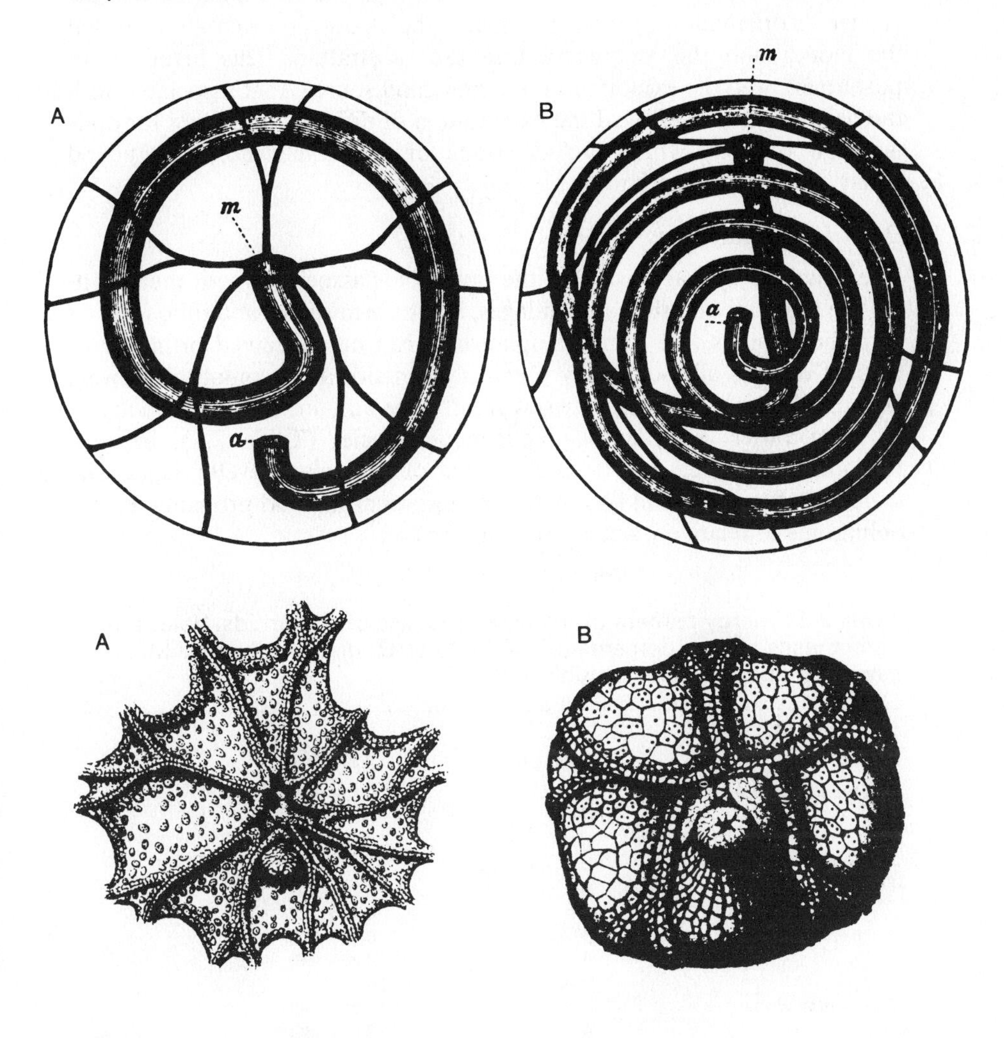

comasterids (A.H. Clark 1915). A generally richer food has been noted in the gut contents of comasterid species than antedonid species (Gislén 1924). The gut of crinoids thus shows a long, undifferentiated gut that is associated with primitive filter feeders. The length of the gut should be longer with a poorer diet, either in terms of quality or quantity (Sibly 1981).

2.1.3. Asterozoa

One of the major evolutionary developments in the echinoderms was the change in orientation of the asterozoans: the change in orientation so that the mouth and the ambulacra face the substratum. This inversion of posture allowed the exploitation of a new food source: that associated with the benthos. It also allowed the exploitation of different strategies for optimisation of diet, which include traits of functional morphology and behaviour, to increase fitness.

2.1.3.1. Ophiuroidea

The structural characteristics of the ophiuroids associated with their feeding are the arms and their appendages, i.e. the arm spines and tube feet. As with the crinoids, the arms of ophiuroids can be conceived primarily as feeding devices although they obviously function in locomotion as well. Because the size of the arms varies greatly, the amount of energy allocated to them varies, even among congeneric species (Table 2.2), being a function of number, length, and degree of muscular development. The structure of the arm and its functioning in procuring food probably restrict ophiuroids to relatively small food particles.

Table 2.2: Energy content of the arms and disc of ophiuroids. Values in parentheses are the percentages of the total. D: disc; A: arms. (J.M. Lawrence and A. Guille, unpublished)

Species	Body part	Energy content (kJ. ind^{-1})
Amphiura antarctica	D	0.138 (21)
	A	0.511 (79)
	total	0.649
Amphiura chiajei	D	0.197 (49)
	A	0.209 (51)
	total	0.406
Amphiura filiformis	D	0.138 (32)
	A	0.293 (68)
	total	0.431
Ophiacantha vivipara	D	1.640 (40)
	A	2.503 (60)
	total	4.143
Ophionotus hexactis	D	2.653 (69)
	A	1.167 (31)
	total	3.820
Ophiura ambigua	D	0.473 (49)
	A	0.485 (51)
	total	0.958
Ophiura brevispinus	D	1.256 (68)
	A	0.598 (32)
	total	1.854

Small size may be adaptive as a means of increasing efficiency of feeding on small prey and of exploiting spatially restricted habitats (P.J. Miller 1979). Ophiuroids are highly susceptible to predation. They are often either cryptic, exposing themselves only when the probability of predation is low or when prey are available, or expose only their arms to capture prey. In the latter situation, predation on the arms occurs and would be a major cost of feeding. Both behavioural patterns would be trade-offs between survival or maintenance costs and feeding effectiveness.

The arms are long, slender and flexible. The ophiuroid arm can bend and twist. The arms of some euryalids branch at considerable distances from the disc. Unlike crinoids, these branches are progressively smaller (Lyman 1877). Because the arms are involved in feeding, one might anticipate an increase in arm number to increase feeding efficiency. However, almost all ophiuroids have five arms; six or seven arms occur in some ophiurids. This limit is probably because of constraints imposed by the use of the arms in locomotion or in the support provided by the disc.

The arm spines provide accessory structures for participation in feeding. The tube feet, like those of crinoids, lack suckers. The change in the structure of the wall and the development of ampullae are probably associated with the greater capacity needed for burrowing. The tube feet do not occur in triads like those of crinoids, but as pairs. Epithelial cilia are absent in most ophiuroid species except around the bursal slits, the undersurface of the disc, and the arm bases (Gislén 1924). Bands do occur on the lower surface of the arms of *Ophiopsila*, and Gislén stated that they produce a current towards the mouth. Other than in these exceptions, food cannot be transferred down the arm to the mouth by ciliary currents. The use of the arm and arm spines as well as the tube feet in feeding is an evolutionary advance and makes the ophiuroids more versatile in feeding modes than the crinoids which rely on the tube feet alone. Although most species of ophiuroids can show a variety of feeding behaviours, most use one predominantly, and ophiuroids can be separated into two major feeding types, each with distinct structural adaptations (Warner 1982). One type includes primarily predatory and carrion-feeding carnivores. Individuals in this category are characterised by short arm spines and tube feet. The second type includes microphagous feeders on particles on the substratum or in the water column. Individuals in this category have long arm spines and tube feet. Although a species may feed primarily in one way and be a specialist, it may feed in alternative ways when appropriate and be a generalist. Thus *Ophiocomina nigra* shows a variety of feeding mechanisms and can be either carnivorous or microphagous. This diversity of feeding habits is no doubt a major factor in the success of the ophiuroids (Fontaine 1965).

Carnivores and carrion feeders fall more into the category of foragers. Microphagous feeders are more sedentary, although euryalids do move to exposed positions during the feeding period. Application of optimality theory to the feeding behaviour of either group has been minimal.

The carnivores primarily use the arm itself to capture food by the so-called 'arm-loop' method (Figure 2.9). Particles 1–2 cm in diameter are captured by *Ophiocomina nigra* (Fontaine 1965). The arm is flexed laterally around the particle which is held in place by the tube feet. The particle is brought to the mouth by sharp curvature of the arm while the disc is raised by the other arms. The tube feet of the arm tip of *Pectinura maculata* grasp food, the arm coils in an oral/aboral direction to the mouth cavity, and the buccal tube feet remove the food and pass it into the mouth (Pentreath 1970). The arm of *Ophiomyxa brevirima* coils laterally to bring the food to the mouth (Pentreath 1970). There the food is grasped by the jaws and pushed into the mouth by the buccal tube feet. Quite obviously, long arm spines would interfere with this process, and long tube feet are not necessary. Only feeble arm loops are formed by the suspension-feeding *Ophiothrix* species which have long arm spines (Warner 1982).

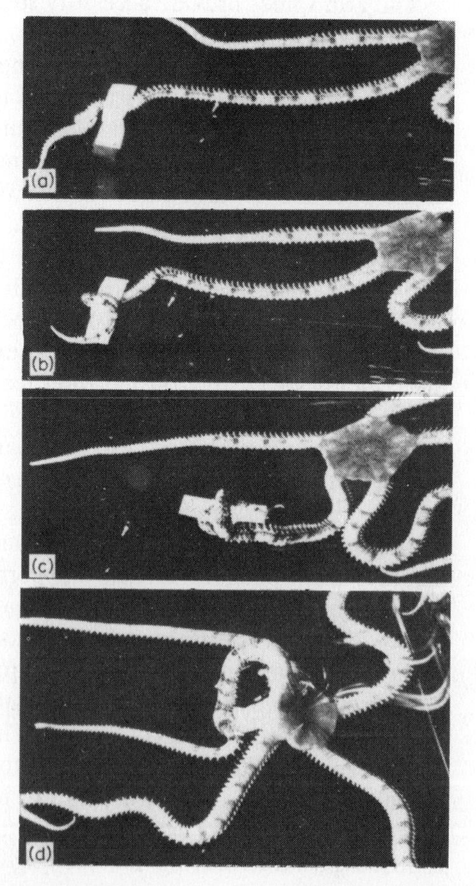

Figure 2.9: *Ophioderma apressum* showing sequence (a to d) of transfer of a piece of paper to the mouth by a coil of the arm (from Reimer and Reimer 1975)

The euryalids have vertebrae which allow oral/aboral coiling of the arms. In the unbranched *Asteroporpa annulata*, the disc is oriented with the aboral surface facing the current with three or four arms extended vertically (Hendler and Miller 1984). The tips of the arms bend into the current or the arms form a sinusoidal pattern, as seen also in *Astrotoma agassizii* (Figure 2.10). It is presumed that prey are captured by loops of the distal part of the arm or by the hooks which occur there.

Extensive branching of the arms occurs in the basketstars. Like the crinoids, euryalids are dependent upon a current flow for feeding. The highly branched arms may be held up singly or in the form of the parabolic filtration fan similar to that of crinoids (Figure 2.11). The oral side of the arms faces into the current (Tsurnamal and Marder 1966), not the aboral side as in crinoids. In contrast, the parabolic feeding fan of *Astrophyton muricatum* is directed towards the current and the ambulacra are down current (Hendler 1982). No arm twisting or variants of arm posture to

Figure 2.10: *Astrotoma agassizii* with unbranched arms held in the sinusoidal feeding posture. (Smithsonian Oceanographic Sorting Center, National Science Foundation)

Figure 2.11: *Astrophyton muricatum* with arms in feeding posture (from Meyer and Lane 1976)

place the ambulacra down current as found in crinoids (Meyer 1982) seem to occur. The ability to hold the tube feet directly into the current is probably the result of transfer of the food particles to the mouth directly rather than by ciliary–mucoid currents in an ambulacrum as occurs in crinoids.

The filtration surface area of basketstars is not as complete as that of crinoids, and large gaps occur in the mesh which, unlike that of crinoids, is not filled by tube feet (Meyer and Lane 1976). Meyer and Lane pointed out that the prey size of basketstars would be larger than that obtainable by crinoids. Movement of the terminal fine branches occurs and would increase the probability of prey interception (Wolfe 1982).

However, instead of the tube feet being used to capture prey, arm tips and arm spines modified into hooks are used as in the unbranched forms (Wolfe 1982). An arm loop is formed around the prey by rapid flexion of the large intervertebral muscles and held by the hooks (Wolfe 1978). In gorgonocephalids, the tentacle spines are modified into strongly curved girdle hooks that occur in double rows on the dorsal side of every joint (Gislén 1924). In *Astroboa*, the divisions of the distal arm are long and slender with girdle hooks, whereas the proximal divisions are short and have tentacle hooks and lack tube feet (Döderlein 1912). Gislén suggested a regional specialization, with the distal hooks being used for feeding and the proximal hooks for anchoring.

The captured prey are transferred to adjacent armlets, and when sufficient in amount the arm coils towards the mouth. As in the macrophagous carnivores, the arm is periodically inserted into the mouth and the food is removed, here by scraping the food against the oral shield (Patent 1968).

Microphagous deposit feeding by ophiuroids involves the tube feet alone. Burrowing amphiurids extend arms out of the burrows, sweep them over the surface of the substratum, and pick up particles with the mucus on the tube feet (Buchanan 1964; Woodley 1975). Cryptic *Ophiocoma* species similarly pick up particles with their tube feet (Sides and Woodley 1985). *Microphiopholis gracillima* selectively ingests large particles and particles coated with organic material or bacteria in conformity with optimal feeding theory (Clements and Stancyk 1984). *Ophiothrix lineata* lives with sponges. It feeds by extending its arms out of the osculum and sweeping them across the inhalent surface of the sponge, using the tube feet and arm spines to collect particles too large to enter the osculum (Hendler 1984).

Microphagous suspension-feeding ophiurid ophiuroids use the arm spines and the tube feet to capture food. *Ophiocomina nigra* raises its arms in a feeding posture (Figure 2.12) and produces a net of mucous threads

Figure 2.12: *Ophiocomina nigra* in the feeding posture (from Fontaine 1965)

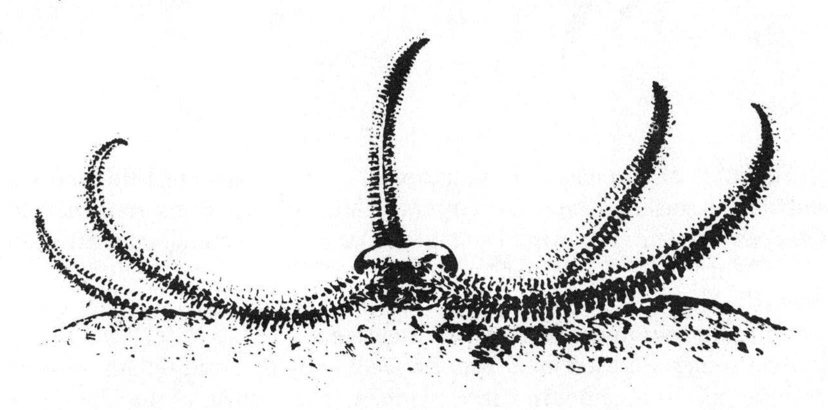

between the arm spines to capture particles (Fontaine 1965). The burrowing amphiurid *Amphiura filiformis* raises its arms above the surface into the water column when there is a current and extends the tube feet (Woodley 1975). Because organic detritus is more easily suspended than particulate substrata, this increases the capture of organic material above that obtained when feeding directly from the substratum surface. The tube feet of *Ophiothrix fragilis* are extended to between two to three times the length of the distal arm spines (Warner and Woodley 1975). In high currents the tube feet project stiffly from either side of the ambulacra in two comb-like rows, but in low currents a four-way orientation occurs in the distal part of the arm so that alternate tube feet are deflected aborally. These changes in tube foot posture are identical to those of crinoids under similar differences in current velocity. The tube feet are also extended into the current flow in the suspension-feeding *Ophiactis resiliens* (Figure 2.13) (Pentreath 1970).

Figure 2.13: (a) Mid-arm of *Ophiactis resiliens* showing arrangement of spines (Sp) and tube feet (Pod). (b) Tube feet held into the current (arrows). (From Pentreath 1970)

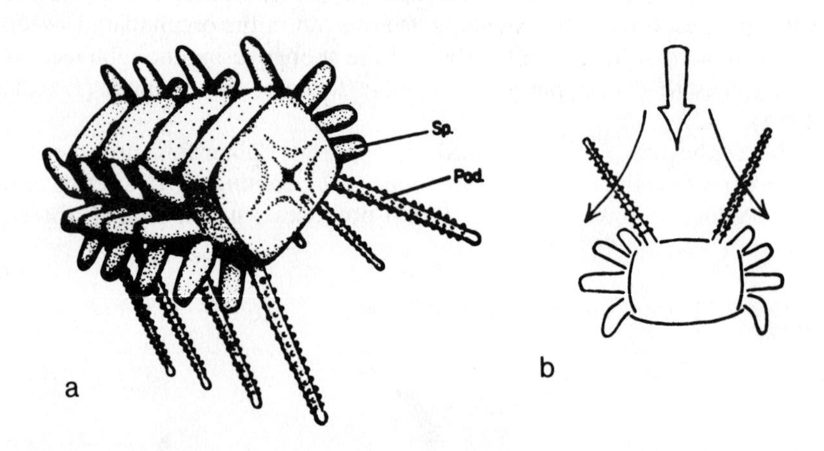

The arms of suspension-feeding ophiurid ophiuroids are held with the ambulacral surface facing the current. Those of *Ophiactis resiliens* and *Ophiothrix fragilis* are twisted so that this posture is maintained (Pentreath 1970; Warner 1971; Warner and Woodley 1975) if the current changes. Similarly, this orientation may be possible because the food is not transferred to the mouth by ciliary–mucoid currents.

Arm spines are used in suspension feeding by ophiurid ophiuroids in a distinct contrast to crinoids. Particles adhere to the mucus of the arm spines of *Ophiopteris antipodum* (Pentreath 1970). *Ophiocomina nigra* produces sufficient mucus to form a mucous net of strands between adjacent strands (Fontaine 1965). Although the arm spines of *Ophiocoma* species capture particles, no mucous net occurs (Sides and Woodley 1985). The capture of prey by these species, which hold their arms vertically into the water column, is facilitated by their occurrence in extremely dense populations (Figure 2.14) (Warner 1971). This has a baffling effect and causes particles to settle out. The arms of *Ophiotholia* can be raised vertically to form a palisade around the disc and place the parasol-shaped hooks into the water column (Figure 2.15) (Lyman 1877).

The tube feet of *Ophiopteris antipodum* (Pentreath 1970) and *Ophiocoma wendti* (Sides and Woodley 1985) are used to remove particles that adhere to their spines (Figure 2.16). In *Ophionereis fasciata*, the tube foot curls upon itself to compact the adhering particles and the mass is then scraped off on to the tentacle scale at the base of the tube foot (Figure 2.17) (Pentreath 1970). Large masses are formed by combining the accumulations of adjacent tube feet. In *Ophiocomina nigra*, particles accumulated at the base of the tube feet are compacted into boluses by the tips of the tube feet (Fontaine 1965).

Figure 2.14: Aggregation of *Ophiothrix fragilis* (*c.* 2000 individuals.m^{-2}) with arms extended into the water column in the feeding posture (from Brun 1969)

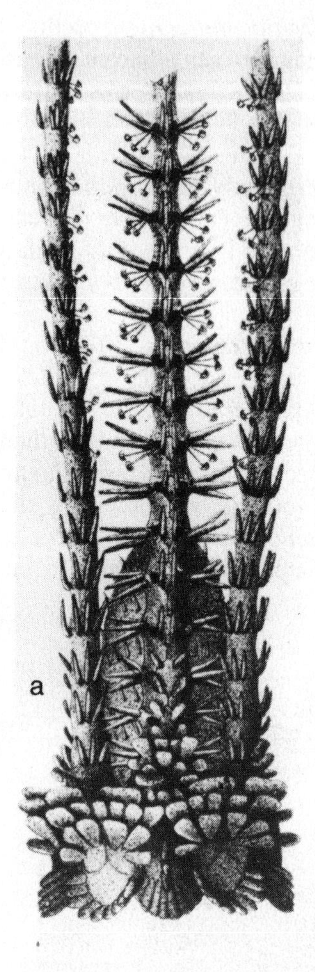

Figure 2.15: (a) Arms of *Ophiotholia* with spine hooks. (b) Enlargement of portion of arm showing tube feet, spines and spine hooks. (From Lyman 1877)

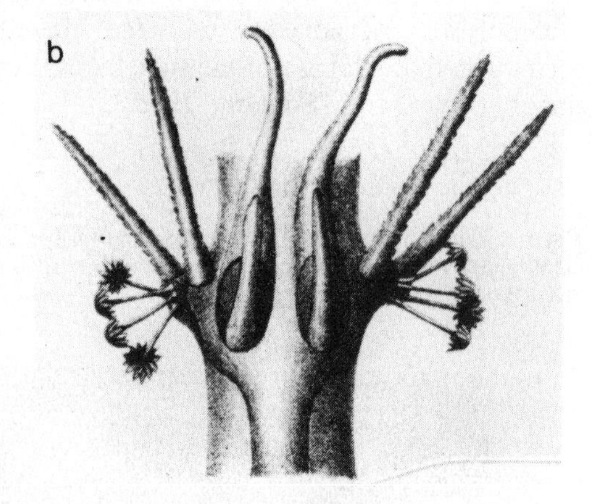

Of all the ophiuroids studied by Gislén (1924) only *Ophiopsila* had cilia on the undersurface of the arms. These cilia form transverse bands, but Gislén stated that they produce a current towards the mouth. In the ophiuroids that lack such cilia, the boluses are not moved towards the mouth by ciliary–mucoid strands as is usual in suspension feeders, but by the tube feet. The boluses formed by *Ophiocomina nigra* are moved by the tips of the tube feet, with periodic halts along the way for the addition of more particles (Fontaine 1965). Some ciliary currents do occur in the proximal part of the arm which aid in transfer. *Ophiothrix fragilis* uses the entire tube foot to grasp the bolus and roll it along the ambulacrum (Figure 2.18) (Warner and Woodley 1975). A similar transfer occurs in *Ophiocoma wendti*, the number of tube feet involved depending on the size

Figure 2.16: A: Tube feet of *Ophiopteris antipodum* wiping spines by stroking spine from tip to base (a) and then straightening to curl upon itself (b, c, d) to compact the particles (e) (from Pentreath 1970). B: Tube feet of *Ophiocoma wendti* wiping spines in groups (a) or singly (b) (from Sides and Woodley 1985)

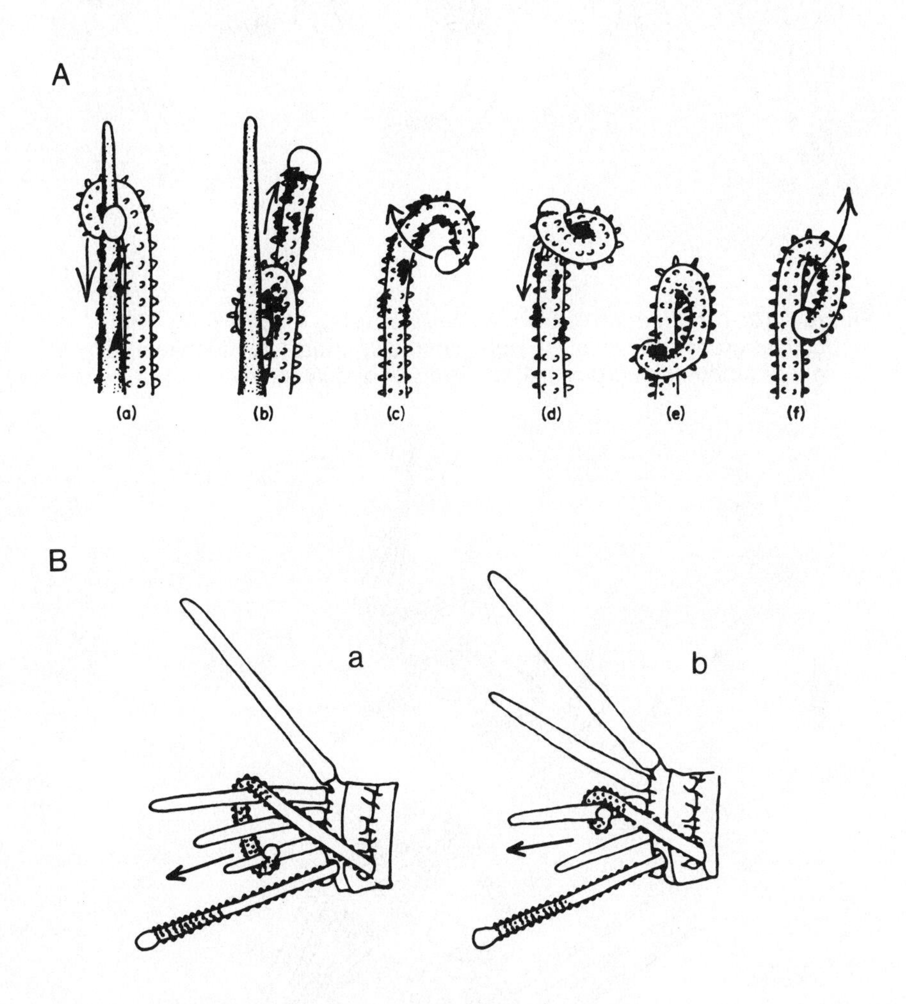

of the bolus (Sides and Woodley 1985). The tube feet of *Ophiactis resiliens* transfer the particles adhering to them by a sequential collapsing action in a proximal direction, so that the tube feet are rubbed against each other and particles are concentrated on the proximal ones (Figure 2.19) (Pentreath 1970). In this instance, boluses are formed only on the most proximal tube feet.

The gut of ophiuroids is a sac in the disc. The lack of an anus has not

Figure 2.17: (a) Tube foot (tf) of *Ophionereis fasciata* curling upon itself (b) to compact particles into a mass that is transferred to the tentacle scale (ts) by scraping (e) (from Pentreath 1970)

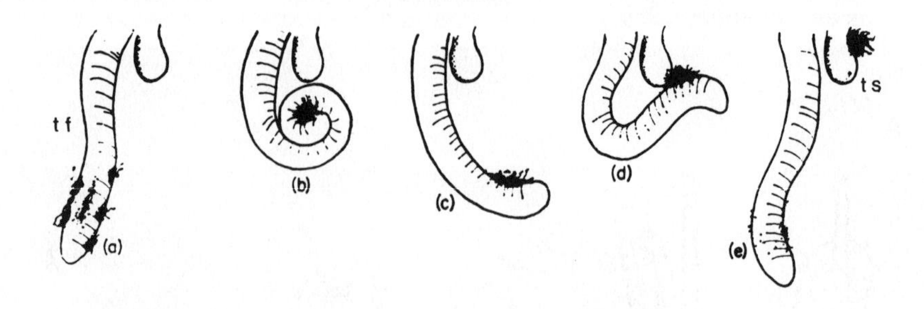

Figure 2.18: Tube feet of *Ophiothrix fragilis* passing a bolus down the ambulacral groove (from left to right, spines omitted). A: Side view, B: view of ambulacral groove. (From Warner and Woodley 1975)

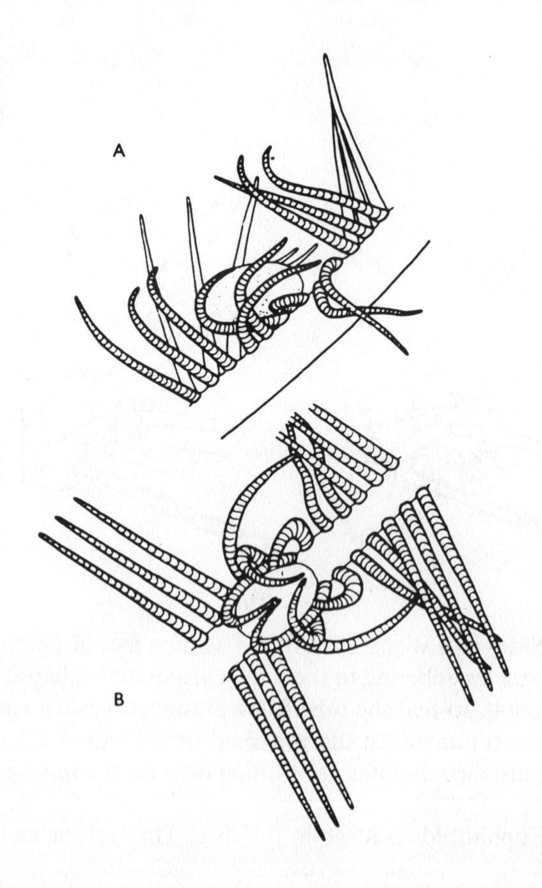

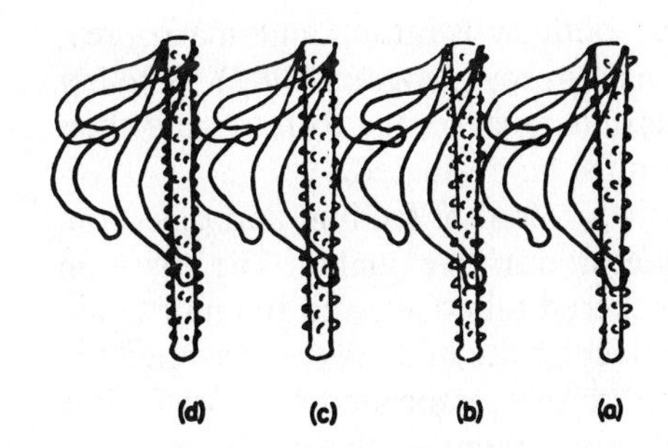

Figure 2.19: Tube feet of *Ophiactis resiliens* collapsing in sequence (a to d). As each tube foot collapses, it bends proximally and rubs against the adjacent tube foot which in turn collapses. The sequence of collapse is towards the mouth resulting in an accumulation of particles on the more proximal tube feet. (From Pentreath 1970)

been satisfactorily explained, but associated with the restriction of the perivisceral coelom to the disc has major consequences for digestion and, consequently, for material and energy acquisition. Because of these characters, regional specialisation is limited and digestive efficiency is probably low. Ingestion may be continuous in microphagous feeders, but food in various states of digestion would be mixed, and incompletely digested food could be egested along with non-digestible material. Macrophagous feeders are intermittent in feeding, and seem to have periods of ingestion, digestion and egestion. They may ingest so much food when it is abundant that subsequent regurgitation occurs (Hendler 1982). These ophiuroids probably feed until the gut is full and then undergo a digestive period. Even though digestion may be complete, particularly that of animal prey, the uptake of material and energy is probably inefficient. Two consequences of the development of the interradial pouches are a space conflict with the interradial gonads as well as the shortening of the nutrient transfer route from the gut to the gonads.

2.1.3.2. Asteroidea

Feeding by the asteroids is profoundly affected by the absence of an arm capable of rapid movement or appropriate for suspension feeding. This is because the skeletal support is not internal, but in the body wall. In a few groups the arms are attenuated and are raised into the water column for feeding. However, by contrast with crinoids and some ophiuroids, holothuroids and echinoids, asteroids generally cannot use water currents to bring food to them and must move in order to obtain food. Consequently, optimal foraging behaviour is an important aspect of their feeding biology. Except in these groups with the arms specialised for feeding, it is not possible to segregate clearly the allocation of energy to a body component for food collection from that of locomotion.

Most of the earliest asteroids are thought to have been microphagous feeders on particulate substrata (Spencer 1951). A major evolutionary development in the asteroids was the change from particulate food to macroprey, leading to a complete forsaking of ingestion of sediment.

Within the paxillosids, luidiids ingest both substratum and macroprey, whereas astropectens primarily ingest macroprey (Jangoux 1982). Whether ingesting substrata or macroprey, asteroids must forage to feed, and their functional morphology and behaviour must be evaluated.

The change to macroprey still involves intraoral feeding and digestion, but involves consumption of food higher in nutritive quality. The ingestion of macroprey thus does not require suckered tube feet, and the nutritional significance of a lack of suckered tube feet by the luidiids and astropectens is simply that these groups are restricted to prey associated with particulate substrata. The exploitation of solid substrata required the development of suckered tube feet. Although locomotion on a hard substratum is possible without suckered tube feet, asteroids cannot maintain position in currents or move up a slope without them.

Another major evolutionary advance associated with feeding in asteroids was the development of extraoral feeding (Jangoux 1982). In extraoral feeding, the stomach is extruded out of the mouth and at least the initial phases of digestion occur outside the body. The transition from intraoral feeding to extraoral feeding opened up many new food resources. Examples of these are surfaces of the substrata (hard or soft) or of prey too large to be ingested. Extraoral feeding eliminates the utilisation of body space to hold food but still restricts digestion to a surface phenomenon as in the intraoral feeders.

The ingestion of particles by the paxillosids (e.g. *Luidia, Astropecten, Ctenodiscus*), which live on soft substrata, is simply a matter of opening the mouth and using the tube feet to push in the sediment or infauna. Analysis of the particles found in the substratum and in the gut of *Ctenodiscus crispatus* indicates that there is no selection of particles in this process (Shick, Edwards and Dearborn 1981); the tube feet seemingly cannot discriminate. This should mean that the composition of the substratum is sufficiently homogeneous to eliminate the potential for selection in the microenvironment, although the possibility of selection of the particular locale exists. Consequently, *Ctenodiscus* might spend energy locating areas high in nutrient level but not in subsequent selection during ingestion.

Capture of particles by ciliary–mucoid and suspension feeding has been suggested, primarily on the basis of the demonstration of external ciliary currents (e.g. Gemmill 1915; Gislén 1924). *Henricia sanguinolenta* can remove suspended prey at experimentally high concentrations, and in the field extends its arms into the water (Rasmussen 1965). *Henricia leviuscula* only occasionally assumes this position in the field (Mauzey, Birkeland and Dayton 1968), suggesting that this mode of feeding may not usually be important. The cribriform organs (Figure 3.8, 3.9) of cribellinids have been considered to produce currents for ciliary–mucoid feeding (Gislén 1924), but seemingly may be important in feeding in *C. crispatus* only in recycling mucus (Shick *et al.* 1981).

Feeding on macroprey opens the potential for optimal diets, in which the costs of foraging and handling the prey affect the energy return (Townsend and Hughes 1981). The costs of foraging involve the availability of the prey. *Luidia clathrata* feeds disproportionately on more abundant prey, and will change preferences even if the prey differ qualitatively (McClintock and Lawrence 1985). Such 'ingestive conditioning', which leads to increased consumption of available prey, also occurs in *Pisaster ochraceus* (Landenberger 1968), *Asterias rubens* (Castilla 1972), and *Acanthaster planci* (Ormond, Handscomb and Beach 1976). As appropriate for a predator, *L. clathrata* is chemosensory to low concentrations of low-molecular-weight compounds, more to amino acids than sugars (McClintock, Klinger and Lawrence 1984). This qualitative and quantitative discrimination would be an important component of search behaviour and food discrimination which *L. clathrata* as well as other asteroids show in the field. Foraging movements by *L. clathrata* are directional until patches of prey are located (McClintock and Lawrence 1985). Similarly, *Astropecten auranciacus* moves directionally but with periodic changes in direction (Burla, Ferlin, Pabst and Ribi 1972) until prey patches are reached (Jost 1982).

Ingestion of macroprey occurs primarily in the primitive paxillosids. The tube feet of *Astropecten irregularis* push food into the mouth (Christensen 1970). *Luidia foliolata* digs into the substratum to capture prey (Figure 2.20) (Mauzey *et al.* 1968). When feeding on the surface, *Luidia clathrata* uses both the tube feet and dorso-ventral bending of the arms to transfer macroprey to the mouth.

Figure 2.20: Digging by asteroids in search of prey (from Sloan and Robinson 1983)

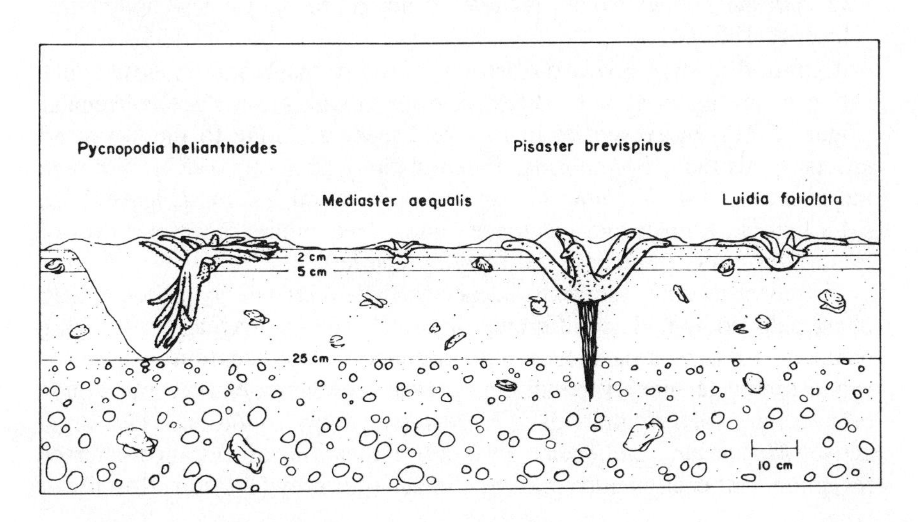

The mouths of paxillosids are capable of considerable distension as very large, often irregularly shaped prey are ingested. *Luidia clathrata*, with an arm length of 140 mm, can ingest a sand dollar with a diameter of 50 mm (Downey and Wellington 1978). However, size selection of prey occurs. Adult *L. clathrata* choose smaller prey and ingest them more rapidly, presumably because these prey are more easily manipulated and can be packed more tightly in the stomach (McClintock and Lawrence 1985). This would indicate that the cost–benefit ratio of manipulating and ingesting large prey items is high relative to that of ingesting small prey. In addition, *L. clathrata* increases the number of prey ingested and the period of activity with an increase in prey density. This suggests an optimisation strategy to increase food acquisition in appropriate areas. Starved *L. clathrata* ingest greater amounts of prey and spend more time foraging than fed individuals.

The paxillosids *Astropecten bispinosus* and *A. auranciacus* overlap at some depths in resource-limited areas of the Mediterranean (Ribi, Schärer and Ochsner 1977). Although their prey are the same, the larger *A. auranciacus* consumes larger-sized prey. This would not completely reduce competition between the two species, because the consumption of juvenile prey by *A. bispinosus* would reduce availability of adult prey for *A. auranciacus*, and consumption of adult prey by *A. auranciacus* would reduce the production of new juveniles for prey of *A. bispinosus*.

Ingestion of live macroprey also occurs in the asteriids, primarily but not exclusively in the multiarmed species. This is probably a consequence of the disc and mouth diameter increasing with the addition of arms. Thus *Pycnopodia helianthoides* ingests sea urchins intact with spines (Fisher 1928). The potential for ingestion is clearly greater for *Crossaster papposus* which has an expanded mouth diameter of several centimetres compared with *Asterias rubens* which has one of not more than a few millimetres (Hancock 1955).

Extraoral feeding involves extrusion of the stomach and its close application to the material to be digested, even though the surface is irregular (Figure 2.21). More species of extraoral feeders belong to the advanced groups (Valvatida, Spinulosida, Forcipulatida) (Jangoux 1982), although the paxillosid *Luidia clathrata* also shows extraoral feeding (Figure 2.22) (McClintock, Klinger and Lawrence 1983). This implies that the variety of food resources is greater for this form of feeding.

Extraoral-feeding asteroids occur in high-water-energy areas, where paxillosids do not. The difference in water energy may affect foraging behaviour. Whereas paxillosids have distance chemoreception in the field where low-water-energy conditions prevail, contact chemoreception may be necessary for forcipulatids in high-water-energy conditions. The large, active *Myenaster gelatinosus* gives no indication of distance chemoreception, but cruises until making contact with prey (Dayton, Rosenthal,

Figure 2.21: *Crossaster papposus* enclosing a mussel (from Hancock 1974)

Mahen and Antezana 1977).

Anderson (1954) proposed that extrusion of the stomach involved opening of the asteroid's mouth, relaxation of the stomach and the intrinsic and extrinsic retractor strands, and contraction of the muscles of the body wall to push out the stomach by hydrostatic pressure; retraction of the stomach involves relaxation of the muscles of the body wall to decrease hydrostatic pressure, contraction of the stomach wall muscles and of the retractor system, and closing the mouth. Hydrostatic pressure applied experimentally to *Asterias* first swells the dermal papillae and then becomes excessive and causes them to burst (Binyon 1980). Binyon stated that individuals with perforated body walls can evert their stomach. This difference in opinion about the role of hydrostatic pressure in extrusion of the stomach needs to be resolved.

The advance of this system over intraoral feeding is immense. Food

Figure 2.22: *Luidia clathrata* with everted stomach (from McClintock 1984)

resources are available by this mode of feeding which are not available with intraoral feeding. Extraoral feeders can be separated into two broad categories (Jangoux 1932). The first category is apparently comprised only of asteriid asteroids that can feed on large and well protected prey such as bivalve molluscs. The second category includes those asteroids that can feed upon surfaces with bacterial films or encrusting organisms, corals and other cnidarians, or echinoderms. In all of these categories, the prey cannot be physically ingested.

The basis for the divergence into feeding categories may be evolutionary pressures not directly involved with feeding. The valvatids include few genera that prey on active, solitary invertebrates, although members of other orders do (Blake 1983). The valvatids dominate tropical, shallow-water asteroid faunas where predation is high. Blake concluded that the protective structures of the asteroids restrict their predatory abilities.

The valvatid *Oreaster reticulatus* is an omnivorous microphagous grazer on sand and seagrass bottoms in the tropics, ingesting benthic micro-organisms and particulate detritus by extraoral eversion of the stomach (Scheibling 1980). Selection occurs by the asteroid's raking the upper layer of the particulate substratum, in which the microphytes are concentrated, into a mound that is then enveloped by the stomach. The degree of individual mound building varies directly with the chlorophyll concentration and grain size of the surface sediment. Foraging by *O. reticulatus* differs from that of *Luidia* and *Astropecten.* The paxillosids must search for prey that are discretely separated, either as individuals or as patches. In contrast, the distribution of the microphytes on which *O. reticulatus* feeds is relatively homogeneous. Consequently, the distance between mounds generally approximates the diameter of the asteroid, i.e. the minimum distance necessary to prevent overlapping of feeding sites (Scheibling 1981b). This optimises foraging efficiency. A directional foraging path prevents regrazing of substrata and extends the foraging range.

In the asteriids, the stomach may be inserted into thin slits no more than 0.1 mm wide (Feder 1955) or force can be used to produce a gap into which the stomach can be inserted (Christensen 1957). The process of forced entry varies with attached or unattached prey (Figure 2.23). Unattached bivalves are oriented so that the bivalve's hinge is opposite to the asteroid's mouth; with attached bivalves, the asteroid's mouth is placed over the free-moving valve. The asteroid assumes a humped posture over the bivalve and the tube feet elongate and then retract. Asteriids are adapted for forcible entry because their tube feet are arranged in four rows rather than in two as in most other extraoral feeders (Figure 3.77). The use of force exerted by the tube feet in this matter would require a rigid body (Christensen 1957). The mechanism to convert the flexible body wall of *Asterias forbesi* to a rigid one seems to be met by a locking mechanism in part of the muscles in the reticular joints and by the collagenous catch mechanism in the connective tissue (Eylers 1976). The whole body acts as

Figure 2.23: Attachment of tube feet of *Evasterias troschelii* to the valves of (A) unattached and (B) attached bivalves. a: Attachment, b: stretching, c: pulling. (From Christensen 1957)

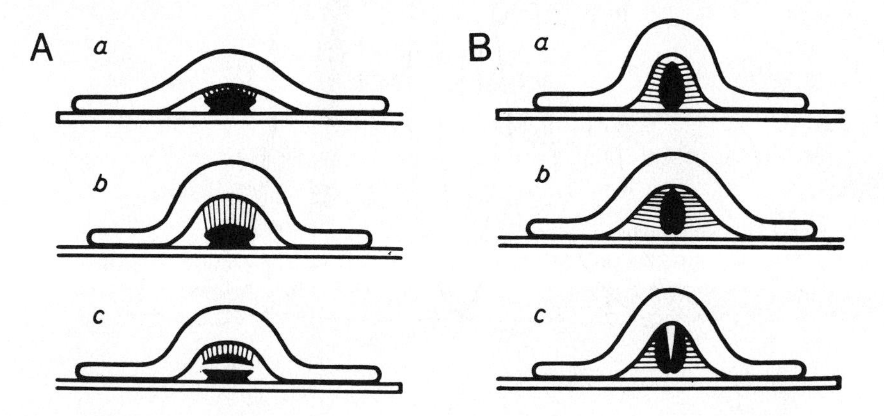

a unit that sustains compressive forces across its undersurface and tensile stresses across its upper surface (Figure 2.24). Only the forcipulatids have an ambulacral mouth frame (Figure 2.25) which supports the centre of the dome in the process.

Only a slight gape of a mussel's valves is necessary for the stomach of *Pisaster ochraceus* to be inserted (Feder 1955). However, penetrating the gape can contribute greatly to the cost of feeding. Increasing the gape of mussels significantly reduces the time required for *P. ochraceus* to insert its stomach, open the valves and digest the mussel (McClintock and Robnett 1986).

The size of the everted stomach can vary considerably, being very large in *Acanthaster* and limited to the area of the mouth in *Henricia* (Jangoux 1982). The size of the individual and that of the stomach that can be extruded can both have an effect. This can result in interaction in co-occurring species, such as *Leptasterias hexactis* and *Pisaster ochraceus* in the north-east Pacific Ocean. Medium- to large-sized prey are eaten by the larger *P. ochraceus*, larger than those eaten by *L. hexactis* (Menge 1972). When *P. ochraceus* is absent, *L. hexactis* begins to ingest larger prey. In

Figure 2.24: Compressive (→ ←) and tensile (⟷) stresses on the ambulacral and adambulacral ossicles and on the mid-dorsal ridge ossicles of *Asterias forbesi* opening a mussel. The body wall has been removed from the right arm to show the ossicles. (From Eylers 1976)

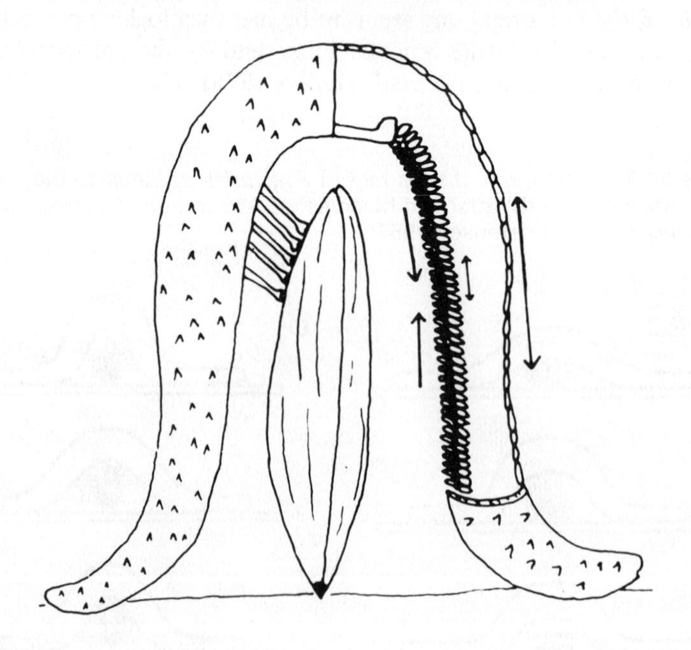

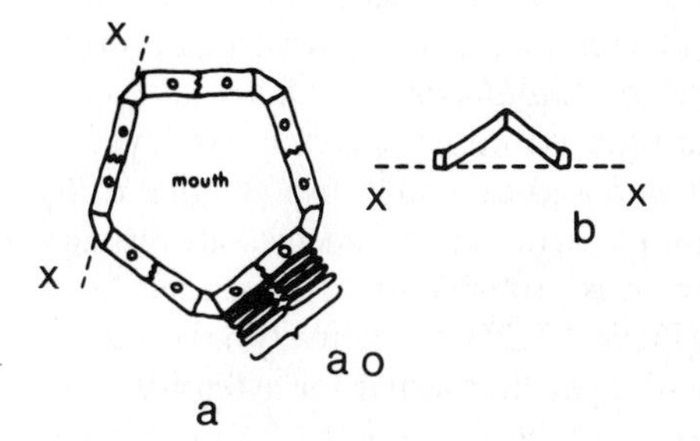

Figure 2.25: (a) Ambulacral mouth frame of *Asterias forbesi.* Ambulacral ossicles (ao) abutting the mouth frame. (b) View of ambulacral ossicles (x---x). (From Eylers 1976)

contrast, *Asterias vulgaris* and *Asterias forbesi* co-occur on the north-eastern North American coast. Both species overlap greatly in the time and intensity of feeding, body size, diet composition, and size of prey consumed (Menge 1979). These overlaps are interpreted as resulting from the occurrence of the two species in densities below carrying capacity so that resource limitation would not be a strong selective pressure.

Because *Acanthaster planci* extrudes its stomach to approximately the same extent at all sizes, the amount of coral that can be covered is proportional to the asteroid's diameter (d) (Lucas 1984). Lucas found that the increase in body mass with growth exceeds the amount of food gained per meal by a factor of $d^{0.32}$. He concluded that the feeding capacity is greatly reduced with the great increase in body size (0.5 to more than 300 mm diameter) and that other factors such as frequency of feeding and quality of food must make up the difference. Lucas proposed that this may be the basis for determinative growth.

The rate of consumption of mussels by *Marthasterias glacialis* increases only slightly with body size, but the size of mussel consumed increases to an extent that energy intake is a relatively constant proportion of the asteroid body size (Penney and Griffiths 1984). The energetic costs to *Asterias forbesi* of opening mussels is such that there is an optimal mussel size which varies with asteroid body size (D.B. Campbell 1984). The optimal mussel size is preferred in feeding except for the largest *A. forbesi.* When simultaneously presented with equal numbers of different sizes of clumps of mussels, *Pisaster ochraceus* chose medium sizes of prey, the choice of size being positively correlated with predator size (McClintock and Robnett 1986). When presented with different sizes of clumped mussels with the gape artificially widened, medium-sized mussels were still chosen, suggesting that the stomach insertion, valve opening and digestion time were not the important constraints determining size-selective predation. Because the force required for *P. ochraceus* to dislodge mussels increases exponentially with mussel size, the constraint involved in the trade-off between energy-maximising and time-minimising strategies may be the force required to dislodge mussels.

Capturing prey by use of the arm is rare in asteroids because the broad arm, its internal skeleton, and lack of muscular development make rapid movement difficult. However, the arm of *Labidiaster annulatus* has a high degree of flexibility associated with a reduced upper skeleton. Its rays are capable of considerable and rapid torsion and can catch fish (Figure 2.26) (Dearborn 1977). Several of the multiple arms of this species are usually elevated and sweep through the water. It is probable that the slender distal portions of the arms of brisingids (Figure 2.27) are actively involved in food capture. These arms have spines longer than usual for asteroids. The tips of the spines are enlarged into blunt hooks (Figure 2.28) (Sladen 1889) and may be used in capturing suspended food.

Figure 2.26: *Labidiaster annulatus* with arms elevated in the feeding posture. (Smithsonian Oceanographic Sorting Center, National Science Foundation)

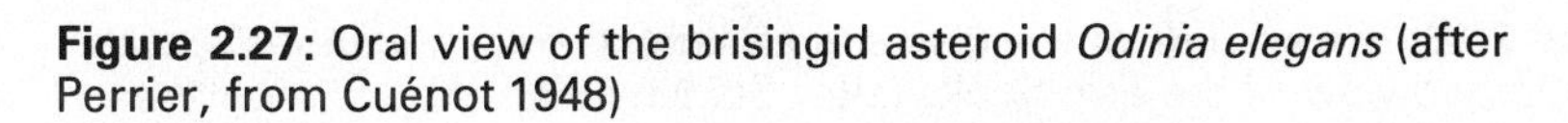
Figure 2.27: Oral view of the brisingid asteroid *Odinia elegans* (after Perrier, from Cuénot 1948)

Capture of macroscopic prey buried in soft substrata can involve the tube feet. The circumoral tube feet of *Pisaster brevispinus* can be extended to 17 cm by large individuals, and prey items in at least the upper 15 cm of substratum are susceptible to being captured and brought to the surface (Figure 2.29) (Van Veldhuizen and Phillips 1978). More often, capture by the tube feet involves digging pits so that the mouth, stomach or tube feet come into contact with the prey. This is done not only by paxillosids but also by asteroids with suckered tube feet (Birkeland 1974; Mauzey *et al.* 1968; Sloan and Robinson 1983). The depth to which the burrows are dug and the posture of the feeding asteroid varies (Figure 2.20). This mode of feeding should increase considerably the cost of obtaining food in terms of time and effort over that of feeding on epibenthic prey.

2.1.4. Echinoidea

Echinoid evolution involved the development of the Aristotle's lantern as a means of ingesting food. Material and energy must be allocated to the

Figure 2.28: Upper view of the disc and an arm of *Freyella dimorpha,* with an enlarged view of the arm hooks (after Sladen 1889)

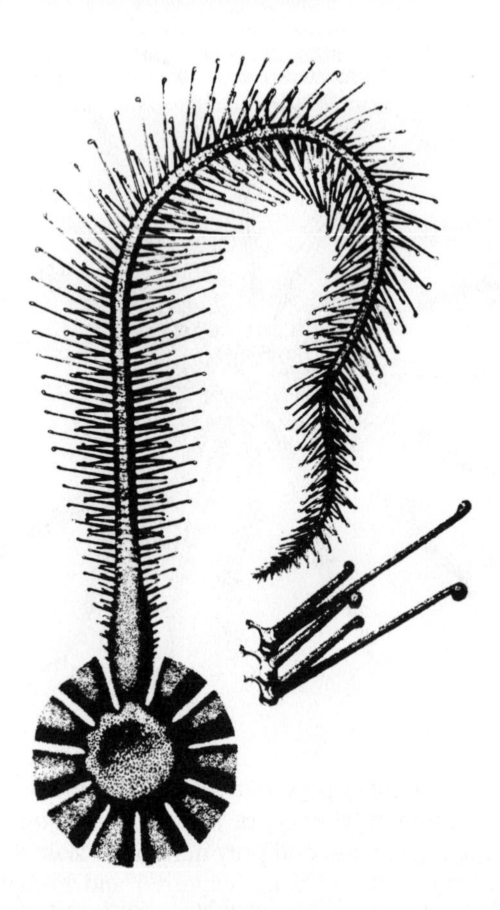

development of this food-gathering component in all echinoids except the atelostomatids. The amount can represent a considerable portion of the total energy of the body (Table 2.3). It is greatest in juveniles and shows considerable variation among species.

The basic functioning of the lantern is the spreading apart and then bringing together of the tips of the five teeth. This requires the support of the lantern in the peristome and muscles to protrude and retract the lantern and to move the teeth. The early lantern was simple and weak and is thought to have functioned only as a scoop and in biting (A.B. Smith 1984). Because the lantern could be used as a scoop, it is possible that the early echinoids lived on soft substrata and ingested it as echinothurids do today (see Lawrence 1975). Alternatively, the early echinoids may have developed suckered tube feet and lived on hard substrata and used the

Figure 2.29: Oral view of the forcipulate *Pisaster brevispinus* showing extension of the tube feet (arrow). The sigmoidal configuration of the arms indicates that the disc had been in an excavated pit (from Van Veldhuizen and Phillips 1978)

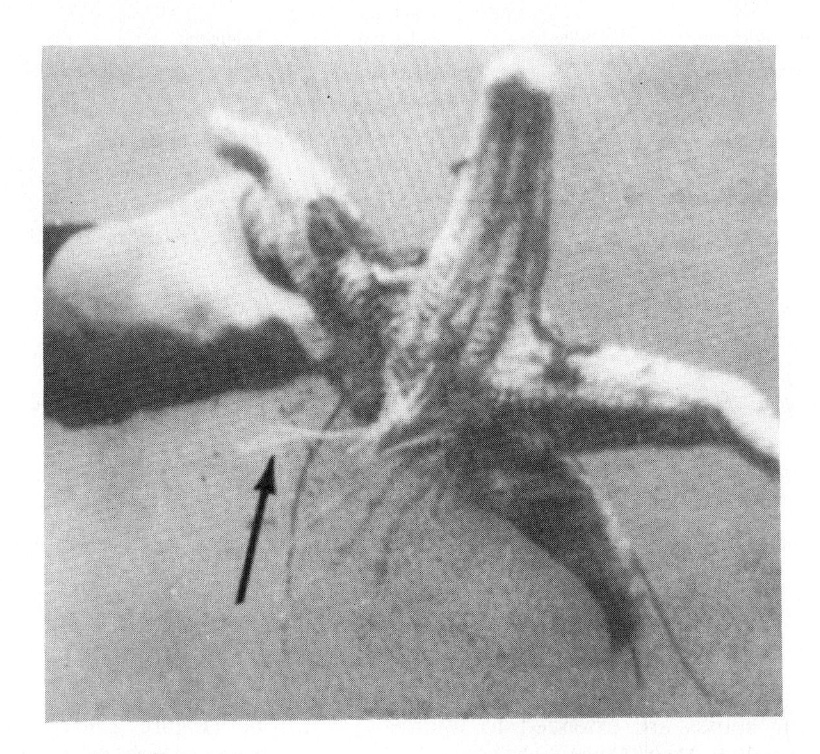

lantern for biting. The evolutionary history of the regular echinoids involved the development of the lantern into a stronger and more effective apparatus for feeding.

The body form and structure of echinoids had potential for being adapted to much different ways of life and feeding types. Most of the differences in body form and function between the regular and irregular echinoids are related to the differences in mode of feeding (Kier 1982). No other extant class of echinoderm has shown such great modification of their basic body form and function.

2.1.4.1. Regular echinoids

The spines and tube feet are both involved in the capture of drift food (see Lawrence 1975). It is difficult to separate adaptations of the spines for food capture distinct from their other functions, but effectiveness in feeding is primarily a function of spine length. *Echinostrephus molaris* is dependent on drift material for food. It excavates a cylindrical burrow from which its

Table 2.3: Energy allocated to the test (T), spines (S) and Aristotle's lantern (AL) of echinoids. Values in parentheses are percentages of the total

Species	Body part	Energy (kJ. ind^{-1})	Reference
Abatus cordatus	T	4.0 (82)	J.M. Lawrence and A. Guille unpublished
	S	0.9 (18)	
	total	4.9	
Colobocentrotus atratus	T	10.9 (47)	J.M. Lawrence unpublished
	S	11.3 (49)	
	AL	0.8 (3)	
	total	23.0	
Echinometra lucunter	T & S	79.5 (85)	J.M. Lawrence unpublished
	AL	14.2 (15)	
	total	93.7	
Echinometra mathaei	T	134.3 (78)	J.M. Lawrence unpublished
	S	31.0 (18)	
	AL	6.2 (4)	
	total	171.5	
Echinostrephus aciculatus	T & S	4.6 (74)	J.M. Lawrence unpublished
	AL	1.6 (26)	
	total	6.2	
Echinothrix diadema	T & S	17.2 (58)	J.M. Lawrence unpublished
	AL	12.6 (42)	
	total	29.8	
Sterechinus diadema	T	34.7 (61)	J.M. Lawrence and A. Guille unpublished
	S	15.5 (27)	
	AL	6.3 (11)	
	total	56.5	

long spines are extended to intercept drift food (Figure 2.30) (A.C. Campbell, Dart, Head and Ormond 1973). Other species such as *Arbacia, Strongylocentrotus, Psammechinus, Paracentrotus* and *Echinometra* catch plant drift and have spines several centimetres in length. By contrast, *Tripneustes* and *Lytechinus* have short spines and use the tube feet primarily for feeding. The spines around the peristome may be spatulate and are probably used to manipulate the food to the mouth. The actual contribution of the drift food to the total amount consumed has not been documented.

The tube feet can also be important in firmly anchoring the individual while rasping hard substrata. The strength of attachment is determined by the tube foot strength and the number of tube feet (A.B. Smith 1981). Phyllodes, areas of enlarged or increased numbers of tube feet in the adoral portion of the test (Figure 2.31), occur in many regular echinoids and increase the ability to counteract the stress of the lantern in rasping food (Kier 1974). Regular echinoids lacking oral suckered tube feet, such as the deep-sea *Phormosoma placenta* which feeds on mud, are probably unable to rasp food.

Figure 2.30: *Echinostrephus molaris* at the entrance to its burrow, collecting algal particles with the spines and tube feet (from A.C. Campbell *et al.* 1973)

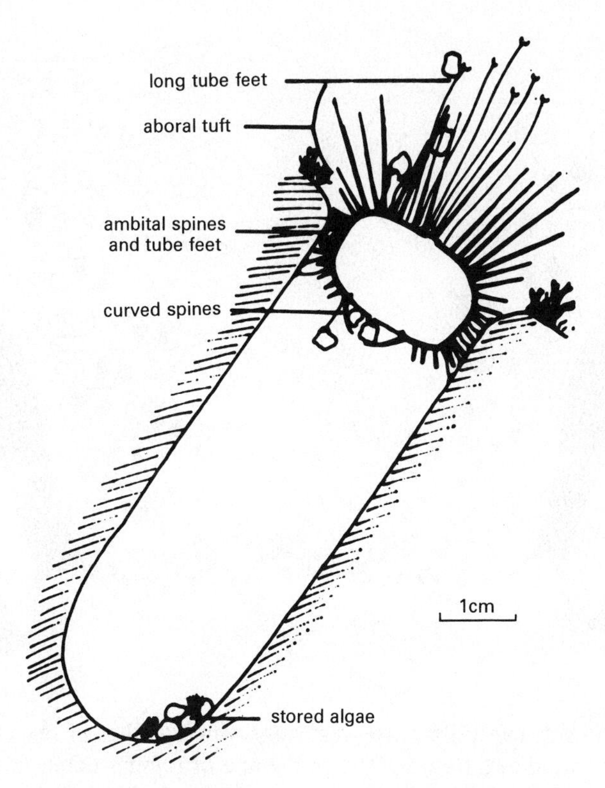

The post-Palaeozoic adaptive radiation of regular echinoids involved major changes in the structure of the lantern which increased its strength and ability to function (see De Ridder and Lawrence 1982; Kier 1974; A.B. Smith 1984). The pyramids of the primitive cidaroids are complete, whereas there is a progressively deeper notch (the foramen magnum) in the advanced euechinoids making the structure lighter (Figure 2.32) (A.B. Smith 1984). Progressive changes in the attachment of the muscles to the test improved its functioning (Kier 1974). The cidaroids developed projections, apophyses, from the interambulacral plates to which both the retractor and protractor muscles attached (Figure 2.33). This perignathic girdle developed further in the euechinoids with the formation of projections, auricles, from the ambulacral plates as well. The protractor muscles remained attached to the apophyses, but the retractor muscles became attached to the auricles. This spread apart the attachment points of the retractor muscles and increased the mobility of the lantern. The lantern

Figure 2.31: Adoral part of ambulacrum of *Toxopneustes pileolus* showing phyllode and adjoining gill slits (from Mortensen 1943a)

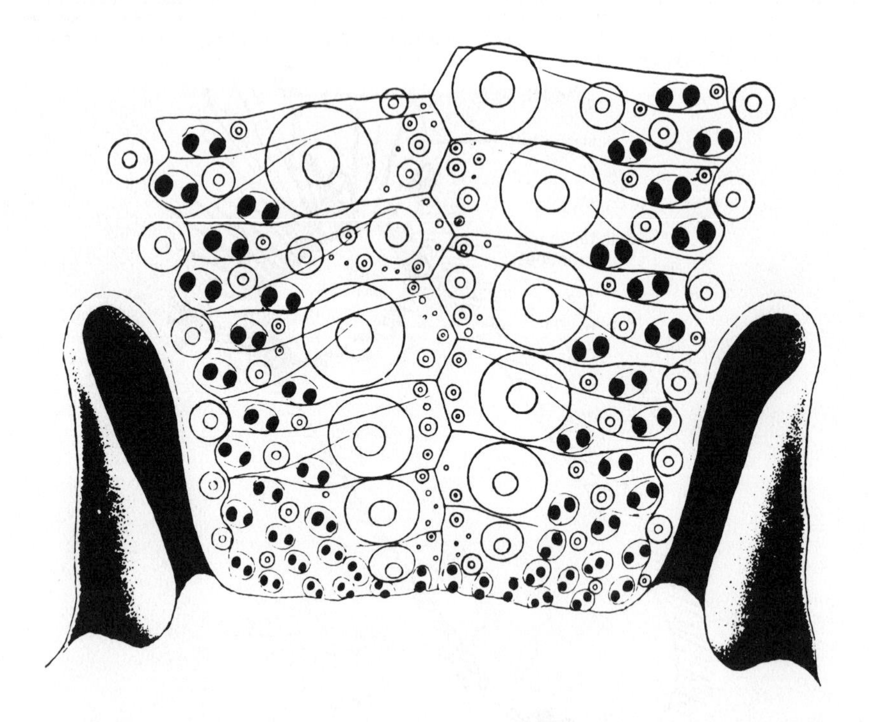

of the cidaroids can only move vertically, whereas that of the euechinoids can be extended at an angle. The peristome of regular echinoids is usually large and has a flexible peristomial membrane to facilitate protrusion of the lantern.

The teeth of regular echinoids are subject to bending stress when they scrape over a hard substatum. The teeth scrape inwardly only, so the direction of stress is always the same. The teeth have a shape and a complex substructure that meets the compressive and tensile forces resulting from the stress (Märkel, Gorny and Abraham 1977). The teeth are either U-shaped (aulodont: cidaroid, echinothuriid and diadematoid), T-shaped (stirodont: arbacioid, camarodont: echinoid), or wedge or diamond-shaped (clypeasteroid) (Figure 2.34). The abaxial part of the tooth (the flange: the bottom of the U, the cross-bar of the T) resists the compression, and the adaxial part (the keel of both) resists the tension. The tooth is composed of two rows of stacked calcareous elements (Figure 2.35). The elements are laminated together in the shaft by polycrystalline calcareous discs to form a strong composite material. The flange, which resists compression, contains only primary plates. The keel, which resists tension, contains the long prisms. The tensile fibres extend into the compression zone and are

Figure 2.32: Possible evolution of the Aristotle's lantern (centre), epiphysis (left) and tooth cross-section (right). A: Archaeocidaroid (*Archaeocidaris*), B: cidaroid (*Eucidaris*), C: phormosomid (*Phormosoma*), D: echinothuriid (*Calveriosoma*), E: early 'pedinoid' (*Diademopsis*), F: diadematoid (*Diadema*), G: stirodont (*Arbacia*), H: camarodont (*Strongylocentrotus*), I: *Eodiadema*, J: holectypoid (*Holectypus*), K: cassiduloid (*Echinolampas*), L: clypeasteroid (*Echinocyamus*). (From A.B. Smith 1981)

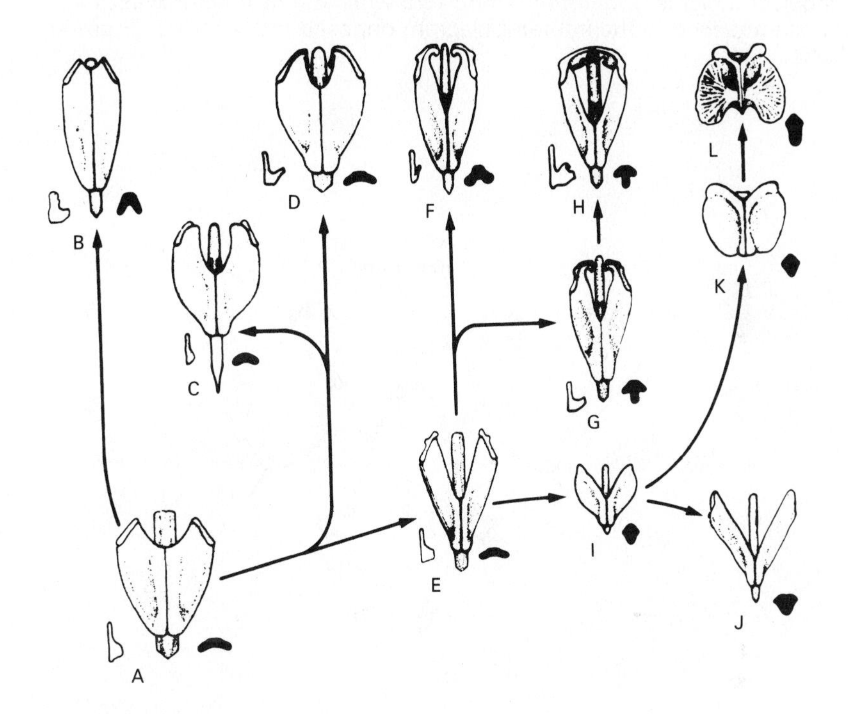

fixed to the primary plates by the calcareous discs.

Although the toughness of the teeth varies with species (Märkel *et al.* 1977), the scratch hardness and indentation hardness do not (Klinger and Lawrence 1985). This is true whether the teeth are U-shaped or T-shaped. The elements of the chewing part of the tooth can be chipped off in the rasping process; continual proliferation of calcium-depositing cells (odontoblasts) in the plumula maintains the tooth (Märkel *et al.* 1977). Echinothuriids ingest soft substratum. Their lantern functions as a scoop and the teeth are exposed to a tensile stress but not a bending stress. As a consequence, the primary plates are small and the prisms are flattened (Märkel 1970).

The ultimate criterion for evaluating the effectiveness of the echinoid feeding apparatus is the rate at which feeding occurs. The feeding rate is affected by physical (salinity, temperature, light) and biological (size and physiological state of the individual, food preferences, quality and quantity

Figure 2.33: Positions for the attachment of the protractor (black) and retractor (stippled) muscles of the Aristotle's lantern to the test. In cidaroids (A) the retractor muscles that attach to the same pyramid come from paired processes of the interambulacrum directly opposite the pyramid. The protractor muscles come from the same processes. In the non-cidaroid and Mesozoic irregular echinoids (B) the retractor muscles from the same pyramid are attached to processes that are outgrowths of ambulacral plates of different ambulacra while the protractor muscles remain attached to the interambulacrum opposite the pyramid. (From Kier 1974)

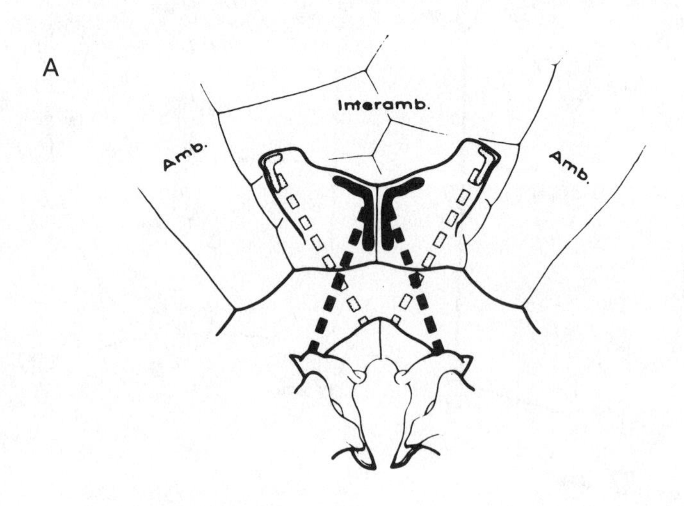

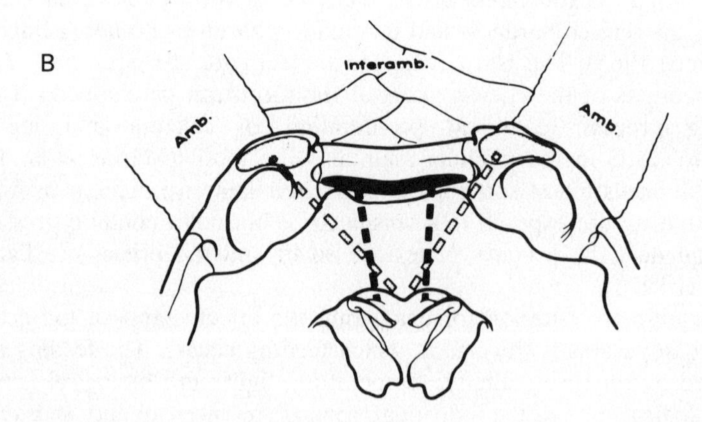

Figure 2.34: Possible evolution of the tooth (cross-sections). A: *Diademopsis*, B: cidaroid (*Stylocidaris*), C: diadematoid (*Centrostephanus*), D: stirodont (*Stomopneustes*), E: camarodont (*Paracentrotus*), F: *Eodiadema*, G: clypeasteroid (*Echinocyamus*), H: clypeasteroid (*Encope*). (From A.B. Smith 1981)

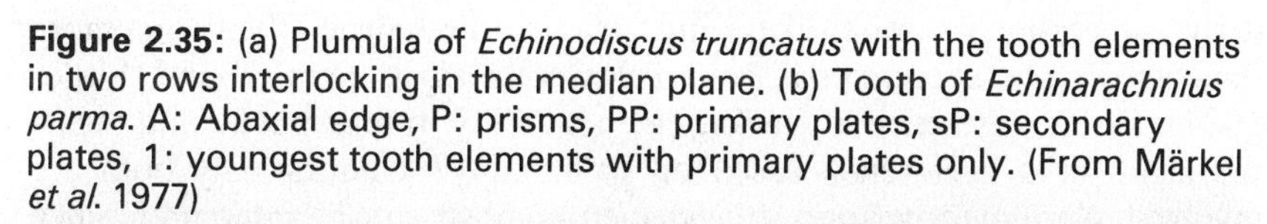

Figure 2.35: (a) Plumula of *Echinodiscus truncatus* with the tooth elements in two rows interlocking in the median plane. (b) Tooth of *Echinarachnius parma*. A: Abaxial edge, P: prisms, PP: primary plates, sP: secondary plates, 1: youngest tooth elements with primary plates only. (From Märkel *et al.* 1977)

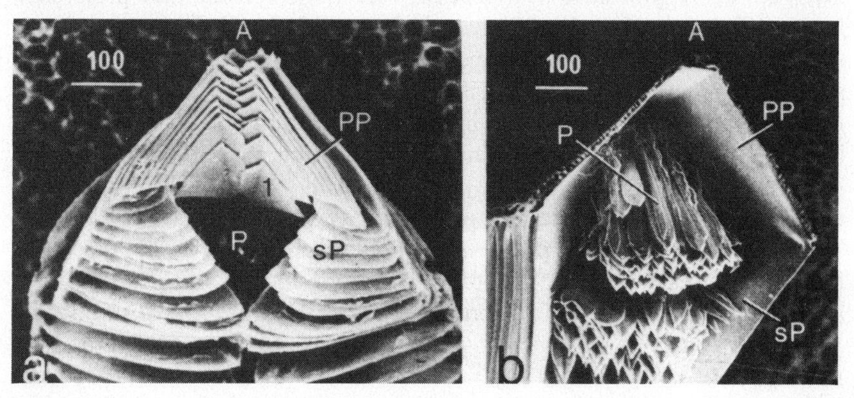

of food) variables (Lawrence 1975; Lawrence and Lane 1982). When the physical factors are held constant, feeding rates vary greatly with the algal food (Table 2.4). Conclusions on the causes of differences in feeding rates on foods in which several biological factors co-vary are difficult.

Table 2.4: *Strongylocentrotus intermedius.* Feeding rate (g wet weight ind^{-1} day^{-1}), absorption efficiency (%), somatic and gonadal growth (mg ind^{-1} day^{-1}), and net growth efficiency (%) of individuals fed different species of algae. (From Fuji 1967.)

Species	Feeding rate	Total absorption efficiency	Growth Somatic	Growth Gonadal	Net growth efficiency
Laminaria japonica	2.71	57	10	10	15
Agarum cribrosum	0.65	73	3	0.2	5
Alaria crassifolia	0.97	67	24	15	29
Sargassum tortile	0.60	63	7	7	17
Scytosiphon lomentaria	1.95	83	15	23	19
Ulva pertusa	0.45	74	12	6	28
Pachymeniopsis yendoi	0.66	68	0.4	7	10
Rhodymenia palmata	0.93	74	12	7	26

The effectiveness of the feeding apparatus may determine feeding rate (Himmelman and Carefoot 1975). Algae have been placed into functional-form groups that transcend phylogenetic and life-history affinities according to their external morphology, internal anatomy, texture and hypothetical abilities to persist in undisturbed 'mature' communities (Littler and Littler 1980). *Echinometra vanbrunti* feeds more on delicate, early successional, sheet-like, and filamentous algae than on more structured, late successional, and calcareous algae (Littler and Littler 1984). Is this difference in feeding rate an ability to manipulate and ingest the algae or a preference based on palatability? Undoubtedly both are involved. *Strongylocentrotus droebachiensis* shows considerable preference for specific algae (Larson, Vadas and Keser 1980). The use of uniform food prepared in different shapes indicates that *Lytechinus variegatus* shows a distinct difference in ability to feed on food of different shapes. Food prepared in a large, block-like form is eaten more rapidly than food in terete, stalk-like forms, and this in turn is eaten more rapidly than flat, blade-like forms (Klinger 1982). This reflects feeding behaviour. The thinner items require constant repositioning while the larger items do not. The feeding structures are primarily adapted for rasping. This also accounts for the greater feeding rate on attached than on loose items. Both observations indicate that the feeding structures are adapted primarily for rasping. It seems that most plant foods do not vary sufficiently in hardness to affect feeding rate.

Optimality theory predicts that the more valuable food should always be accepted when encountered until its availability falls to a level at which it becomes profitable to include a less valuable food (Townsend and Hughes 1981). For *Strongylocentrotus droebachiensis* this occurs in the laboratory when the preferred, more valuable algae are present in equal concentrations. *Strongylocentrotus droebachiensis* consumed its preferred, most rapidly eaten algal food (*Lamiharia longicauda*) when mixed with four other species at 36% the rate when the alga was supplied alone (Larson *et al.* 1980). The rate of consumption of the less preferred foods was greater when supplied alone than in combination with the others. In addition, the feeding rate when individuals were supplied with equal quantities of the five algal species was only 77% of that when they were supplied only with the preferred food. Preferences may develop from feeding experience. *Strongylocentrotus purpuratus* fed either red, green or brown algae develop a preference for their food, but this is transient and disappears within several days (Leighton 1968).

Psammechinus miliaris can be found in areas with either attached or drift kelp, *Laminaria* (Bedford and Moore 1985). Young individuals grow more rapidly on the decaying kelp but adults do not. Despite this, adults given a choice between the two quickly develop a preference for the latter.

The relative feeding rate of regular echinoids decreases with size (Fuji 1967; Klinger 1982; Leighton 1968). The basis for this is not known, but the allometry of the feeding apparatus (spines, tube feet, Aristotle's lantern) is probable. Thus the Aristotle's lantern is relatively larger in juveniles than in adults (Figure 2.36), but whether the necessary support for efficient functioning exists in the test is not known.

The relative size of the Aristotle's lantern varies among populations of the same species (*Echinus esculentus*: Hagström and Lönning 1964, *Paracentrotus lividus*: Régis 1978, *Diadema setosum* and *Strongylocentrotus purpuratus*: Ebert 1980, *Echinometra mathaei*: Black, Johnson and Trendall 1982; Black, Codd, Hebbert, Vink and Burt 1984). A functional inverse relation between the relative size of the Aristotle's lantern and the level of food availability has been hypothesised (Ebert 1980). This implies the existence of a physiological set point for the level of nutrients for the organism. Otherwise, the Aristotle's lantern would be as large as possible even with abundant food. The rate of feeding on some but not all species of algae is directly related to the size of the Aristotle's lantern in *E. mathaei* (Black *et al.* 1984). The relative size of the gut of *E. mathaei* does not vary with the level of food available. This implies that the acquisition of food by the echinoid is the limiting factor, and not the processing or absorption of ingested food. However, the absorption efficiency of *Strongylocentrotus intermedius* varies with the rate of feeding (Fuji 1967).

It is difficult to compare the ability to feed with the relative size of the

Figure 2.36: The relations between sizes of structures and of functions and body size in *Strongylocentrotus intermedius.* AL: Aristotle's lantern. The scale for gonad production, total production, and gut contents is the left ordinate; the scale for gut contents and the sizes of the gut and Aristotle's lantern is the right ordinate. (Original, prepared from the data of Fuji 1967)

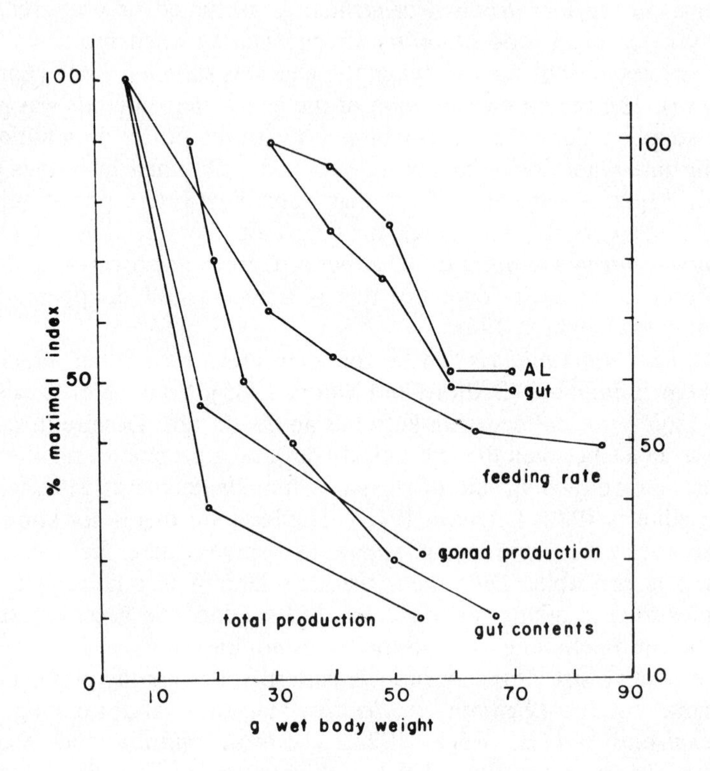

Aristotle's lantern among species because of the differences in the characteristics of foods. However, *Arbacia lixula* feeds by rasping and has a relatively larger Aristotle's lantern than *Paracentrotus lividus* (Régis 1978). The rate of feeding of *Strongylocentrotus franciscanus* is two- to four-fold greater than that of the smaller *Strongylocentrotus purpuratus* fed the same algal species (Vadas 1977). It is probable that this relationship is not isometric.

The gut of regular echinoids is a long, coiled tube with regional differentiation (De Ridder and Lawrence 1982) that should increase the efficiency of digestion of refractory plant foods. Echinoids feed continuously except when constrained by environmental or behavioural phenomena (Lawrence 1975). The siphonal groove or tube is a modification of the basic gut structure (Figure 2.37) (Holland and Ghiselin

Figure 2.37: Occurrence and location of siphon grooves and tubes in echinoids. Si.Gr.: Siphon groove, Si: siphon, Phar.: pharynx, Eso.: oesophagus, Stom.: stomach, Int: intestine, Rec.: rectum. (From Holland and Ghiselin 1970)

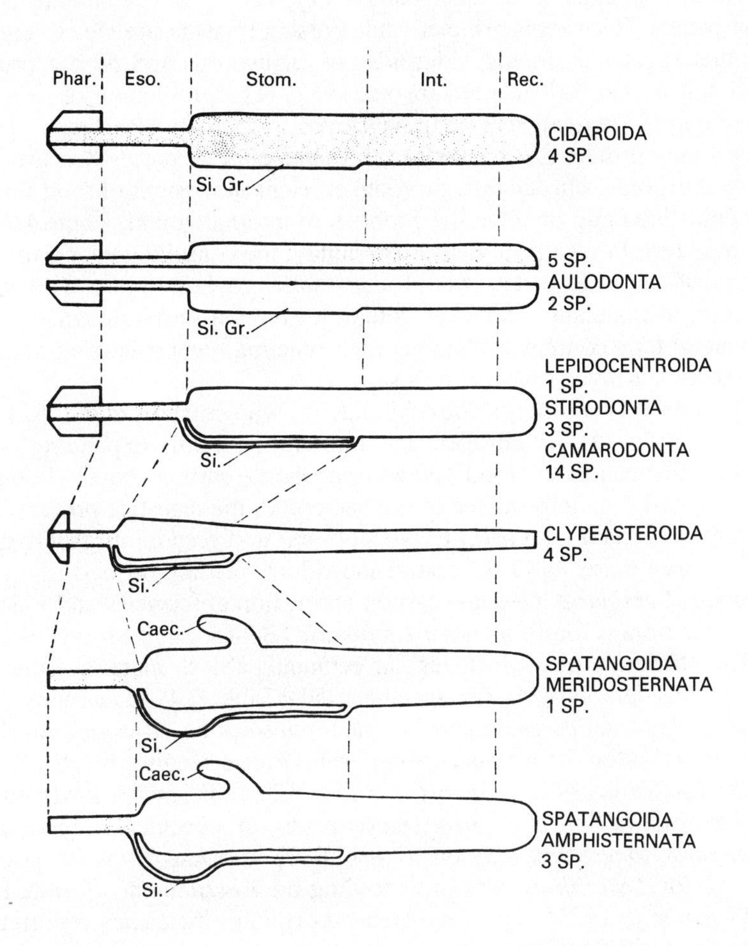

1970). Both the groove and tube are ciliated, and peristalsis occurs as well (Cuénot 1948). A siphonal groove occurs in cidaroids and diadematids, while a presumably more efficient tube occurs in other regular echinoid groups. The function of the siphon in regular echinoids is thought to be provision for a bypass of water around the stomach to prevent dilution of the food digestive enzymes there (Cuénot 1948). A role in respiration has been discounted (Farmanfarmaian 1966). Water may be ingested during feeding, and may also accompany food pellets formed in the oesophagus passing into the stomach (Buchanan 1969). However, water is continually ingested even when the regular echinoid is not feeding (F.C. Stott 1955).

The Aristotle's lantern of *Lytechinus variegatus* is in constant feeding motion (Klinger 1984). Water may be ingested by this process even if food is not.

The oesophagus of regular echinoids typically produces mucus-coated food pellets which remain intact while passing through the gut. In contrast to other regular echinoids, cidaroids are carnivorous and do not produce food pellets (De Ridder and Jangoux 1982). The production of pellets by non-cidaroid, regular echinoids has several functions (Buchanan 1969). Pellets may protect the gut surface, produce discrete faeces that are more easily removed from the test, promote efficient movement of food through the fluid-filled gut, or solve the problem of internal volume control due to the rigid test. In regard to the last possibility, the consolidated and rounded nature of the food pellets would allow water to flow freely. This would function to maintain a constant volume of the gut despite changes in the volume of food contents. Whatever their function, the production of mucus for the pellets involves a cost to digestion.

Bacteria exist in the gut of echinoids (Lawrence and Lane 1982). The gut microflora can fix nitrogen, the amount apparently depending on the level of nitrogen in the food and an appropriate carbon source (Fong and Mann 1980). The importance of the bacteria to the digestive process is not known. Bacteria do not seem to be important in digestion of carbohydrates by *Psammechinus miliaris* because individuals feeding on bacterially rich decaying *Laminaria* have low carbon absorption efficiencies and a limited microbial flora is found on fresh *Laminaria* (Bedford and Moore 1985).

The efficiency with which regular echinoids absorb material from food varies considerably with the specific food (Table 2.4) (Lawrence 1975, 1987a). The total (organic and inorganic) absorption efficiency can be as high as 91% for *Strongylocentrotus franciscanus* feeding on *Nereocystis luetkeana* (Vadas 1977). The organic absorption efficiency is lower but can still be high (77% for *Strongylocentrotus droebachiensis* feeding on *Laminaria longicruris*, Larson *et al.* 1980) and can even be negative (−35% for *Lytechinus variegatus* feeding on *Euecheuma isiforme*, Lowe and Lawrence 1976). The very high absorption efficiencies reported for many foods would require the absorption of most of the inorganic material as well as the organic material.

The amount of nutrients absorbed is the product of the absorption efficiency and the amount of food eaten. To maximise the amount absorbed, as much as possible of the preferred food should be eaten and absorbed. *Strongylocentrotus droebachiensis* feeds most rapidly on preferred food, but absorption efficiencies are not correlated with preference (Larson *et al.* 1980). In contrast, although the rate of feeding on the preferred food decreased when less preferred foods were available in equal amounts, the total rate of feeding was the same in both situations for *Strongylocentrotus franciscanus* (Vadas 1977).

If the availability of food becomes below the potential rate of ingestion, the retention time should increase to increase the digestive and absorptive processes (Sibly 1981). *Diadema setosum* feeds nocturnally and retains food in the intestine during the day (Lawrence and Hughes-Games 1972). Food may be held longer in the gut when echinoids are fed low rations in the laboratory. *Strongylocentrotus intermedius* showed an increase in absorption efficiency in such a situation but not sufficiently to compensate for the decrease in amount eaten (Table 2.5). *Lytechinus variegatus* decreases its feeding rate and increases its food retention time when exposed to low temperatures, but although the activity of the digestive enzymes does not change, the absorption efficiencies and amounts of organic material absorbed decrease (Klinger, Hsieh, Pangallo, Chang and Lawrence 1986). *Psammechinus miliaris* has a longer gut retention time when fed fresh *Laminaria* than decaying material, which seems to be associated with the degree of difficulty in digestion of intact cells (Bedford and Moore 1985). In this instance, a trade-off between the gut retention time and the absorption efficiency for protein sources results in equivalent growth rates of adults on the two food types.

Table 2.5: The effect of feeding rate on the absorption efficiency and amount of food absorbed by *Strongylocentrotus intermedius* fed *Laminaria japonica*. (From Fuji 1967.)

Feeding rate (mg ind^{-1} day^{-1})	Absorption efficiency (%)	Amount absorbed (mg ind^{-1} day^{-1})
205	66	135
137	72	99
50	80	40
32	82	26

Much of the food of regular echinoids may be carried to them as drift by water currents (Lawrence 1975). The ecological importance of this is that it has freed many echinoids from the necessity of movement to obtain food, and allows them to exist in habitats and at densities which would not be possible otherwise (Lawrence and Sammarco 1982). Another important consequence is that regular echinoids thus have two basic strategies of feeding: sitting-and-waiting and foraging. Many species of regular echinoids are found intertidally in burrows, and their mobility and foraging potential are limited (Lawrence and Sammarco 1982). The strategy provides the echinoid with a supply of food and protection, but costs in that the feeding period is limited to high tide and the individuals are dependent on an environmentally regulated food supply.

The two different modes of feeding also differ in interindividual effects on feeding. No interference between individuals exists in the sit-and-wait mode, but considerable interaction can occur among foraging individuals. *Lytechinus variegatus* frequently aggregates in the field, and the rate of consumption of seagrass is inversely related to density (Greenway 1977). Similar aggregations occur with foraging algal-eating regular echinoids (see Lawrence 1975).

Even burrowing species that are not completely restricted in movement can forage and interact intraspecifically. *Echinometra* species live in burrows on exposed intertidal shores in the tropics. Because of wave activity, foraging is restricted primarily to the burrow and immediate areas. The burrows can completely cover the intertidal so that space is limiting. In such a situation, territoriality might be expected because of limited area for grazing and the requirement for food for survival, growth and reproduction. One expects the value of winning to be highest and costs lowest near the centre of an individual's territory, and that a contest will be won by the individual with the higher value to cost ratio (G.A. Parker 1985). *Echinometra mathaei* occurs in two types, A and B (distinguishable phenotypically) on Okinawan reefs (Tsuchiya and Nishihira 1985). Type B individuals are more aggressive than type A individuals and repel them. Type B individuals also interact, and larger individuals win. In contrast type A individuals interact less and even form aggregations within the burrows.

Populations that are not exposed to wave action do not burrow, but are cryptic when predation is a potential danger. It is in this situation that the alternate feeding strategies are possible. Abundant drift algae can supply sufficient food to eliminate the necessity for foraging. Foraging may be local when algal supplies are sufficient in the vicinity, such as that by *Diadema antillarum* which leads to the production of 'haloes' around patch reefs in the Caribbean (Ogden, Brown and Salesky 1973). Typically shallow-water species have a diel activity rhythm, foraging when predation danger is minimal (Lawrence 1975). A cost associated with this strategy of foraging is again a limitation in time of feeding, but is a trade-off for survival. Under extreme conditions, drift and attached algae may be so low in biomass that regular echinoids abandon their cryptic behaviour and forage continuously. The amount of foraging by *Strongylocentrotus franciscanus* is inversely correlated with the amount of brown-algal drift material (Harrold and Reed 1985). Avoidance of predator behaviour by *Strongylocentrotus droebachiensis* is subordinate to feeding (Vadas, Elner, Garwood and Babb 1986).

Unlike carnivores, regular echinoids may depend more on random encounters than upon directed movement as a foraging strategy. *Lytechinus variegatus* locates food at a very close distance (Klinger and Lawrence 1985). Foraging by *Tripneustes gratilla* seems random in seagrass beds (Nojima & Mukai 1985). However, *Strongylocentrotus droe-*

bachiensis responds quickly to algae in its vicinity (Vadas *et al.* 1986), and various species of regular echinoids are reported to perceive and move towards algae in the field (see Lawrence 1975). *Evechinus chloroticus* moves more when drift kelp is present, but its movement is still non-directional as though it perceives the presence of food but not the precise location (Andrew and Stocker 1986).

2.1.4.2. Irregular echinoids

Although regular echinoids may live on particulate substrata, their food is primarily macroscopic. Exceptions are echinothuriids which scoop up the substratum indiscriminately. A major evolutionary event in echinoids was the development of irregular echinoids that feed on particulate matter (Kier 1974). This led to the massive invasion of habitats and exploitation of a food source not available to the regular echinoids. This development occurred twice, producing two groups which have unique suites of characters. The gnathostomaceans have retained the Aristotle's lantern although its original functioning in food ingestion has been lost. Gnathostomaceans occur on the surface or at minimal depths in the substratum. The atelostomaceans have lost the lantern, and occur at greater depths in the substratum.

2.1.4.2.1. Gnathostomacea. The primitive holectypoids in the superorder Eognathostomata, which includes only two extant genera, *Echinoneus* and *Micropetalon*, had a highly vaulted test with a flat oral surface. The adults lack a lantern, but the juveniles do not. Mortensen (1948b) pointed out that the discovery by A. Agassiz in 1909 that juveniles possess a complete lantern was one of the great events in 'Echinology'. The lantern is never used in feeding, however, as it is completely resorbed even before the mouth opens. *Echinoneus* lives buried in coarse substrata or is attached by its tube feet to the underside of coral slabs. No observations have been made on its feeding mechanisms, but the type of substrata in which it lives and its gut contents indicate that the tube feet alone are involved.

Members of the superorder Microstomata have a very small peristome, a great variety of body forms, and may have a lantern. They use the tube feet and spines for feeding in various ways. The cassiduloids seem to have a feeding mechanism similar to that suggested for the holectypoids. *Cassidulus caribaearum* has a lantern as a juvenile which persists for some time, but it is not known if it functions in feeding (Gladfelter 1978). The tube feet of the anterior ambulacrum are concentrated to form a phyllode (Figure 2.38) (Kier 1962). Here, tube feet are crowded around the mouth to assist in moving particles in feeding. With specialisation of the tube feet, a reduction in the number necessarily occurred. Because these tube feet were not used for respiration, pairing was not necessary, and the unipore which evolved was structurally stronger. A swelling of the interambulacral

Figure 2.38: Oral view of the cassiduloid *Cassidulus pacificus* showing enlarged interambulacral plates (bourrelets) and the concentrated tube feet pores in the ambulacra (phyllodes) (from Lovén 1874)

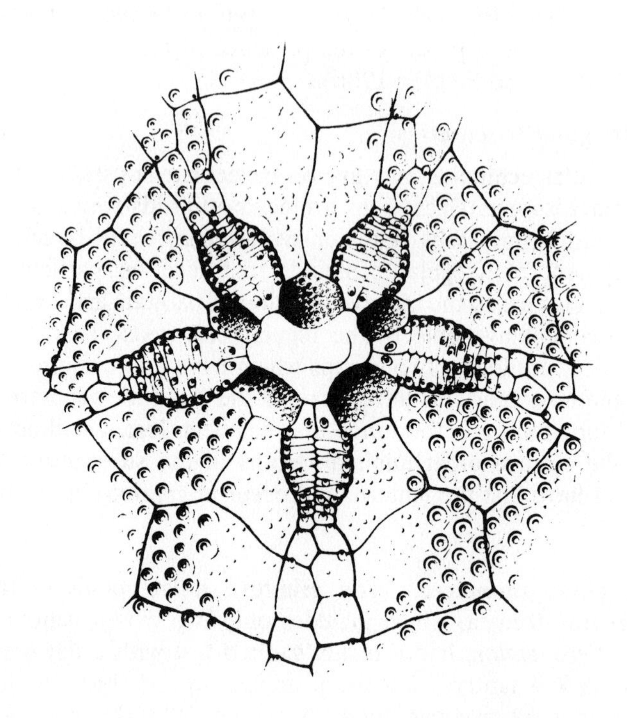

plates at the peristome occurs also to form bourrelets (Figure 2.38) (Kier 1962). In some extinct species, the bourrelets were huge and toothlike. They are covered with many small spines which are used to push food particles into the mouth. These tube feet pass sand particles posteriorly to the mouth where they are ingested through the action of the radial spines there. The smaller sand particles are ingested because of the difficulty in manipulation of larger sizes. This would not be selection in the true sense, but it should result in an increase in the total surface area to volume ratio of the particles ingested and consequently increase organic intake.

The clypeasteroids (sea biscuits) and laganoids (sand dollars and fibulariids) are microstomatids which retain the lantern as adults, probably having evolved neotenously from the cassiduloids (Phelan 1977). The earliest clypeasteroids had a small, high test with the periproct in the primitive dorsal position, a very erect lantern, no food grooves and a large peristome (Kier 1984). The extant forms are low to flattened in profile, and have a flattened lantern and a small peristome. The clypeasteroids have ambulacral lantern supports and the laganoids have interambulacral ones.

The fibulariid *Echinocyamus pusillus* is a small form which uses buccal tube feet and accessory tube feet (Figure 2.39) (Telford 1983) to pass sand particles to the mouth. The accessory tube feet are spread on the surface of *Echinocyamus* by their occurrence at the edges of a broad, lobed, radial water vessel (Figure 2.40) (Phelan 1977). The particles may be passed directly into the mouth. More usually the circumoral spines hold the particles in place and rotate them with the free edge of the peristomial membrane while the teeth rasp off the organic material (Nichols 1959; Telford 1983).

Figure 2.39: Cross-section through the mouth of *Echinocyamus pusillus*. The thickened peristomial membrane acts as a mobile lip that holds particles for the teeth to scrape (from Telford, Harold and Mooi 1983)

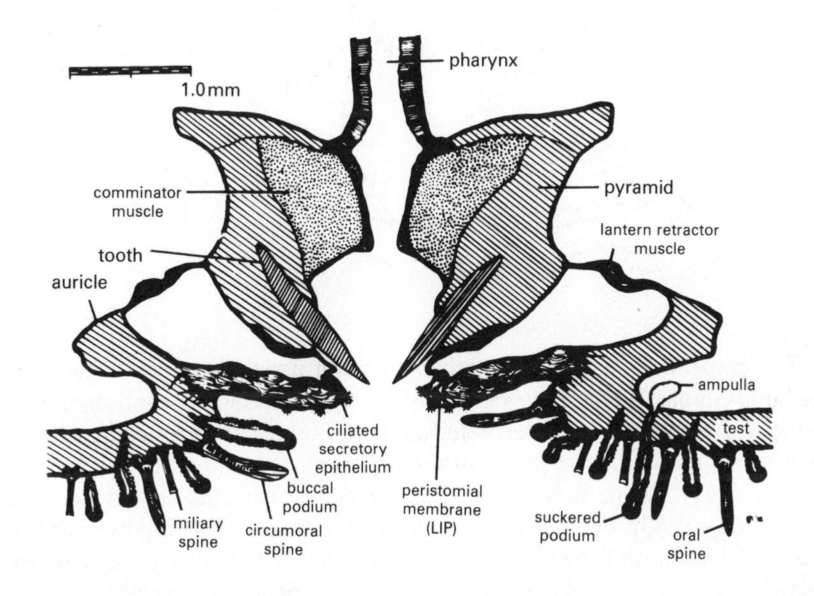

The broad form of the clypeasteroid sea biscuits and laganoid sand dollars has been interpreted as a mechanism to increase the surface area for food collection on the undersurface and, in the latter, on the upper surface. However, the test form must also be appropriate for burrowing and hydrodynamics.

The proportion of the upper surface of the test that could function in food collection would be inversely related to the development of the respiratory petals. The area of the interambulacra, which would be involved in food collection on the upper surface, is relatively greater in the rotulinids and mellitids than in laganids and arachnoidinids (Durham 1966).

Figure 2.40: Radial water canal (r) of *Echinocyamus* with lobes (l) that serve the accessory tube feet ampullae (a). p: Peristome, t: ambitus. (From Phelan 1977)

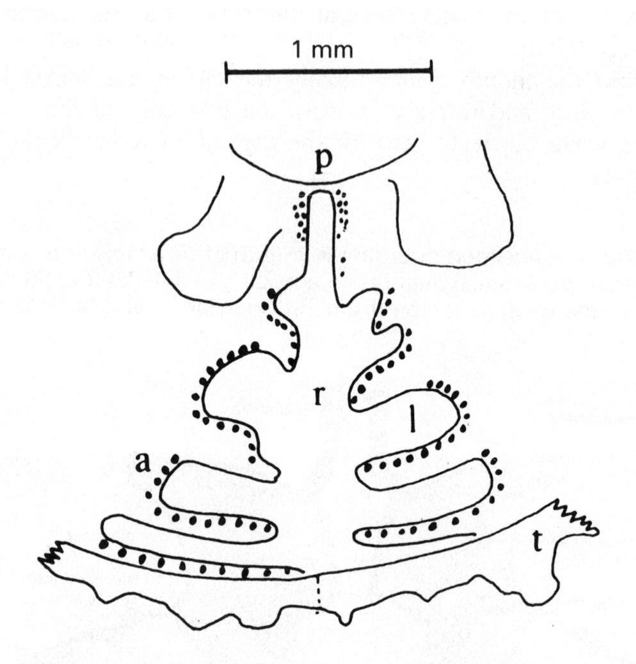

The epibenthic sea biscuits probably do not use their upper surface for feeding (Ghiold 1984). Involvement of the upper surface of sand dollars in food collection has been interpreted as being the result of a sieving action, with small particles dropping between the spines on to the test surface (Figure 2.41) where they are carried by ciliary currents to the ambitus (Goodbody 1960). Seilacher (1979) proposed that this sieving mechanism was associated with forward locomotion within the substratum. The smaller particles reaching the test supposedly would be richer organically and also sufficiently small to be carried by ciliary currents to the ambitus. Transfer of the particles around the sharp ambitus from the upper surface to the undersurface of the test is a major difficulty. Accessory tube feet are abundant around the ambitus of *Mellita quinquiesperforata* and could function with the marginal spines to transfer particles. Only a small portion of the particles collected on the upper surface may be retained by the process in *Echinarachnius parma* (Ellers and Telford 1984).

Conspicuous food grooves converging on the mouth occur on the undersurface of clypeasteroids and laganoids. They differ greatly among groups in development (Figure 2.42). The amount of material carried by the food grooves should consequently vary. The functional consequences of these

Figure 2.41: Shoe and miliary spines on the upper surface of *Echinarachnius parma* showing the relationship between the distance between spines and the size of particles (from Mooi and Telford 1982)

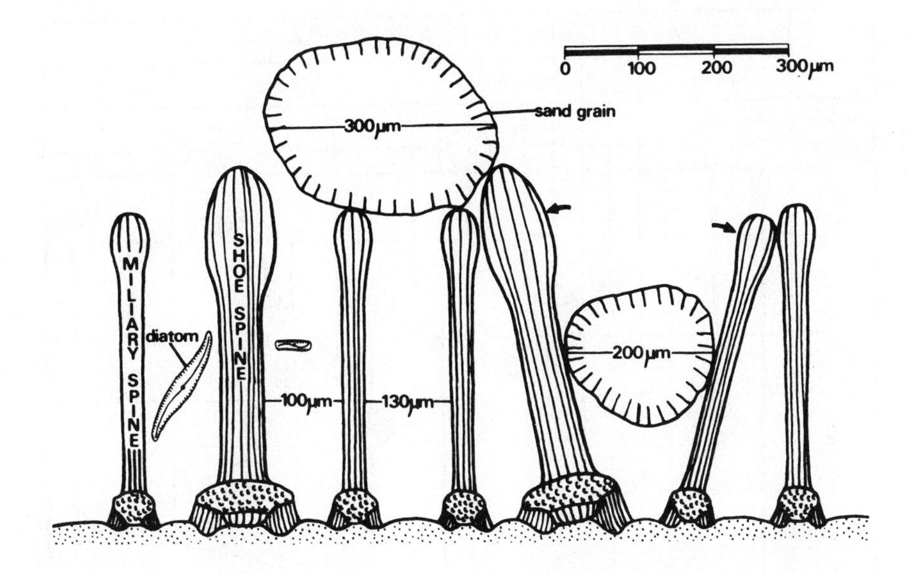

structural differences in degree of development of the food grooves are not known. Thus it is difficult to evaluate Cowen's (1981) suggestion that elaborate, complex feeding networks can be energetically expensive. Cowen suggested that the advective food grooves of sand dollars might be analysed in terms of the effectiveness of transport of food particles to the mouth. He noted that the theoretically most effective 'banana plantation' pattern is not common among sand dollars. *Arachnoides* fits the 'banana plantation' paradigm, but *Echinarachnius* does not (Ghiold 1984). The degree of development of the food grooves may differ with the degree of involvement of the undersurface because the locomotory areas and feeding areas are distinct.

The food grooves have accessory tube feet. These are suckered tube feet that have very small single pores (Phelan 1977). They are found abundantly within and beside the food grooves in some genera, but only beside the grooves in others (Table 2.6) (Ghiold 1984). The basic perradial food groove is a simple, unbranched furrow along the perradial suture between two columns of ambulacral plates. The radial water vessel lies beneath on the interior of the test and separated from it by the internal position of the radial nerve and haemal vessel. The polyfurcating food groove has a very short section on the perradial suture near the peristome, but soon divides into two branches, each of which extends along the growth

Figure 2.42: Development and evolution of food grooves (in black) on the under surface of Clypeasteroida. Locomotory spines occur between the food grooves. (From Durham 1966)

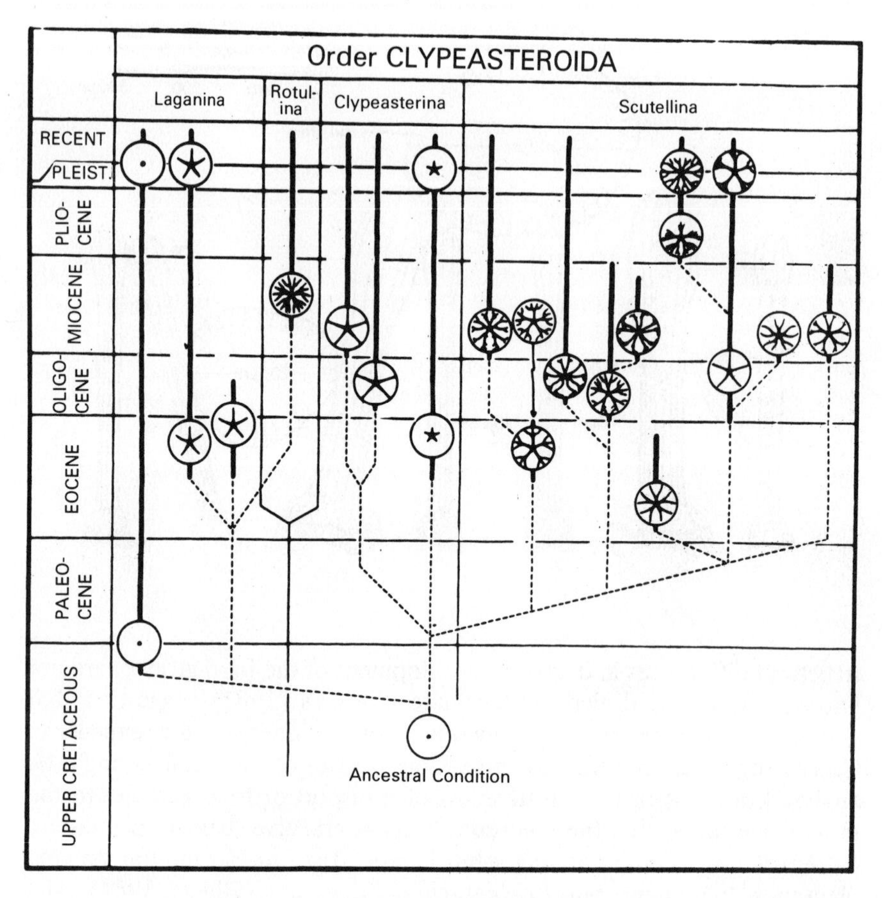

Table 2.6: Distribution of accessory tube feet on undersurface of laganoids and clypeasteroids (modified from Ghiold 1984)

Family	Distribution of accessory tube feet
Clypeasteridae	Food-groove borders
Fibulariidae	Rows parallel to perradial suture, at right angles to suture at ambitus
Laganidae	Food-groove borders
Arachnoididae	At right angles to food grooves
Rotulidae	Ubiquitous
Astriclypeidae	Test margin, food-groove borders
Dendrasteridae	Test margin, food-groove borders
Echinarachniidae	Food-groove borders, within grooves
Mellitidae	Food-groove borders, within grooves

centres of a column of ambulacral plates. These commonly branch also. These lateral water vessels are not obstructed by the radial nerves and haemal vessels from direct access to the test, and the accessory tube feet are consequently found within the food grooves. *Clypeaster* has simple perradial food grooves that almost reach the ambitus, and lateral vessels that do not branch (Figure 2.43). Numerous accessory tube feet occur on either side of the groove until near the ambitus where they occur within the grooves. In the sand dollars, the arrangement of the food grooves and consequently the lateral water vessels and their accessory tube feet is more complex (Phelan 1977).

Figure 2.43: a: Inner surface of ambulacral plates of the undersurface of *Clypeaster subdepressus* showing the lateral water vesels (lwv) extending from the radial water canal (r) with position of accessory tube feet ampullae (a). b: Outer surface of an adoral area showing a portion of the food groove (fg) and perradial suture (ps). The notch on each accessory pore points in an adoral/adradial direction where the lateral water vessels are at right angles to a perradial suture. Accessory pores are absent in the perradial food groove. (From Phelan 1977)

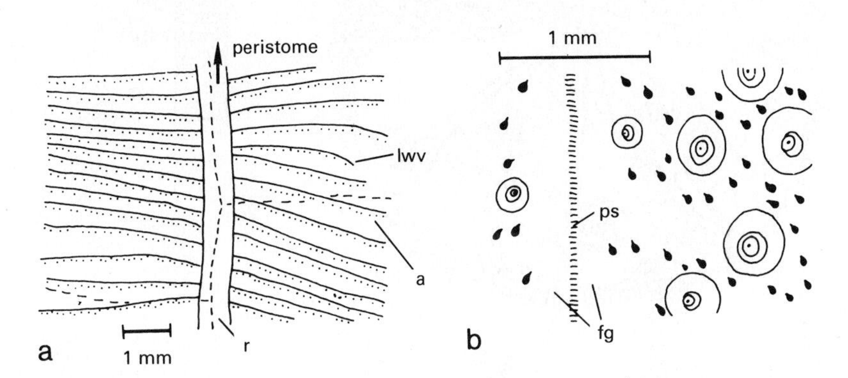

The accessory tube feet on the lower surface also pick up particles from the substratum (Ellers and Telford 1984; Ghiold 1984). Particles are picked up by long, barrel-tipped tube feet that form a narrow fringe around the geniculate spine fields, and are passed by short, barrel-tipped tube feet found among the geniculate spines to the food grooves (Figure 2.44) (Telford, Mooi and Ellers 1985). The particles are incorporated into mucous cords in the food groove and are moved by the tube feet associated with the food groove to the mouth. Buccal tube feet push the particles into the mouth. It is probable that most of the particles ingested by sand dollars are collected by the accessory tube feet on the ventral surface.

The shape of the test of the sand dollars has hydrodynamic properties that stabilise it on the surface (Telford 1983). These hydrodynamic

Figure 2.44: a: Distribution of accessory tube feet on the undersurface of *Mellita quinquiesperforata*. ac: Accessory tube feet, fg: food-groove tube feet, lb: long barrel-tipped tube feet, sb: short barrel-tipped tube feet. b: Cross-section through a lateral ambulacrum of the undersurface of *Mellita quinquiesperforata* showing (on the right) a sequence of collection and transfer of a particle by the barrel-tipped tube feet to the food groove. f: Food-groove tube foot, g: geniculate spine, lb: long barrel-tipped tube foot, m: miliary spine, p: pressure drainage channel spine, pdc: pressure drainage channel, pfg: primary food groove, sb: short barrel-tipped tube foot, sfg: secondary food-groove. (From Telford *et al.* 1985)

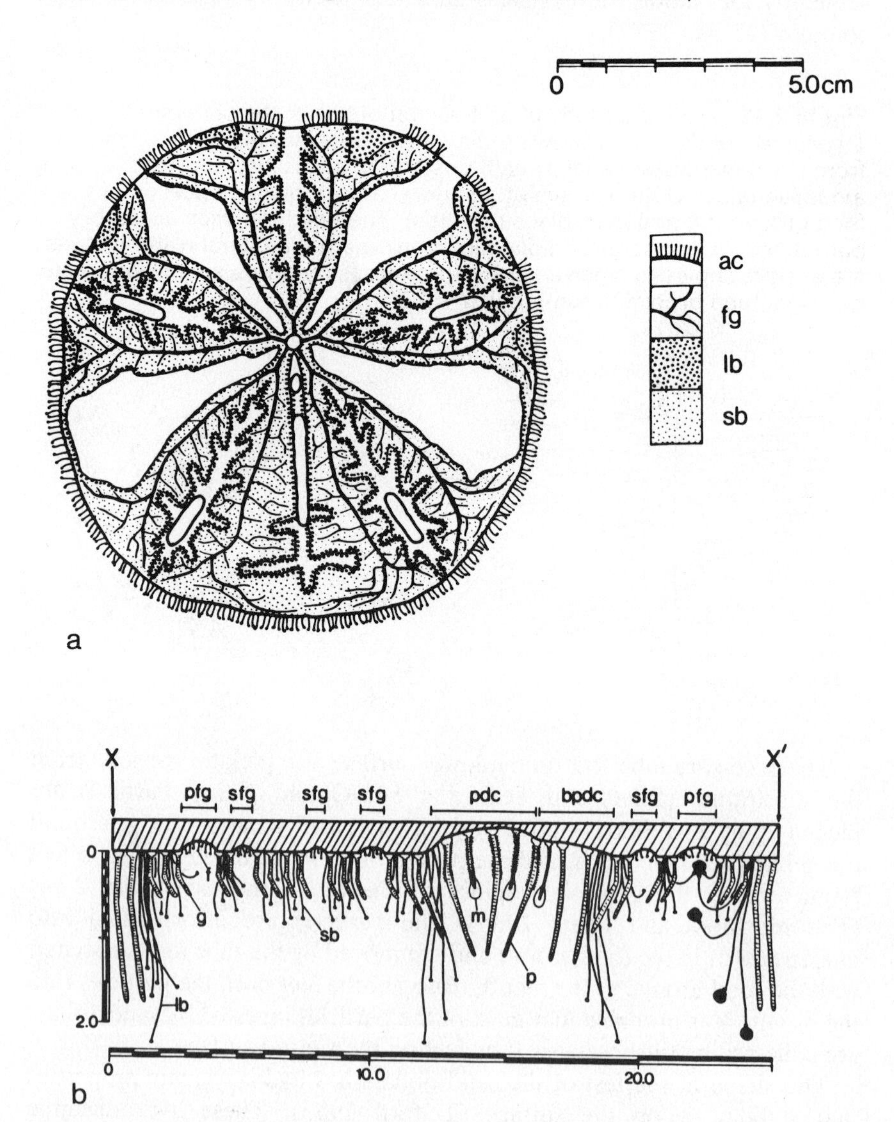

properties are used by *Dendraster excentricus* for suspension feeding (O'Neill 1978). In low currents, individuals lie flat on the substratum and move, supposedly feeding in the manner as described above for other sand dollars. At moderate water currents, however, individuals in denser populations are in an inclined position with the anterior end buried (Figure 2.45) (Merrill and Hobson 1970) and the tube feet on the oral surface capture particles (Timko 1976). In this posture, the test is parallel to the water flow and acts as a lifting body (Figure 2.46) (O'Neill 1978). The streamlined curvature resulting from lift moves particles in the direction of the feeding surface and thus should bring more in reach of the tube feet. The spacing and posture of individuals are a function of the water velocity which may maintain the optimal gap to take advantage of the effect of neighbours. This is thus a third posture in filter-feeding echinoderms: the ambulacra being down current in the crinoids, up current in some filter-feeding ophiuroids, and parallel to the current in *D. excentricus.*

2.1.4.2.2. Atelostomacea. The Atelostomacea lack a lantern throughout their postmetamorphic existence. The development of a sunken, anterior ambulacral groove and a body elongated to varying degrees are general developments of the two extant orders, holasteroids and spatangoids. The holasteroid pourtalesiids are tiny barrel-shaped forms with a funnel at the anterior end (Figure 2.47). Presumably the individual burrows through the upper part of the substratum, shovelling food into its mouth with its oral spines.

The spatangoids show modifications of the body form, spines and tube feet which are associated with feeding. The feeding tube feet are highly modified and found in the phyllodes (Figure 2.48) (A.B. Smith 1980a). These tube feet have a broad disc covered with papillae. The large size of the ampulla is associated with the extensibility of the tube foot. In general, a particle or cluster of particles of the substratum adheres to the oral, prehensile, penicillate tube feet, is tranferred to the mouth, and is scraped off against the spines surrounding the peristome (Nichols 1959). The non-funnel building *Meoma ventricosa,* which lives just beneath the surface of the substratum, uses the numerous penicillate tube feet on the phyllodes around the mouth to collect food particles (Chesher 1969).

Spatangoids that construct respiratory funnels (Figure 3.22) feed on particles from the surface. Particles can fall into the funnel passively or as a result of the activity of the funnel-building tube feet (Chesher 1963; De Ridder 1982; Nichols 1959). Particles that fall into the funnel or are carried by the respiratory currents are transferred to the mouth by the highly specialised anterior ambulacrum.

Particles that fall on to the cordyles at the apex of the body of *Moira atropos* stick to the mucus produced by these ciliated spines and are carried by ciliary currents towards the midline (Figure 2.49) (Chesher 1963).

Figure 2.45: *Dendraster excentricus* in the feeding posture. The tests are parallel to the surge current. (From Merrill and Hobson 1970)

Figure 2.46: The trajectory of food particles past the oral surface of *Dendraster excentricus* and the distance (D) of capture by the tube feet (TF) if (A) no lift were generated, and if (B) lift were generated (from O'Neill 1978)

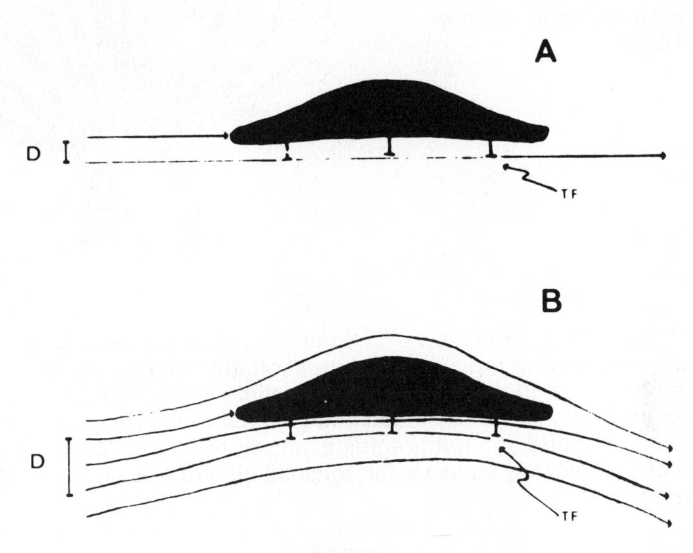

Figure 2.47: Views of the (a) upper, (b) side, and (c) lower surfaces of the atelostomatid *Pourtalesia* (from A. Agassiz 1881)

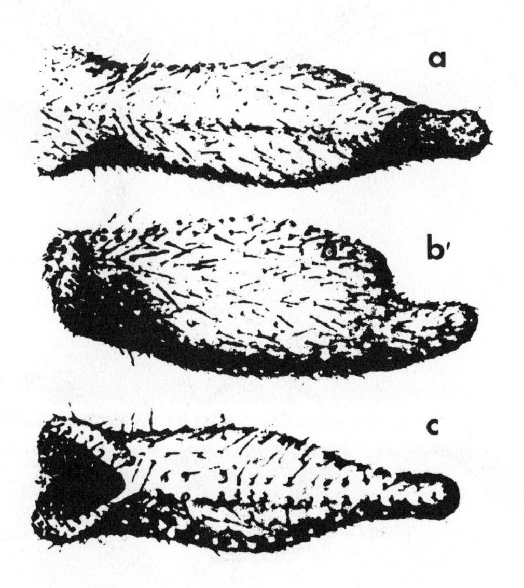

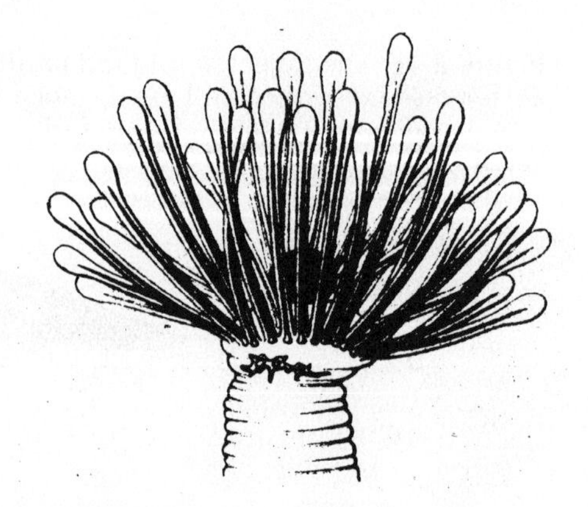

Figure 2.48: Phyllode tube foot of *Palaeostoma mirabile* (from Lovén 1883)

Figure 2.49: A (a): Lateral view of the depressed frontal ambulacrum of *Moira atropos* showing regional differentiation. (b) Orientation of sectioned frontal ambulacrum to the test. B: Cross-sections of the frontal ambulacrum showing the formation and position of the mucous rope (stippled area). (a) Region I, (b) region II, (c) region III. mt: Mucous trough, fw: wall of fasciole, ap: aperture for consolidation of mucous rope. (From Chesher 1963)

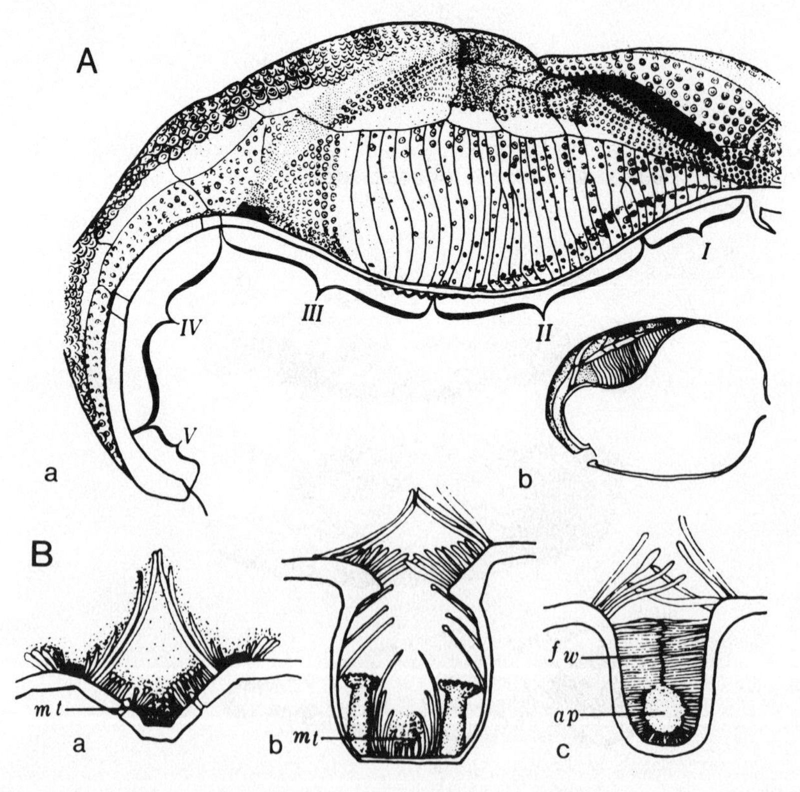

Spatulate spines move the mucus-embedded particles down the depressed ambulacrum. The particles are crammed together and pushed through the fasciole by large spatulate spines to form a mucous rope, which is broken by the oral tube feet. The tube feet with adhering rope sections are inserted into the mouth like the feeding tentacles of holothuroids.

Echinocardium cordatum ingests both surface material and that at the level of the mouth in the burrow (De Ridder and Jangoux 1985). Starved individuals ingest more of the organically richer surface sediment, implying a capacity to regulate intake to nutritional condition.

The general arrangement of the gut of irregular echinoids is similar to that of regular echinoids, but the curves have a broader sweep (De Ridder and Jangoux 1982). The arrangement of the gut of the young echinoid probably reproduces the primitive condition, and is retained in adult clypeasteroids (Cuénot 1948). Siphon tubes occur in both clypeasteroids and spatangoids, two being present in some species of spatangoids (Figure 2.37). Faecal pellets are not produced, but a mucoid covering is associated with the faeces as in holothuroids.

2.1.5. Holothuroidea

Holothuroids use their tentacles to feed on particulate food. These arise from the five radial water vessels which grow anteriorly from the water ring. The difference in the structure of the tentacles is so great that it is used to separate the group at the subclass or order level (Pawson 1966). The tentacular ampullae are lacking in the elasipods, absent or little developed in the dendrochirotes, small in the apodids, and long and narrow in the molpadids and aspidochirotids (Hyman 1955). A calcareous ring provides support for the tentacles in most holothuroids.

Modifications of the tentacular complex are the basis for variation in feeding type and associated body structures and form. The most profound division in the holothuroids in this regard is the separation of those sessile forms whose tentacles function in suspension feeding, and those vagile forms whose tentacles function in deposit feeding or capture of prey. The length of the tentacles is limited by their lack of support and their method of transferring food to the mouth. Just as the characteristics of the food-gathering structures possibly limit the diameter of the calyx of crinoids, the sole use of the oral tentacles in food gathering may limit the body diameter of holothuroids.

The basic arrangement of the holothuroid gut is that of a young echinoid gut (Cuénot 1948). Like the echinoid gut, the holothuroid gut shows regional differentiation (Féral and Massin 1982). The gut is long except in some synaptids and elasipods, indicating a low-quality diet. A 'stomach' which seems to be a triturating organ occurs in many synaptids and some dendrochirotids and dactylochirotids, but not in other orders. The 'intestine' varies greatly among species in length and structure. It is longest in dendrochirotids, which suggests adaptation for digestion of refractory plant material. Similarly, the shorter length of the intestine of aspidochirotids and apodids suggests that the organic matter associated with the particles ingested is less refractory. As in the echinoids, feeding is continuous in holothuroids unless they are under environmental or behavioural constraints. The faeces of aspidochirotids are often enclosed in a mucous casing. This casing may be produced to protect the gut lining as the often irregular particles are moved through it by peristaltic action.

2.1.5.1. Dendrochirotacea

The dendrochirotids are considered to be the most primitive of the living holothuroids, whose richly branched tentacles have evolved from initially simple finger-like oral tentacles by repeated branching (Figure 2.50) (Fell and Moore 1966). As suspension feeders, the dendrochirotids are restricted to areas with suspended food which is primarily small plants and animals or organic particles (Massin 1982). The prominence of dendrochirotids in temperate and sub-tropical regions but not in tropical regions (Pawson 1970) or at great depths (Hansen 1975) may be related to the availability of suspended material. Suspension feeders can be completely sessile if an adequate supply of food is dependable and if they can survive physical or predatory danger. Thus, there are two feeding strategies with dendrochirotids. They can exist both on hard substrata or buried in soft substrata and move relatively little or not at all, or movement may be a part of their feeding activities.

A permanent modification of the body form can be associated with this 'pelmatozoan' habit. In epibenthic species such as *Psolus chitinoides* and *Cucumaria frondosa*, the mouth is displaced upwards to position the tentacles into the water column (Figure 2.51). Burrowing species such as *Sclerodactyla briareus* live with the mouth and anus at the surface of the substratum (Figure 2.52) (A.S. Pearse 1908). Most extreme is *Rhopalodina lageniformis* which has a barrel shape and the mouth and anus close together at the end of the narrow neck (Figure 2.53).

In contrast, other dendrochirotids have behavioural modifications. *Cucumaria curata* lives attached to hard substrata in high-energy areas. They are small in size; large forms probably could not resist the water action. Here the mouth is at the anterior end of the body axis, and it is raised away from the surface to form a U-shape in feeding (Brumbaugh 1965). *Aslia lefevrei* attaches to hard substrata in crevices and extends the anteriorly positioned tentacle crown out into the water column (Figure 2.54) (Costelloe and Keegan 1984). *Neopentadactyla mixta* is a burrower, but its U-shaped posture is behaviourly assumed (Figure 2.54) (T.B. Smith 1983). The tentacles are not passive in feeding, but may actively sweep

Figure 2.50: Tentacles of holothuroids. a: Dendritic (mainly dendrochirotids). b: Pinnate (synaptids). c: Peltodendritic; DE: dendritic end, M: mouth, TS: tentacle shaft (aspidochirotids). d: Digitate; B: buds, D: apical digitations (synaptids). e: Digitate; M: mouth, T: tentacle (molpadids). f: Peltate (aspidochirotids and some elasipods). g: Digitate (elasipods). (From Massin 1982)

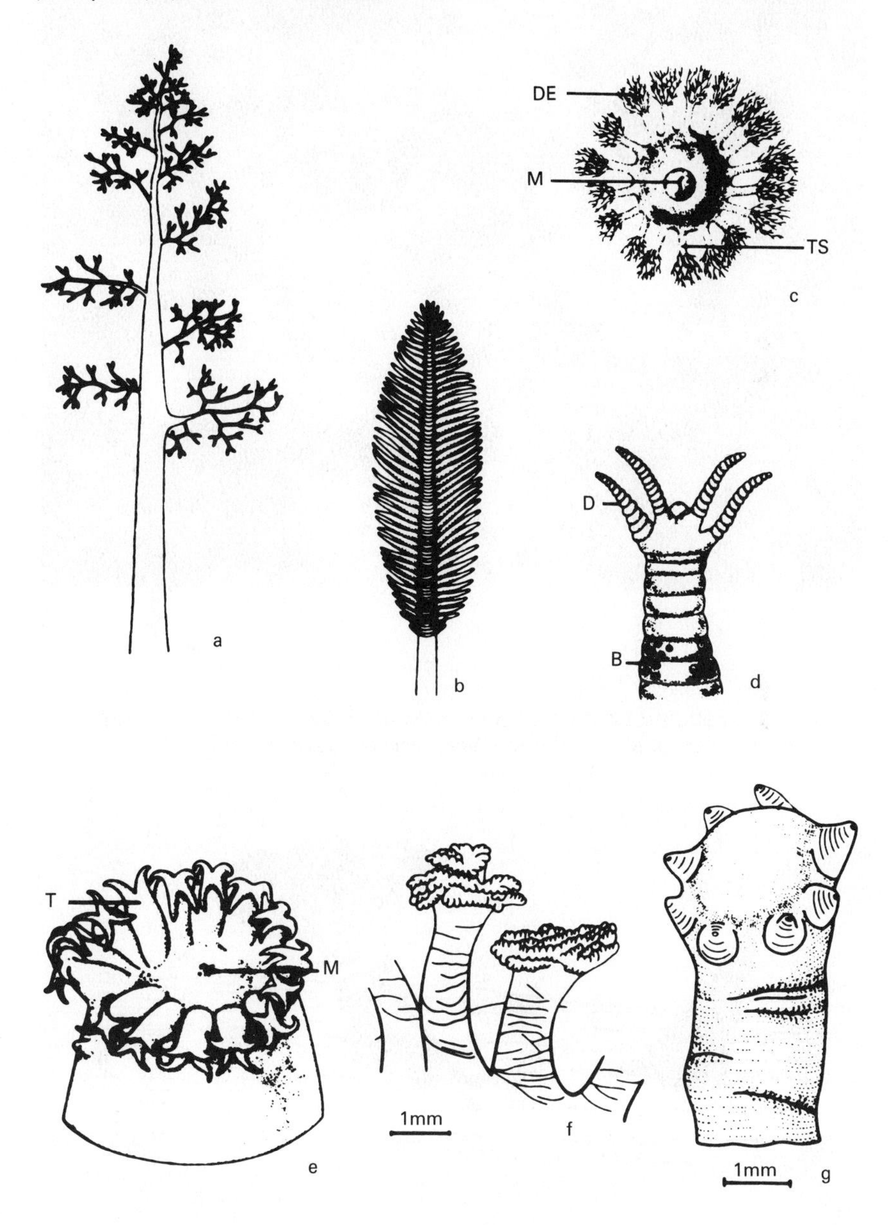

Figure 2.51: Feeding posture of *Psolus chitinoides* (from Fankboner 1978)

Figure 2.52: Feeding posture of *Sclerodactyla briareus* with both tentacle crown and the anus above the surface. (From A.S. Pearse 1908)

Figure 2.53: *Rhopalodina lageniformis,* with the mouth and anus at the end of the narrow portion of the body (from Semper 1868)

through the water (Brumbaugh 1965; Fankboner 1978). The unusual dendrochirotid *Leptopentacta elongata* lives with its anterior end buried, and is a deposit feeder (Figure 2.55) (Fankboner 1981). Members of the order Dactylochirotida also live buried in soft substrata with the body structurally U-shaped. Their eight to thirty tentacles do not branch but are digitiform or digitate (Pawson and Fell 1965) and are probably used for detrital feeding (Pawson 1982b).

As with crinoids, the effectiveness of the dendritic tentacles of the dendrochirotids should be a function of the surface area involved, but this has not been analysed. There are ten to thirty tentacles in various species, and the degree of branching is variable. The branching of the tentacles is like that of the arms of the ophiuroids rather than like the arms of the crinoids, i.e. branching occurs the length of the tentacle rather than proximally only (Figure 2.50). The upper limit of the number of tentacles may be the limit to the effective diameter of the collecting devices of a particulate feeder as with crinoids.

A major difference between the suspension feeding of the dendrochirotid

Figure 2.54: Intermeshing of the tentacles of the feeding dendrochirotids (A) *Aslia lefevrei* and (B) *Neopentadactyla mixta* (from Costelloe and Keegan 1984 and T.B. Smith 1983, respectively)

Figure 2.55: Feeding posture of *Leptopentacta elongata* buried in mud. Faecal pellets are at the left of the burrow entrance (from Fankboner 1981)

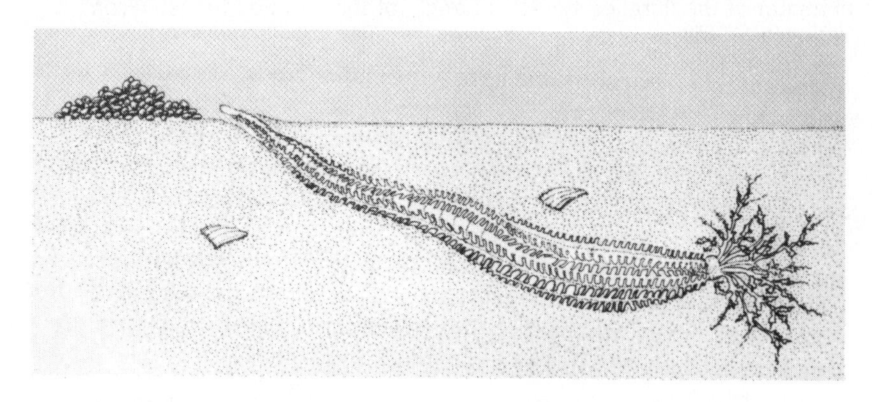

holothuroids and that of the crinoids or ophiuroids is that the tentacles are not structurally supported extensions. This has several important consequences. Lacking skeletal support, the tentacles never reach the length of some crinoid arms. The limited length of the tentacles and an inability to raise them far above the substratum restricts the interception of particles to the lower water column.

The length of the tentacles can be important in feeding. The tentacle is longer and the tentacle length to body length ratio is larger in *Thyone auriata* than in *Pentacta doliolum* (Velimirov 1985). In addition, the tentacles of *P. doliolum* are held at a lower angle and form a greater filtration-funnel diameter than those of *T. auriata* of the same body length (Figure 2.56). These two species occur together, with *P. doliolum* forming a dense layer over *T. auriata*. The differences would allow stratification of feeding.

Orientation of the tentacles in holothuroids may be difficult, and filtration fans have not been reported. The effectiveness of filtration thus would depend on degree of tentacular branching. As in some ophiuroid species, dense populations of dendrochirotids produce a forest of tentacles which would promote particle capture (Figure 2.54). Particles adhere to sticky material secreted by the papillae of the tentacles (Costelloe and Keegan 1984; Fankboner 1978, 1981; T.B. Smith 1983). In addition, particles may be physically trapped. *Psolus chitinoides* traps larger particles in an enclosure formed from expanded tubules which are bent inwardly (Fankboner 1978). The use of adhesion for food capture probably allows the capture of fine particles only (Fish 1967).

Particles collected by the tentacles are not transferred to the mouth by ciliary–mucoid currents as in crinoids. Instead, the tentacles are inserted one at a time into the mouth, and the particles are wiped off into the mucus

Figure 2.56: Filtration funnels of *Thyone auriata* and *Pentacta doliolum*. L: length of tentacle, H: height of tip of tentacle above the substratum, D: diameter of the filtration funnel, α: angle of the filtration funnel. (From Velimirov 1985)

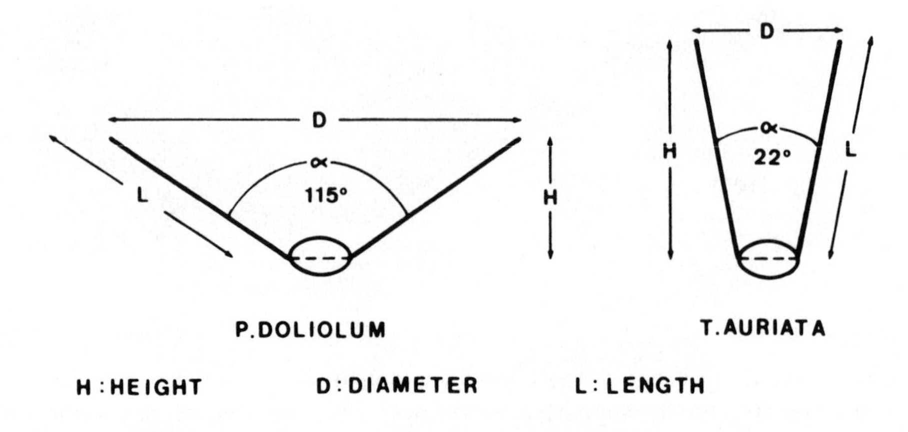

of the pharynx (Figure 2.57) (Costelloe and Keegan 1984; Fankboner 1978; Könnecker and Keegan 1973). Some tentacles are smaller and assist in removing particles from the long tentacles. *Pseudocucumis mixta* has fifteen large tentacles and an inner circlet of five small ones (Könnecker and Keegan 1973). As many as three small tentacles follow a large tentacle into the mouth and probably remove food particles as the large tentacle is withdrawn. Both the large and the small tentacles of *Aslia lefevrei* help push particles into the pharynx, and the small ventral ones can assist in removing particles as the large tentacles are withdrawn from the mouth (Costelloe and Keegan 1984). The removal of most particles from the tentacles seems simply to result from the tightly closed peribuccal lip. The mode of transfer of food particles into the mouth probably limits the length of the tentacles. In addition, this mechanism for transfer of particles from the tentacles to the mouth means that particle collection by a tentacle is not continuous. However, the rate of transfer would be more rapid than by the ciliary–mucoid mechanism.

The exposed tentacles in the water column are susceptible either to physical damage or predation. The introvert retracts the tentacle crown by contracting muscles attached to the radial pieces of the calcareous ring (Figure 3.26).

2.1.5.2. Aspidochirotacea

One of the most important evolutionary events in the holothuroids was the development of a more efficient means of deposit feeding, the development of the shield-shaped tentacles of the aspidochirotids (Figure 2.50). This

Figure 2.57: A: Tentacle crown, mouth and pharynx of *Aslia lefevrei*. BM: buccal membrane, LT: tentacle branch, PH: pharynx, M: mouth, ST: tentacle stalk. B: Feeding individual with a large tentacle inserted into the mouth. (From Costelloe and Keegan 1984)

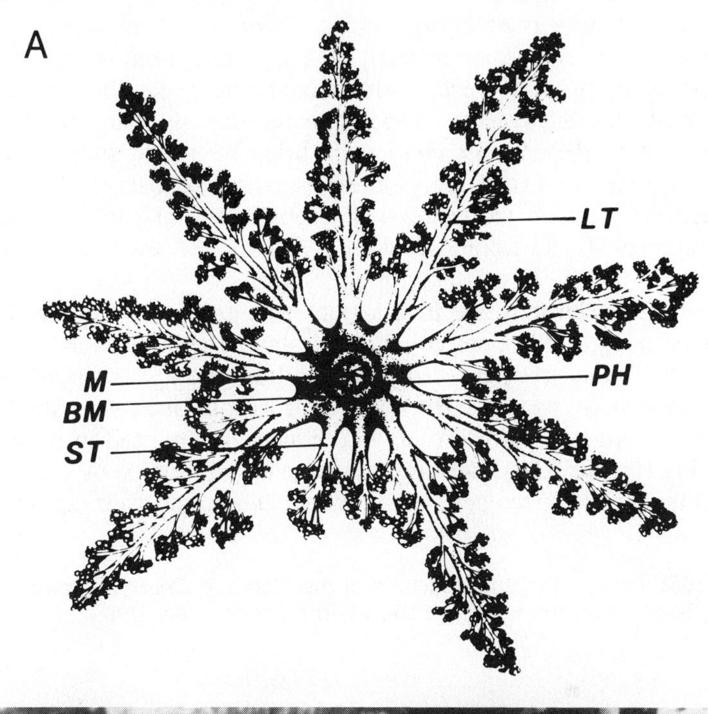

results from branching only at the distal end of the tentacle. For the most part, these deposit feeders can no longer depend upon food being brought to them as can the suspension-feeding dendrochirotids. Consequently the aspidochirotids are vagrant.

The foraging strategy of *Parastichopus californicus* seems simply to be to locomote and feed simultaneously, the apparent random movement pattern resulting from an evenly distributed and renewable source of detrital food (Da Silva *et al.* 1986). The reef-flat dwelling *Holothuria* species to which the currents continually bring new food supplies are a partial exception to this. None of these is completely sessile, however. Being dependent upon particles on the substratum, aspidochirotids are most prominent in the tropics (Pawson 1970) or the deep sea (Hansen 1975).

For the aspidochirotids, the food supply is benthic and below the body rather than being in the water column and above the body as in the dendrochirotids. Many species of aspidochirotids show a morphological change in the body form so that the mouth is positioned ventrally. This occurs in the order Aspidochirotida but is most pronounced in the order Elasipodida (Figure 2.58). Because the tentacles are not exposed into the water column, only the tentacles are retracted and no introvert occurs.

Figure 2.58: View of the undersurface of the elasipod *Deima validum* showing the mouth, the anus and the trivium (from Théel 1882)

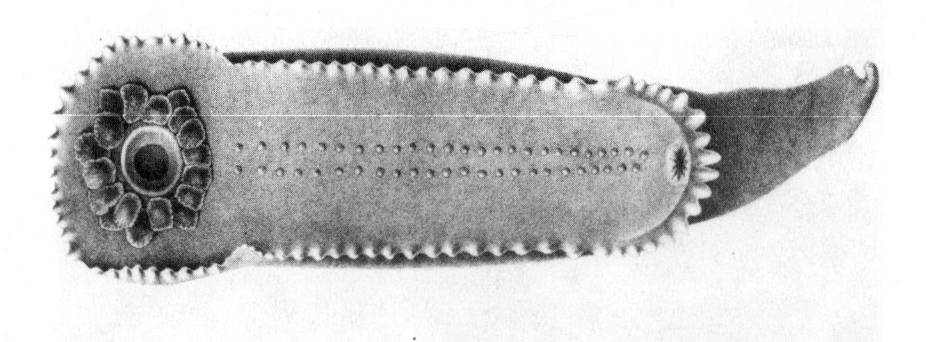

The shield-like end of the aspidochirotid tentacle is used to shovel or push particles directly into the mouth, or to first collect particles by mechanical trapping or by adhesion. A difference in the sizes of the particles ingested should result from these three different mechanisms. Large particles are ingested by shovelling or raking, moderate sized-particles by mechanical trapping, and fine particles by adhesion. The occurrence of coral rubble several centimetres in diameter in the gut of *Holothuria atra* indicates that at least some particles are pushed into the mouth by the tentacles. Mechanical trapping of particles has been reported. The tentacles of *H. atra* are expanded over and into the substratum for particle collection (Figure 2.59) (Roberts and Bryce 1982). The surfaces of the tentacles of *Holothuria forskali* expand similarly over particles which are trapped between its branches.

Figure 2.59: Particle handling by the tentacles of *Holothuria atra* from (a) side and (b) front views. The expansion of the tentacle nodules is indicated in the circles. (From Roberts and Bryce 1982)

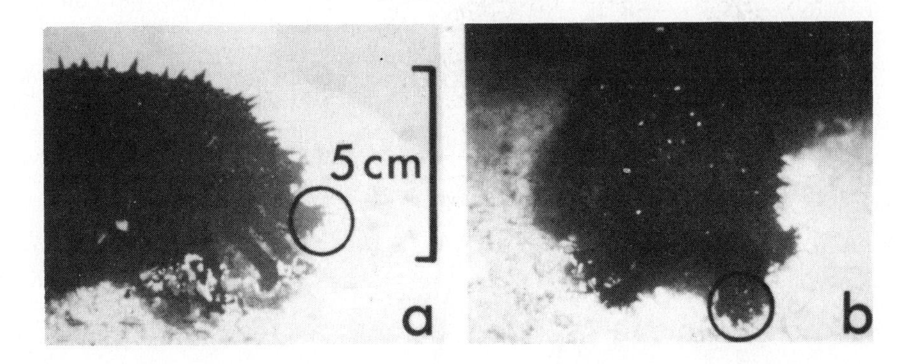

In contrast to these holothuriid species, mechanical trapping by the tentacle of *Parastichopus californicus* is considered secondary to adhesion for food collection (Figure 2.60) (Cameron and Fankboner 1984). Particles adhere to the surface of the tentacles and are also held in the cup formed from collapse of the tentacle as it is pulled from the substratum. The tentacle end is inserted into the mouth and the particles are released within the pharyngeal cavity (Roberts 1979; Roberts and Bryce 1982; Cameron and Fankboner 1984). The aspidochirotid tentacle has been modified for suspension feeding in *Holothuria glaberrima* (Nutting 1919) and *Holothuria cinerascens* (Roberts and Bryce 1982). The surface of the tentacles of deposit-feeding aspidochirotids has irregular nodules, whereas that of the tentacles of *H. cinerascens* has regular nodules with dense papillae which are secretory (Roberts and Bryce 1982).

The aspidochirotid tentacle has the ability to capture a wide range of particle sizes. However, the ability of the aspidochirotids to select particles of certain sizes in feeding in the field has been difficult to document. No evidence for particle size selection has been found for a number of species of tropical aspidochirotids, and it would seem that particle size selection is not a mechanism for resource partitioning among them (Hammond 1982).

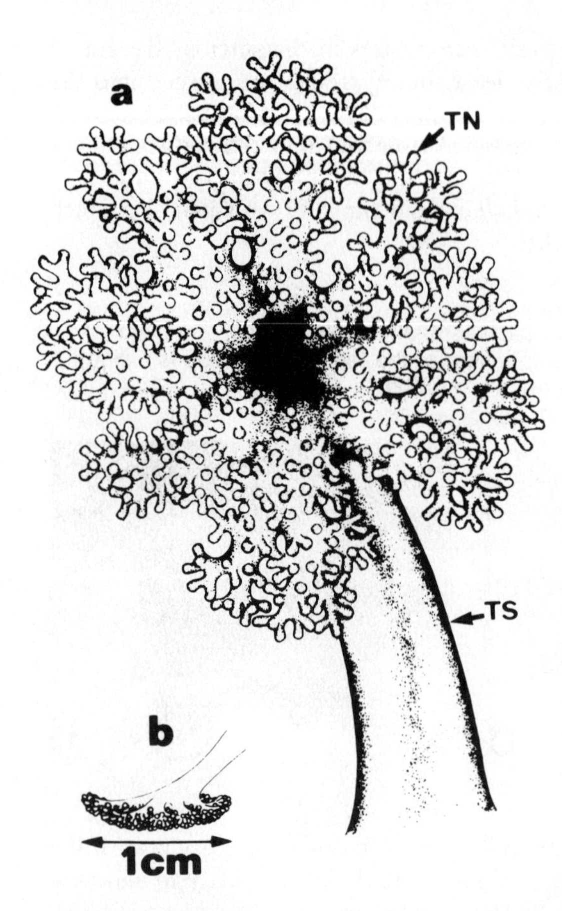

Figure 2.60: Expanded tentacle of *Parastichopus californicus*. (a) Tentacle nodules (TN) forming an adhesive surface on top of the tentacle stalk (TS). (b) Side view of an expanded tentacle being placed against the substratum during feeding. (From Cameron and Fankboner 1984)

Various species do differ in the sizes of particles ingested. Tropical aspidochirotids can be separated into those that ingest fine particles ($<$ 0.15 mm), those that ingest moderate-sized particles (0.25 to 1 mm), and those ingesting large particles ($>$ 1 mm) (Levin 1979). The considerable difference in size of particles ingested by closely related species (Levin 1979) indicates that the size ingested results in large part from differences in the availability of particles and not from a difference in functional morphology. This does not rule out selection of particle sizes, but indicates that it is not a prominent feature of feeding in the field.

Little selection of particle size would result from the use of shovelling or raking in ingestion of food, and probably little from mechanical food collection (although the sizes of the spaces in which particles are trapped offer a possible mechanism). Selection through properties of adhesion would seem to be important primarily for small particles and would depend on their weight and surface area.

Because of the feeding mechanism, the body size of different aspidochirotid species is not related to the sizes of particle ingested (Levin 1979). However, small *Parastichopus parvimensis* feed on fine particulate material whereas large individuals feed on granular deposits (Yingst 1982). The size of the calcareous ring provides support for the pharynx and tentacles, and may be more important in determining the maximal size of particles which can be ingested. The calcareous ring may be greatly reduced or essentially absent in some deep-sea elasipods which ingest very fine particles (Hansen 1975).

The tentacles of the deep-sea macroplanktonic elasipods are greatly modified for suspension feeding, ending in fine subdivisions (Barnes, Quetin, Childress and Pawson 1976; Hérouard 1906). In *Peniagone diaphana* the tentacles are raised, extended, and then laterally expanded to sweep through the water.

Several species of north-temperate aspidochirotids lack guts during winter months. Although a seasonal evisceration of unknown function in some species (Swan 1961), it seems to be associated with a controlled process of visceral atrophy in *Parastichopus californicus* (Fankboner and Cameron 1985). This would be associated with a pronounced seasonal change in food availability, and indicates that the cost of maintaining the gut is greater than that of regenerating it.

2.1.5.3. Apodacea

The apodids have pinnate tentacles (Figure 2.50). The flat surfaces of the tentacles of *Synaptula hydriformis* and *Opheodesoma spectabilis* are applied to the surface of the substratum or plants, and particles that adhere are ingested (Olmsted 1917; Berrill 1966, respectively). However, P.V. Fankboner (pers. comm.) found no adhesive properties associated with the tentacles of *O. spectabilis* that grasp food by means of flat pads on their external surfaces (Figure 2.61). Fankboner pointed out that the manner of feeding contrasts with the use of the inner surface of the tentacles of other groups and that locomotion and food gathering occur simultaneously in *O. spectabilis.* The tentacles are extended and collapsed, which leads to capture of particles on the outer surface. Retraction of the tentacles draws the body forwards, simultaneously removing the particles from the substratum. Food is released on the far side of the buccal cavity instead of on the near side as in aspidochirotids and dendrochirotids.

Leptosynapta tenuis ingests particles while it burrows, but also forms feeding funnels (Figure 2.62) (Powell 1977). Particles that adhere to the tentacles extended out of the funnel and applied to the surface of the substratum, as well as particles which fall down the funnel, are ingested. The tentacles are inserted into the pharynx and withdrawn to remove food particles.

It is assumed that the digitate tentacles of the burrowing molpadids also

Figure 2.61: A: Tentacles of *Opheodesoma spectabilis* being extended for particle collection and being inserted into the pharynx. B: Sequence showing (a) extension of a tentacle, (b) application of the extended tentacle and spread branches to the substratum, (c) retraction of the tentacle and branches, and (d) upward bending of the branches with particles adhering on the outer surface. (P.V. Fankboner, unpublished)

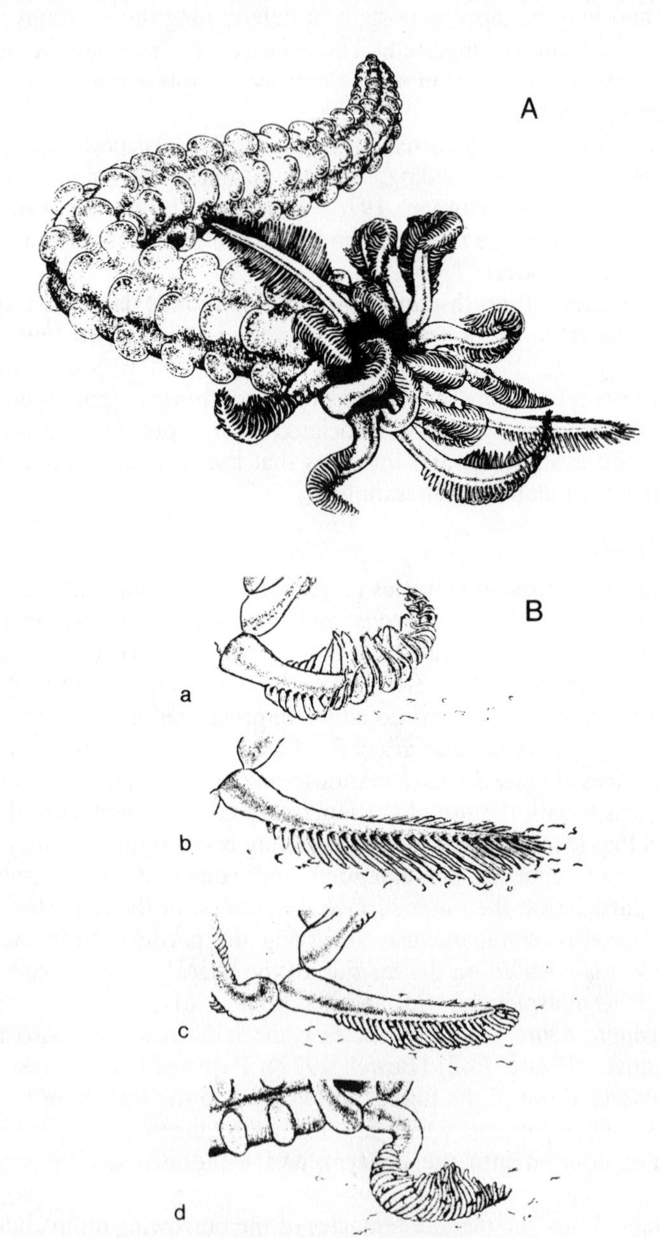

Figure 2.62: *Leptosynapta tenuis* in the feeding posture. A: Funnel-shaped depression, B: faecal mound, C: site of initiation of new burrow. (From Powell 1977)

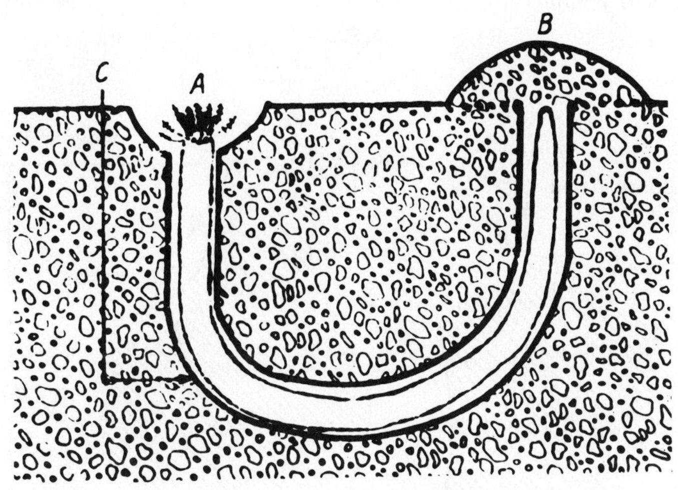

use adhesion to collect food particles (Figure 2.63). *Molpadia* recycles the substratum to bring the organically rich surface particles to the buried mouth (Rhoads and Young 1971). It would thus be restricted to fine substrata in which this type of feeding is possible.

A theoretical analysis of the optimal feeding hypothesis suggests that deposit feeders would ingest small and organically rich particles because it would maximise net energy gain per unit time feeding (Taghon, Self and Jumars 1978). *Leptosynapta clarki* selected small glass beads in feeding although the higher proportion of small particles in the gut than in the substratum was not great (Taghon 1982). However, *Leptosynapta tenuis* ingests all particles from the natural substratum except the smallest and largest (Myers 1977), and the large *Euapta lappa* does not discriminate between different sizes of particle (Hammond 1982).

2.2. UPTAKE OF NUTRIENTS ACROSS THE BODY SURFACE

Parts of the bodies of large echinoderms are distant from the gut and their mode of nutrition has always been a problem. This is particularly true of skeletal structures. Two mechanisms provide for direct uptake of nutrients across the body surface. They certainly could provide for skeletal elements and potentially could have a much larger role. Few analyses of echinoderm energy budgets have considered this source of acquisition.

Uptake across the body surface can be by active transport (Bamford 1982). This has been demonstrated for all echinoderm classes. The entire

Figure 2.63: Feeding posture of *Molpadia oolitica* showing surface faecal cone with population of the polychaete *Euchone incolor* and depression containing unconsolidated faeces produced by subsurface feeding (from Rhoads and Young 1971)

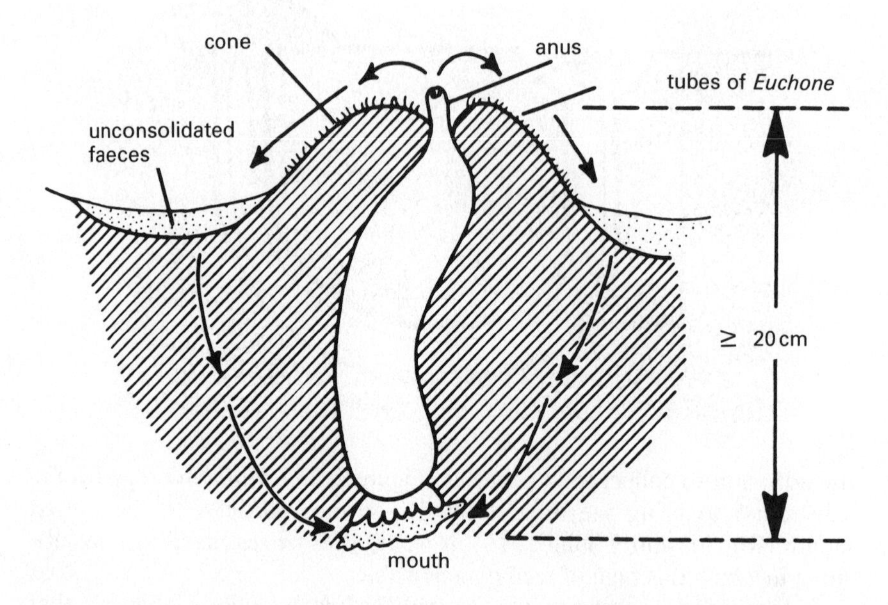

body surface has this capacity in contrast to the limited areas involved in molluscs and crustaceans. Kholodov (1975) calculated that 13% of the food requirements of *Strongylocentrotus droebachiensis* are satisfied by dissolved organic material and that 80% of the material was accumulated in the external body structures. The amount of amino acids taken up by active transport has been calculated to be equivalent to 1% of the aerobic respiration of the entire individual in *Sclerodactyla briareus* (Ferguson 1982c) to 58% in *Dendraster excentricus* (Stephens, Volk, Wright and Backlund 1978). The capacity for active transport of nutrients across the external body surface should free many structures from a dependence on food as a source of energy and material. Because echinoids in particular, and possibly all echinoderms, move water through their guts, it is possible that some of the uptake of dissolved organic material is via the gut in addition to the external body surface.

External digestion of complex organic material and living tissue has been noted in an asteroid, an ophiuroid and two species of echinoids (Péquignat 1970, 1972). This is a consequence both of the release of hydrolytic enzymes and of activity of coelomocytes that migrate out of and back into the external epithelium. The contribution of this process to the energy budget has not been calculated.

3

Maintenance Activities

Organisms must maintain themselves through time. This is survival, and is one component of fitness. Maintenance activities include those processes that counter the consequence of the second law of thermodynamics. Such processes are the maintenance of the ionic imbalance between the cell contents and the medium in which they occur and the repair of structural deterioration from subcellular to organ structures. Phenomena that protect the individual from the environment, whether physical or biological, are maintenance activities. These maintenance phenomena may be either structural or behavioural. Maintenance activities can be considered the 'cost of living' (Calow 1984a) and have been applied to various echinoderm activities (Ebert 1982). The significance of maintenance activities in considering the functional biology of echinoderms is that these activities utilise part of the acquired nutrients and make less available for somatic growth and reproduction. It is essential that the characteristics of maintenance activities be understood in order to evaluate their costs and their effects on somatic growth and reproduction.

3.1. UPTAKE OF OXYGEN

3.1.1. Oxygen requirements

Echinoderms are aerobic organisms although some may withstand periods of low levels or absence of oxygen for considerable periods of time (Shick 1983). The high dependency on oxygen may be due to the strongly aerobic nature of the body wall (Farmanfarmaian 1966) because the internal organs seem to have some capacity for anaerobiosis (Ellington 1982; Shick 1983).

Although aerobic, echinoderms do not have a high absolute requirement for oxygen. Rates of oxygen consumption calculated on a wet-weight basis show the effect of amount of skeletal material present in asteroids, being *c.* 1 ml oxygen ind^{-1} h^{-1} in *Asterias rubens* (Miller, Mann and Scarratt 1971) and 0.002 ml oxygen ind^{-1} h^{-1} in *Linckia laevigata* (Webster 1975). Similarly, ophiuroids have a low rate, e.g. 0.003 ml oxygen ind^{-1} h^{-1} in *Amphiodia occidentalis*, because of the great development of the skeleton (Webster 1975). Rates of oxygen consumption of species of other echinoderm groups are of similar magnitude (see Lawrence and Lane 1982).

The allometric relationship between the rate of oxygen consumption and respiration varies with species, season, food and temperature. The allometric constant varies among the classes (Shick 1983). On an individual basis, the allometric constant is greatest in holothuroids (mean = 0.85), in which the respiratory trees deliver oxygen to the interior of the body, and in asteroids (mean = 0.84), in which the body is relatively flattened, and specialised surface exchange areas (papulae) occur. It is smaller (mean = 0.64, near the expected 0.67 which would occur if the rate were proportional to body surface area) in the echinoids, probably related to the globoid shape and calcified test. It is lowest in the ophiuroids (0.54), in which the proportion of metabolically inert inorganic material increases with size.

The respiratory rate of echinoderms does not seem to be affected by latitude, indicating that the species are adapted to the environmental temperature at which they live (see Lawrence and Lane 1982). Thus the necessary adaptations at the molecular level have occurred to allow the appropriate rate of energy production over the entire range of temperatures at which echinoderms are found (−2 to 35°C). Species similar in structure and function should be expected to have similar rates.

In general, there seems to be little or only partial seasonal acclimatisation of respiratory rate of echinoderms to temperature (see Lawrence and Lane 1982). However, the occurrence of different respiratory rates at different seasonal temperatures does not necessarily mean a lack of acclimatisation as the physiological state of echinoderms typically changes seasonally.

Either because of the low rate of aerobiosis or because the basic body structure does not have the potential for appropriate modification, mechanisms to increase the uptake of oxygen from the environment are not highly developed except in infaunal groups and little regulation of oxygen consumption occurs.

3.1.2. Crinoidea

Undoubtedly oxygen diffuses across the tube feet of the arms into the radial water canals, but the characteristics of the canal make it unlikely that it supplies oxygen to other parts of the body. The anal tube of comatulids shows rhythmic pulsation and may have a respiratory function (Hyman 1955). The tegmen is perforated with ciliated hydropores which provide communication between the coelomic fluid and the surrounding medium (Figure 3.1) (Breimer 1978). The number of hydropores varies between 500 and 1500 in adults, and they are thought to be a part of the water vascular system. They may provide a route for oxygen into the body cavity. Because the water vascular system in crinoids is completely in the soft tissue external to the skeletal system, supply of oxygen to the skeletal tissue must be solely by diffusion. This would also be the case for the stalk of stalked crinoids. Oxygen demand must be low there either because of the small number of cells or an anaerobic capacity.

Figure 3.1: Section through the tegmen of a crinoid showing ciliated hydropores. c: Coelom; nf: nerve. (From Hamann 1889)

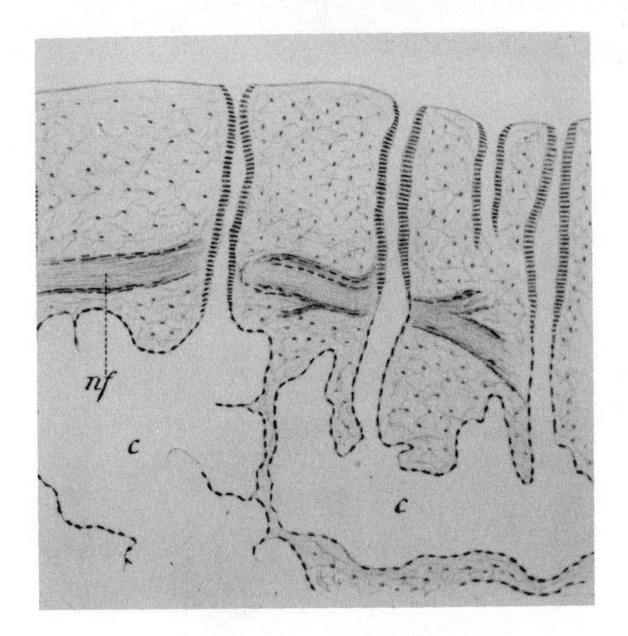

3.1.3. Ophiuroidea

Although much of the body surface of ophiuroids lacks an epithelium and consequently cilia, there are ciliated epithelial surfaces along parts of the arm and the ventral surface which create surface water currents (Gislén 1924). The bursae are sac-like invaginations of the oral body wall at both sides of each arm (Figure 3.2) which presumably arose to meet the requirements of life in burrows (Spencer and Wright 1966). If so, their retention

Figure 3.2: *Ophiopteris antipodum*. 1: Vertical section of disc showing complex bursal extension into the external interradial muscle. 2: Horizontal section of disc at level A of 1. Vertical section of disc in plane C of 2 showing bursal extensions under the stomach and over the proximal portion of the arm. Ext. inter. m: external interradial muscle; Interamb. pouch: interambulacral pouch of the stomach; Interv. m: intervertebral muscle; Prox. arm seg.: proximal arm segment. (From Pentreath 1971)

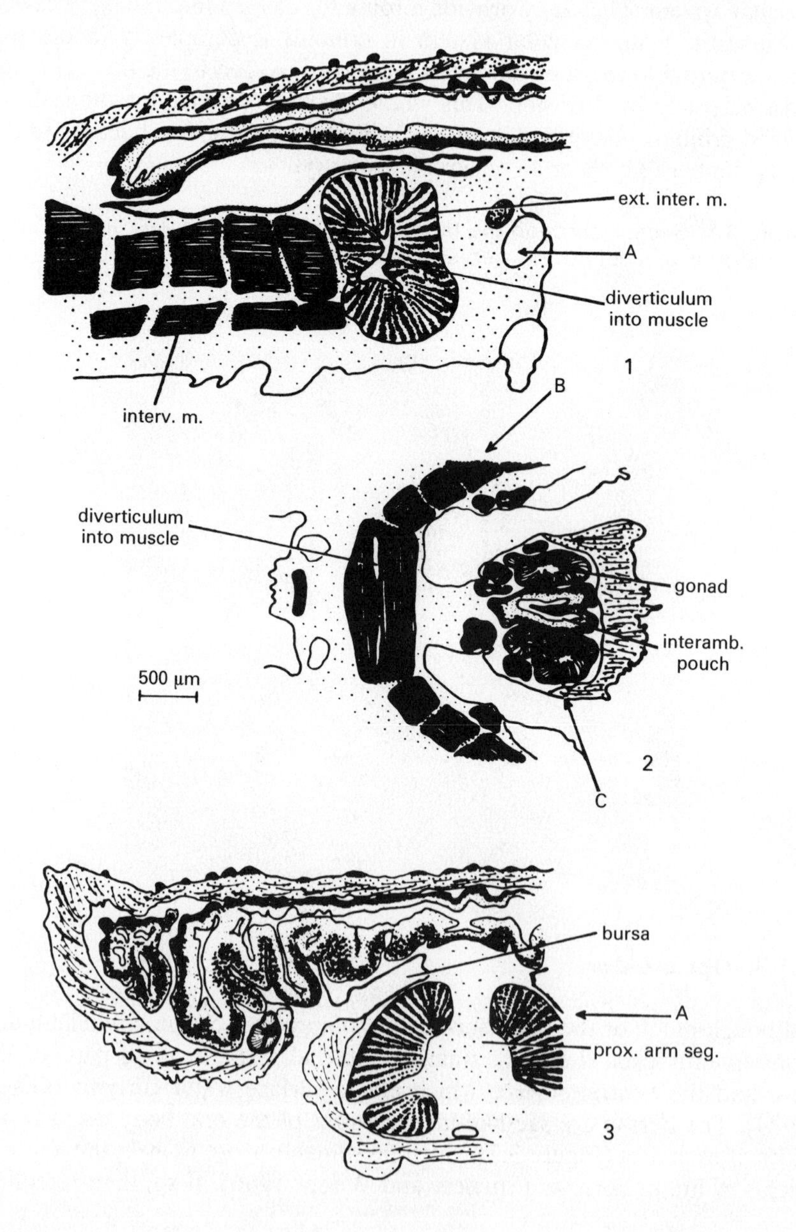

in extant, non-burrowing forms may be associated with their role in reproduction. The bursae are usually well-developed, but are reduced or lacking in some forms, indicating that they are not always essential for respiration. Most species that lack bursae are less than 2 mm in diameter (Austin 1966), near the limit which Farmanfarmaian (1966) calculated to be maximal for an individual to obtain oxygen by diffusion alone.

The bursae open by slits which are supported by the genital shields. Cilia associated with the bursal slits produce currents which enter and leave the bursae on the lower and upper sides (Austin 1966; Gislén 1924). Pumping movements by the soft lower body wall of *Amphiura filiformis* while the bursal slits are open are interpreted to be respiratory movements to increase water exchange in the bursae (Gislén 1924). The gnathophiurine jaw of some families (Amphiuridae, Ophiactidae, Ophiotrichidae, Ophionereididae, Ophiocomidae), which has greatly enlarged interradial and radial muscles and perhaps greater mobility at the adradial articulation, may have evolved to pump water through the bursae (Wilkie 1980).

The bursae of *Ophiopteris antipodum* and *Ophiactis resiliens* have few bursal muscle fibres but possess diverticula which penetrate the large external interradial muscles (Figure 3.2) (Pentreath 1971). In contrast, *Ophionereis fasciata* has well developed bursal muscles but lacks diverticula in the interradial muscles. These may be alternate methods for ventilatory movements (Pentreath 1971). Certainly the bursae are sites of oxygen uptake in these species because oxygen consumption is reduced when their openings are blocked. The bursal slits of *Ophiothrix spiculata* also widen in low concentrations of oxygen although there is no increase in the disc pumping rate (Austin 1966).

Burrowing forms must have a mechanism for obtaining oxygenated water from the surface. The burrows of amphiurids are ventilated by undulations of the arms (Woodley 1975). Waves pass down the undulating arm(s) from the surface to the disc chamber and then pass up the channels past the inactive arm(s). The rate of undulations increases with reduced oxygen concentration. The undulatory movements of the arms are made more efficient by closing the gaps between the spines by means of the tube feet (Figure 3.3). The tube feet may be held out sideways between the spines and lightly anchored to the burrow wall. They may have this posture but hold particles from the substratum to form a continuous fringe, or the particles may be transferred to the arm spines to form the fringe. The capacity of the tube feet to function in respiration or feeding would be affected by which mechanism was employed. The tube feet would be effectively non-operative for other functions when forming a respiratory fringe in *Ophionephthys limicola.*

Figure 3.3: (a) The posture of tube feet of *Acrocnida brachiata* on an arm engaged in respiratory undulation. The tube feet are extended laterally and flexed towards the disc. (b) The respiratory fringe formed of mucus-bound particles attached to the tube feet and adjacent spines of *Ophionephthys limicola*. (c) The respiratory fringe of mucus-bound particles attached to the spines of *Amphiura filiformis*. (From Woodley 1975)

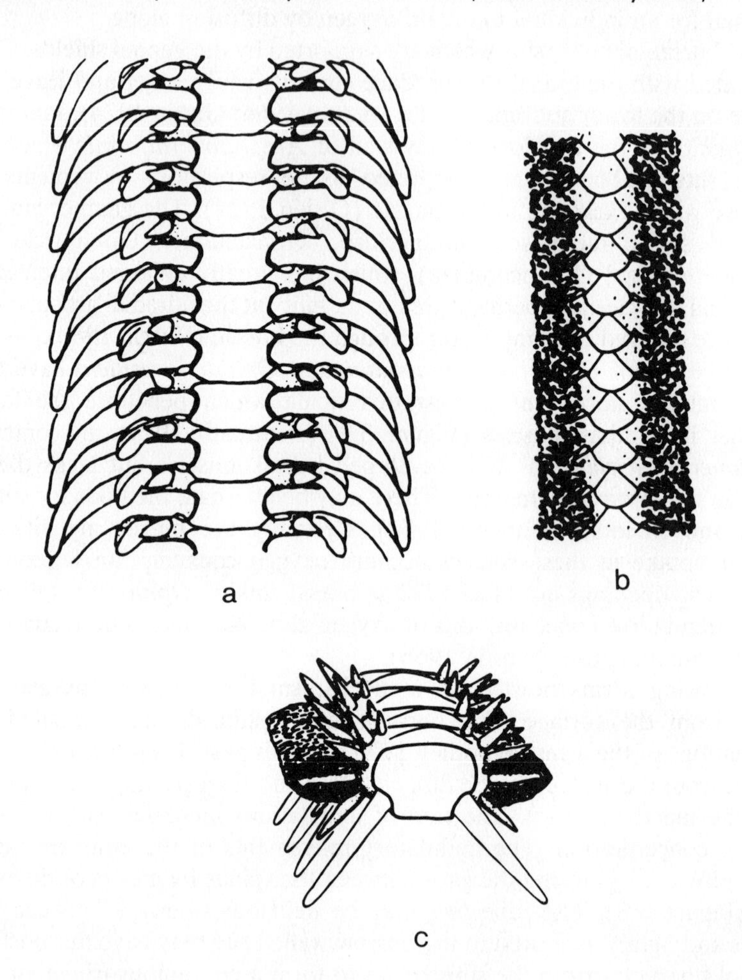

3.1.4. Asteroidea

Oxygen uptake is by the tube feet and across the body wall. The latter is made possible by development of the ossicular skeleton of the body wall. This permitted the development of papulae, portions of the body wall in which the dermal layer is very thin (Figure 3.4). These can be swollen,

Figure 3.4: Cross-section of the body wall of an asteroid showing a papula. c: Connective tissue; ci: cavity within the dermis; cp: cavity around the papula; g: mass of particles within the papula; m: sub-peritoneal muscle layer. Cilia cover the external and internal surfaces of the papula. (From Cuénot 1948)

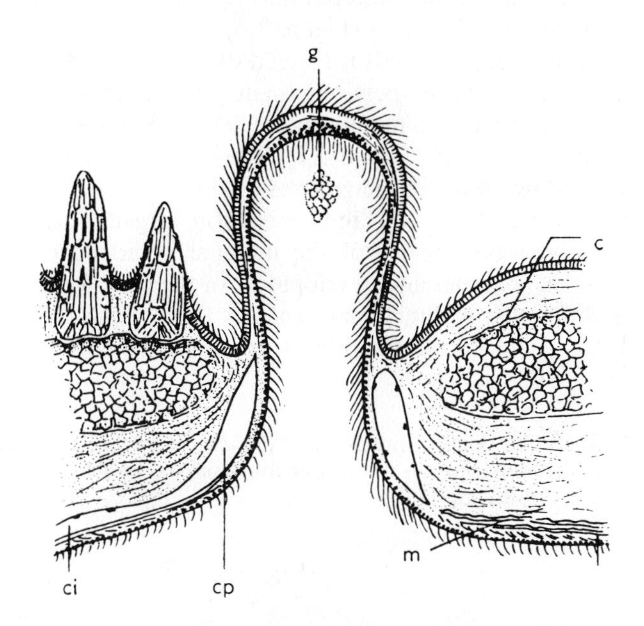

presumably by hydrostatic pressure of the coelomic fluid, and retracted by muscles found on the inner side of the dermis. The papulae are lined with flagellated epithelial cells. External ciliated currents in asteroids are produced by cilia associated with the papulae and the base of the spines, and are thought to function in respiration and cleansing (Gislén 1924). No major current directions occur on the upper surface because of the diverse arrangement of the papulae and spines, but particles are moved to the side. Ambulacral currents seem to be absent or weak towards the mouth, and ventral interradial currents are away from the mouth.

The occurrence of the papulae is variable with development and among taxonomic groups (Sladen 1889). The paxillosids have papulae only on the upper surface, only at the base of the ray in some but scattered over the entire upper surface in others. The lack of papulae on the lower body surface is probably because of the shallow-burrowing habit and association with particulate substrata of members of this group. The forcipulatids have papulae on both the upper and lower surfaces. Newly metamorphosed asteroids have no papulae initially. Papulae first appear on the upper surface near the base of the ray and only subsequently on the lateral and lower surfaces. The evolutionary and developmental difference in the extensiveness of papulae coverage implies a functional significance, as does their form. The papulae can be bush-like as in *Pycnopodia* (Figure 3.5), long and vermiform as in *Rathbunaster*, or small and numerous as in *Astrometis*. The papulae are clustered into papularia in the proximal mid-radial upper surface of *Pontaster* (Figure 3.6).

The body frame of paxillosids is formed of two series of large marginal plates with vertical grooves between them. In *Luidia* and *Astropecten*, these grooves are edged with minute spines, which A. Agassiz (1877) compared to echinoid fascioles (Figure 3.7). The small spines are ciliated and create currents down through the grooves.

Elaborate ciliated grooves, the cribriform organs, occur between adjacent superomarginal plates of the infaunal goniopectinids and porcellanasterids. The number and development of the cribriform organs vary, implying a difference in functional capacity (Figure 3.8) (Sladen 1889; Madsen 1961). The cribriform organs are partially enclosed and have

Figure 3.5: Bush-like compound papulae (P) with surrounding dwarf and giant straight pedicellariae on the upper mid-surface of *Pycnopodia helianthoides* (from Fisher 1928)

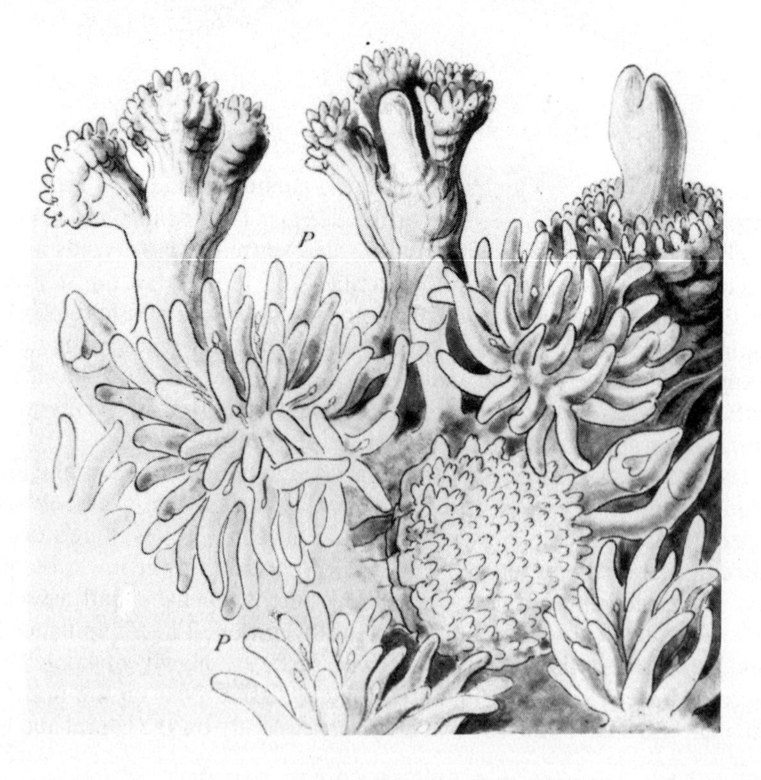

Figure 3.6: Papularia on proximal mid-radial upper surface of *Pontaster limbatus* (from Sladen 1889)

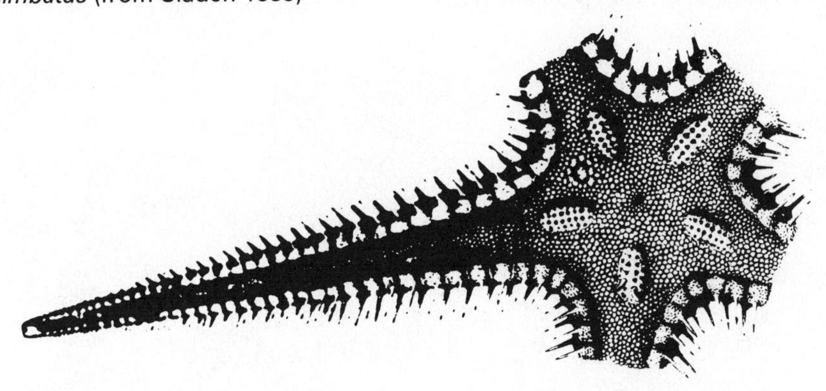

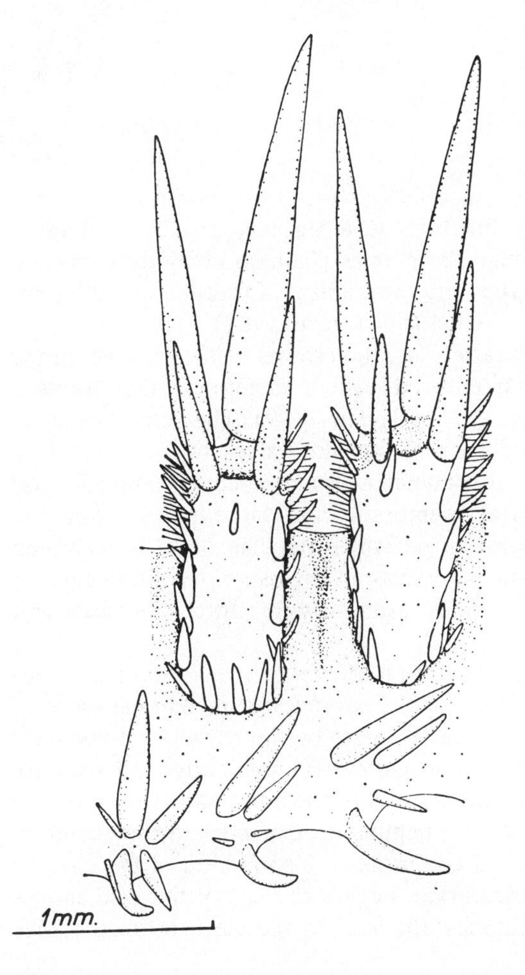

Figure 3.7: View of the lower surface of the arm of *Astropecten spiniphorus* showing the inferomarginal plates with long protective spines on the outward edge and minute spines on the lateral surface (from Madsen 1950)

Figure 3.8: Cribriform organs in the interradius and apical cone of *Abysdaster tara* (from Madsen 1961)

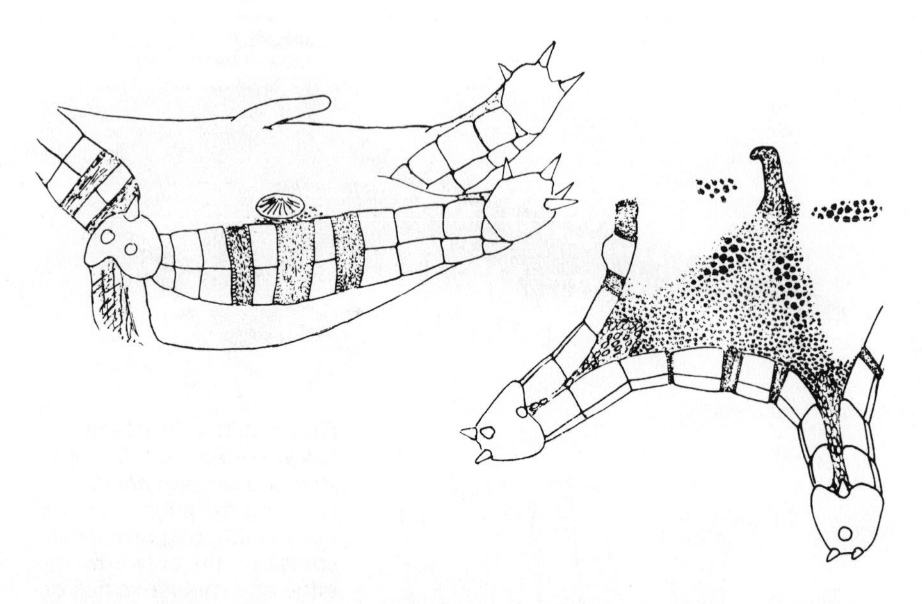

lamellae (Figure 3.9). They function as a turbine, producing a high-pressure downward flow through the burrow channels along the arms and across the papulae (Figure 3.10). This mechanism of ventilation is adaptive to life in an easily disturbed sediment (Shick *et al.* 1981).

In the spinulosid family Pterasteridae, the crowns of the paxillae on the upper surface are united and covered with a membrane that forms a supradorsal membrane (Figure 3.11) (Sladen 1889). The central opening over the anus may be closed by five fan-like valves or by webbed or papillose spinelets. There are also numerous small openings (spiracles) on the surface of the supradorsal membrane and large pores along the ambulacral grooves. The body wall on which papulae occur is separated from the supradorsal membrane and forms an enclosure (the nidamental or respiratory chamber). This system is respiratory in nature (Johansen and Petersen 1971).

Muscular contractions of the body wall of *Pteraster tesselatus* and relaxation of the circular muscle fibres of the supradorsal membrane increase the volume of the chamber, and water enters by the ambulacral openings (Nance and Braithwaite 1981). The ossicles in the body wall overlap, which allows greater than usual contraction of the body wall. The increased hydrostatic pressure also causes the papulae to expand into the chamber. Deflation occurs by a reversal of the process. Contraction of the circular muscles of the supradorsal membrane begins at the ray tips and moves towards the disc. Thin membranes attached to muscle-controlled spines

Figure 3.9: Cross-section of cribriform organs taken from an interradius of *Ctenodiscus crispatus*. e: Ciliated epithelium; m: marginal ossicles; s: skeletal support of ciliated lamella. (From Cuénot 1948)

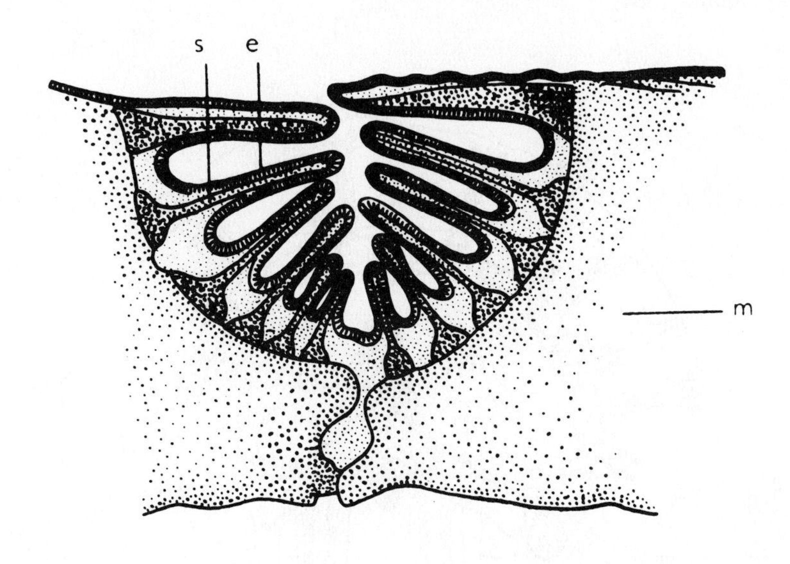

Figure 3.10: Irrigation current flows of buried *Ctenodiscus crispatus*. (From Shick *et al.* 1981)

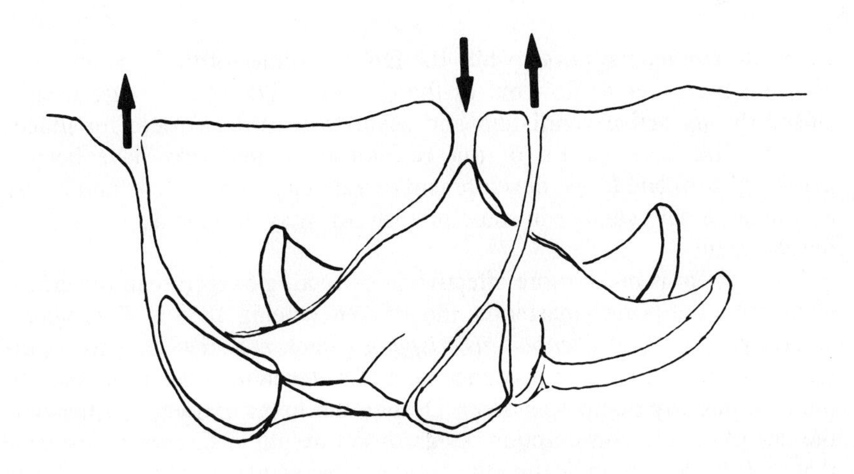

Figure 3.11: Cross-section of the arm of *Pteraster tesselatus*. AMP: Ambulacral pore; AP: ampulla; CO: coelom; DB: dermal branchia; NDC: nidamental chamber; OS: ossicle; PX: paxilla; PYC: pyloric caecum; SDM: supradorsal membrane; SPI: spiracula; TF: tube feet. (From Nance and Braithwaite 1981)

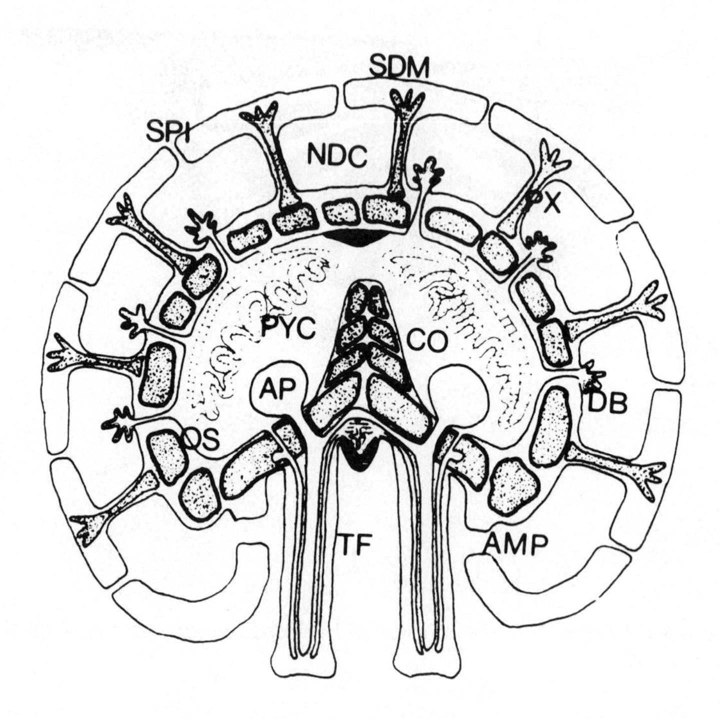

cover the ambulacral pores, while the fan-like spines of the osculum relax allowing the water to flow out of the chamber. The spiracles are usually closed during active ventilation and seem to provide a route for mucus release. The development of this respiratory system may have been a necessary correlate to the development of a defence mechanism, and its use as a nidamental chamber in some pterasterids may be seen as a secondary development.

This system is much more effective in extracting oxygen than the cribriform system of porcellanasterids, the efficiency being 18% in *P. tesselatus* and only 7 to 8% in *Ctenodiscus crispatus* (Shick *et al.* 1981). This is perhaps related to the longer residence time of water in the tidal as opposed to the continuously pumping system. Despite the lower extraction efficiency, the rate of oxygen consumption per gram wet weight in *C. crispatus* is twice that of *P. tesselatus*, indicating the greater effectiveness of the continuously flowing system.

Many infaunal asteroids also have an apical cone (anal cone, epiproctal cone) (Figure 3.10, 3.12) (Sladen 1889). In the porcellanasterids, the cone is a morphological feature. The apical cone of the goniasterid *Ctenodiscus crispatus* is extendable and is used by buried individuals to maintain a connection with the water column. The degree to which the apical cone of *C. crispatus* is extended is dependent upon oxygen concentration (Figure 3.12) (Shick 1976).

Figure 3.12: Changes in the apical cone of *Ctenodiscus crispatus*. Individual at rest (a) or exposed to hypoxia on surface of substratum (b), or burrowed (c). (From Shick 1976)

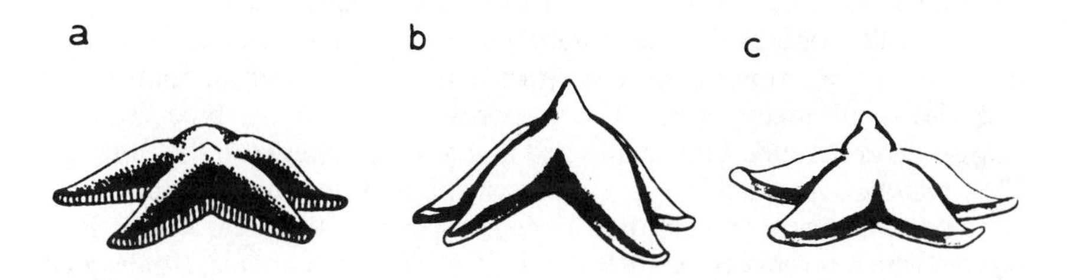

3.1.5. Echinoidea

The echinoids have tube feet on both the upper and lower body surfaces. Although some tube feet on the upper surface function in feeding, others have been freed for modification for respiratory purposes. That the tube feet are the primary site of oxygen uptake was demonstrated by the use of bacteria which fluoresce only in the presence of oxygen (Farmanfarmaian 1966). The capacity for oxygen uptake by the tube feet has been increased in three ways: an increase in the number of the tube feet, modifications in their structure, and an increase in the movement of the medium over their inner or outer surfaces.

Tube feet open through the ambulacral plates of the echinoid test except for the accessory tube feet of clypeasteroids and spatangoids which open through the sutures between the plates. Because there is only one tube foot for each ambulacral plate, the evolutionary trend of increased number of tube feet involved an increase in the number of ambulacral plates (Kier 1974). This has occurred in two ways. The evolution of non-cidaroid Palaeozoic echinoids involved an increase in the number of columns in an ambulacrum from two to more than twenty (Kier 1966). This may have been possible because of their imbricating plates which made a non-rigid test. The extant cidaroids have only two columns of plates in each ambulacrum. The ambulacral plates remained vertically aligned in cidaroids and consequently became very narrow in height. Consequently, the evolutionary development of compound plates (a single primary tubercle extending over several adjacent ambulacral plates), which occurred in the euechinoids in a variety of forms (Figure 3.107) was not necessary to increase the number of tube feet. The adaptiveness of compound plates is associated not only with the functions of the tube feet, but also with the development of wide ambulacra and the large spines associated with them. The evolutionary development of pore arcs, which is associated with compound plates, would result in the increase in the number of tube feet possible along an ambulacrum.

The tube feet, pores and ampullae have all been modified for respiration (A.B. Smith 1978, 1980a). The respiratory tube feet of regular echinoids have relatively larger isopores than those of adhesive tube feet (Figure 3.13). These isopores diverge inwardly and open at a distance from each other on the interior of the test, producing long, flattened, thin-walled ampullae with many septa. The isopores of respiratory tube feet are elongate and the tube foot is flattened along the axis between the isopores. They may be separated by a partition (partitioned isopores) or joined by a furrow (conjugate isopores). Partitioned respiratory tube feet have a long septum which is reduced or absent in non-respiratory tube feet. Conjugated tube feet have two thin-walled cylinders joined by a convoluted area which may have extensions at each end. This convoluted area greatly increases the surface area of the tube foot and divides the inflowing deoxygenated ambulacral fluid into many streams. This system also separates the incurrent from the outcurrent to a greater degree than in the partitioned tube feet.

The respiratory tube feet on the upper surface of irregular echinoids can have either isopores or anisopores (A.B. Smith 1980a). Additional pores are associated with the very wide tube feet of clypeasteroids (Figure 3.14) (Durham 1966). The tube feet of irregular echinoids have two large tubes on either side connected by a thin-walled central area. This central area is partitioned into narrow passageways which are stacked vertically in cassiduloids, most clypeasteroids and some spatangoids.

The tube feet of spatangoids can show the greatest complexity, being exceedingly convoluted (Figure 3.15) (A.B. Smith 1980a). The walls of the ampullae of the respiratory tube feet of irregular echinoids are extremely broad and flat with septa partitioning the lumen into passageways (Figure 3.15). The ampullae may even have channels on opposite sides. The respiratory tube feet are thin with only a single layer of epithelial cells on either side of a very thin layer of connective tissue fibres, and sparse muscle fibres. The central, partitioned area of the respiratory tube feet of irregular echinoids maintains the narrow–elongate shape of the tube foot during contraction and relaxation. The septa also separate the incurrent into many streams and lead to full use of the central region of the tube foot. The

Figure 3.13: A: Longitudinal sections of the tube feet of *Centrostephanus nitidus* (P1), *Echinus esculentus* (P2), and *Arbacia lixula* (P4) with associated isopores. B: Partitioned isopores of type P1 of respiratory tube feet (slight development of attachment area around pores); type P2 (with attachment area continuous around the pores except for the neural canal); and type P4 (with small pores that are separated by a large partition, and with a broad attachment area around most of its circumference). (From A.B. Smith 1978)

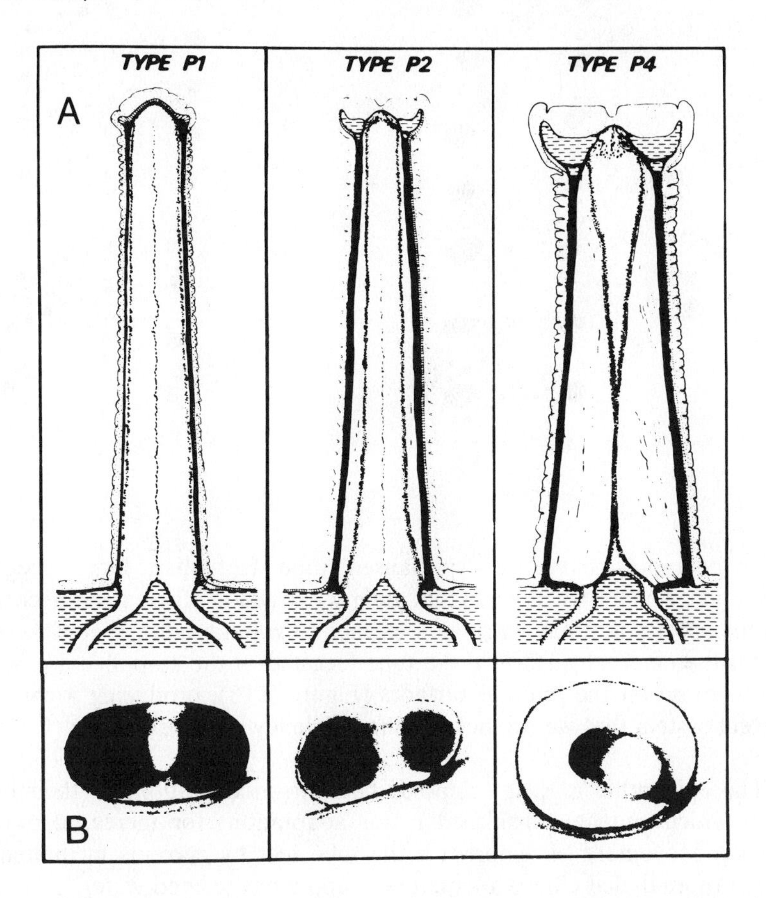

central region increases the surface area of the tube foot and isolates the incurrent deoxygenated and outcurrent oxygenated ambulacral fluid. The lateral extension and reduced height of respiratory tube feet of the irregular echinoids are probably an adaptation to the infaunal way of life. The more elongate tube feet are typically skewed so that the incurrent is in the taller part, presumably causing the incurrent to be better distributed through all the pathways of the central partitioned region. The development of side branches allows for increased length without increased height. Small irre-

Figure 3.14: Portion of the petals of A *Rotula orbiculus* with a single outer pore, and B *Rotula augusti* with multiple division of the outer pore (from Mortensen 1948b)

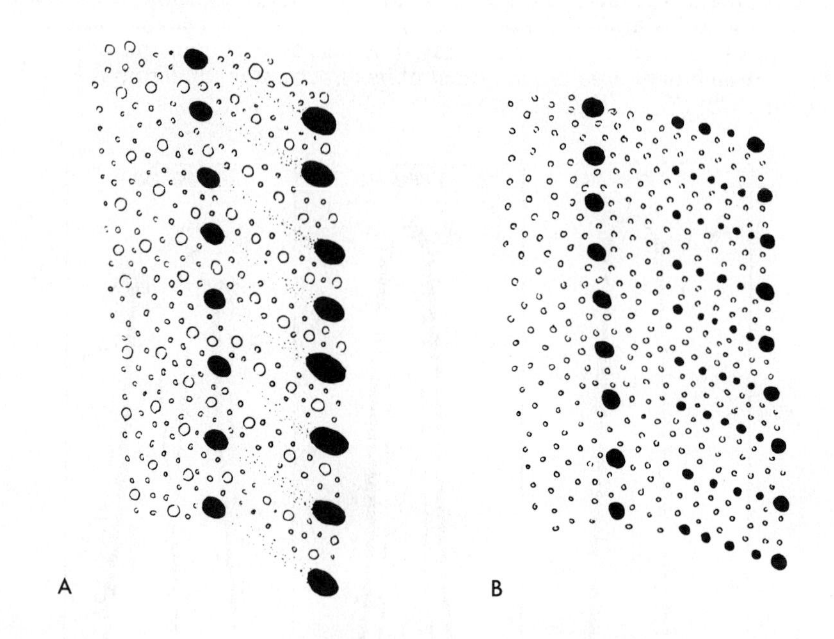

gular echinoids have less specialised tube feet than large irregular echinoids, which may indicate that greater extraction efficiency is necessary for the species that have a smaller surface to volume ratio. Water currents across the external surface of the tube feet and of the ampullae are opposite to those on the internal surfaces (Figure 3.16), producing a counter-current system that would increase the efficiency of gas exchange (Chesher 1969).

The epibenthic regular echinoids have no modifications of their body form which can be considered major adaptations for increased oxygen uptake. Movement of water over the tube feet by currents facilitated by action of epithelial cilia is adequate to supply oxygenated water. This contrasts with the irregular echinoids which have modifications of the body form (petals) and of the appendages (spines) to facilitate oxygen uptake. The irregular echinoids which lack these modifications, the holectypoids and pygasteroids, are the most primitive (Kier 1974). Cassiduloids have petals that are less developed than those of spatangoids and clypeasteroids.

Spatangoids, which inhabit the fine substrata in which water circulation is difficult, have depressed petals protected by spines and special arrangements for directed and higher velocities of current flow (Figure 3.17). Spatangoids have only four petals as the anterior ambulacrum is modified

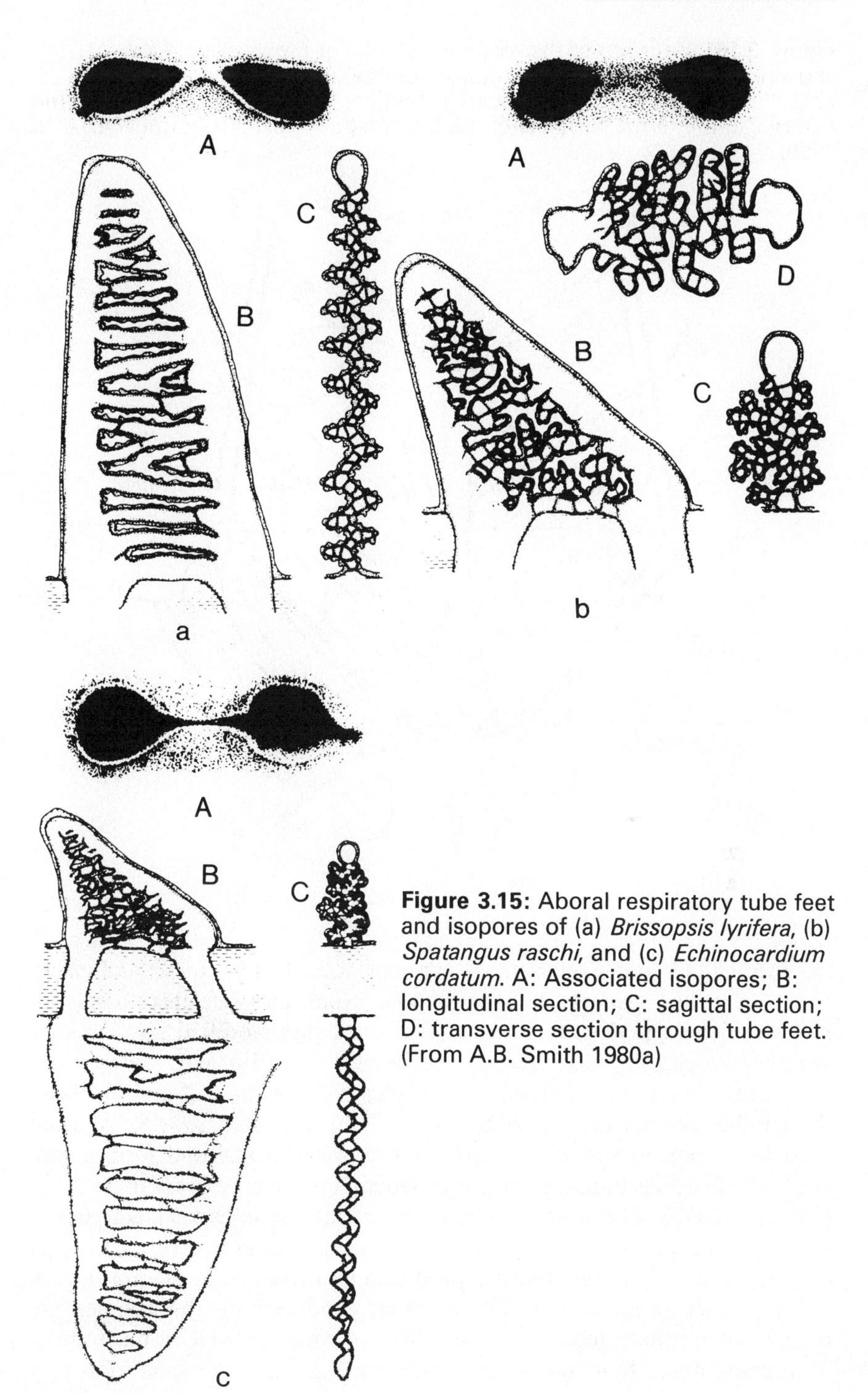

Figure 3.15: Aboral respiratory tube feet and isopores of (a) *Brissopsis lyrifera*, (b) *Spatangus raschi*, and (c) *Echinocardium cordatum*. A: Associated isopores; B: longitudinal section; C: sagittal section; D: transverse section through tube feet. (From A.B. Smith 1980a)

Figure 3.16: A portion of the respiratory petal of *Meoma ventricosa* showing the two roles of respiratory tube feet (g) and the convoluted ampullae (a). The solid and dashed arrows represent water currents on the external and internal surfaces of the test, respectively. (From Chesher 1969)

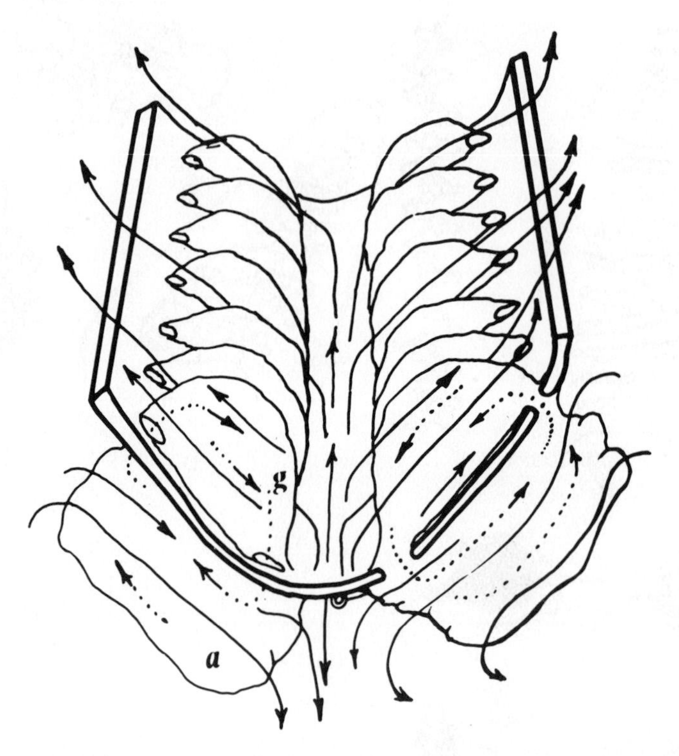

for feeding. Currents over the respiratory tube feet in these ambulacra result from the general body cilia and from the fascioles, bands of numerous clavulae (Figure 3.18). Clavulae are thin modified spines with a slightly enlarged top that secretes mucus and a band of cilia on opposite sites. This arrangement extends the cilia away from the test and increases the number set at right angles to the direction of water flow. Substratum particles adhere to the mucus produced by the clavulae and form a wall above the fascioles to make current production more efficient (Figure 3.19) (Chesher 1969). The number and shape of the fascioles vary among the spatangoids, and the strength and direction of the currents vary accordingly (Figure 3.20). High pressures are produced by these arrangements (Figure 3.21) (Foster-Smith 1978). The currents produced by the facioles are thought to facilitate feeding and removal of waste as well as respiration. The respiratory function seems essential. *Abatus cordatus* normally

Figure 3.17: Postero-lateral view of the body surface of *Echinocardium cordatum* without spines showing direction of water currents drawn down the respiratory funnel, over the depressed petals, and through the respiratory tube feet or out of the anal exhalant tube (from Nichols 1959)

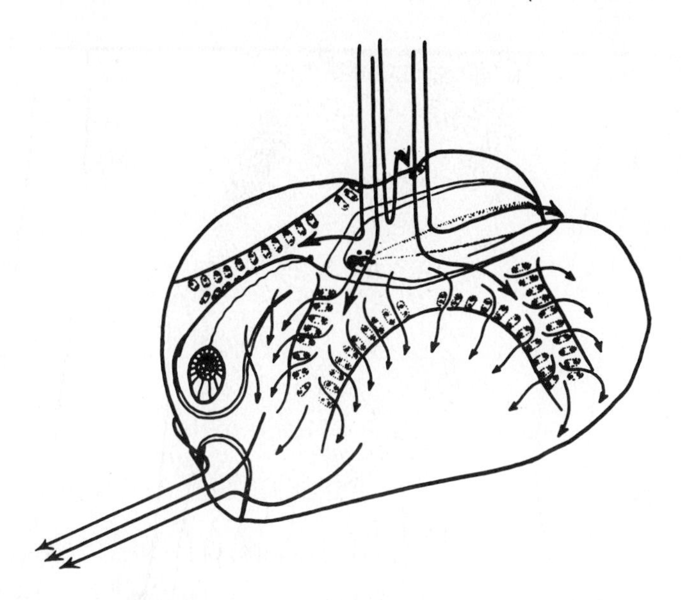

Figure 3.18: (a) Dense, minute band of tubercles which bear the clavulae and form the fasciole of *Echinocardium cordatum*. A pore pair, which bears a respiratory tube foot, is present in each plate. (b) Side view of two clavulae showing the bands of cilia on opposite sides. (From Nichols 1959)

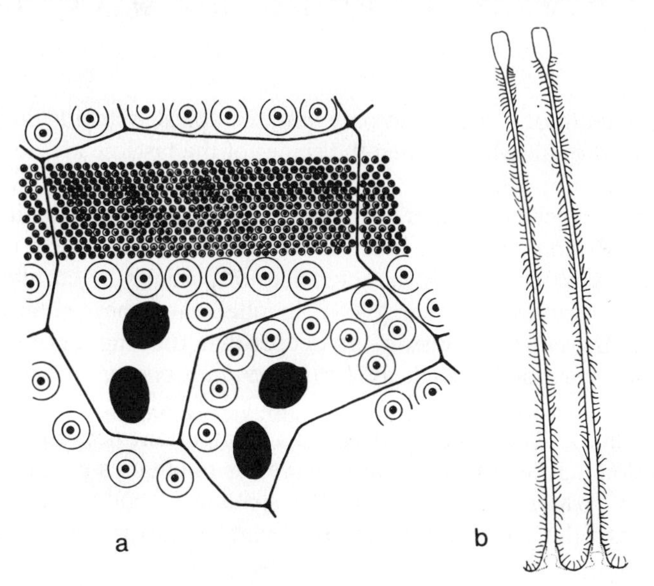

Figure 3.19: Mucous-sand wall (MS) produced by the peripetalous fasciole of *Meoma ventricosa*. The respiratory section (R) of the burrow on the left and the drainage area (D) on the right. Dashed arrows indicate the direction of water currents. (From Chesher 1969)

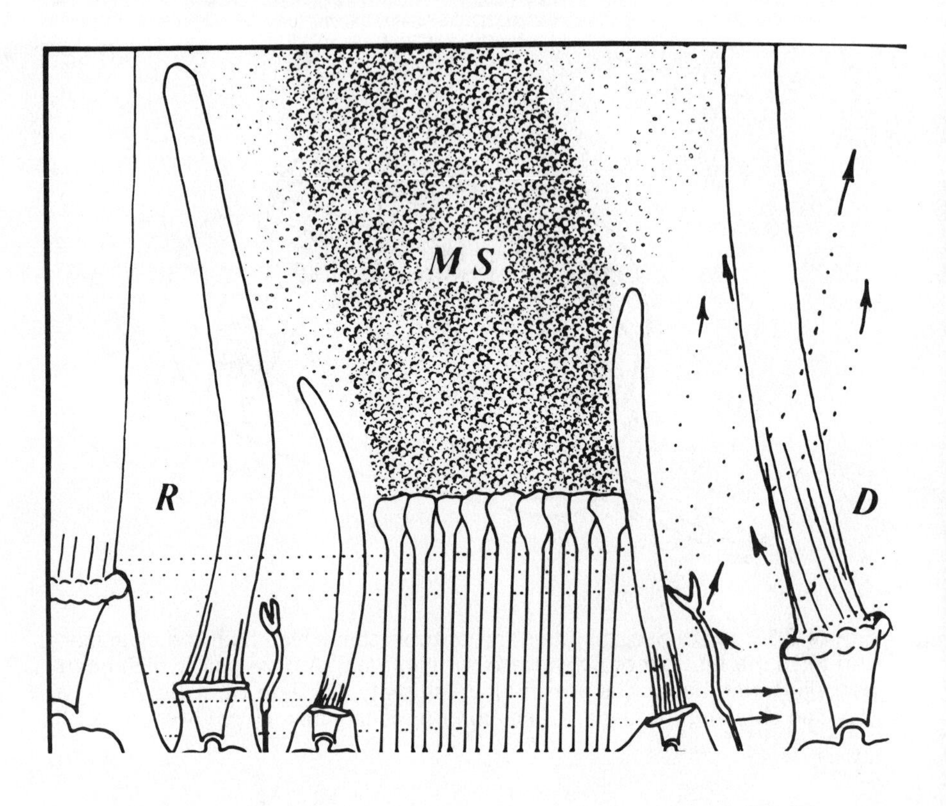

burrows to a depth of *c.* 5 cm, but only until the apical system is flush with the surface of the substratum when the spines of the fasciole are destroyed (J.M. Lawrence, unpublished.).

Access to oxygenated water is obviously necessary for the infaunal spatangoids. *Meoma ventricosa* burrows to depths of up to 10 cm in clean sand and draws water through the interstices of particles by currents produced by the fascioles (Chesher 1969). In silted sand the system cannot function and *M. ventricosa* burrows no deeper than the length of the apical spines. Other spatangoids such as *Echinocardium cordatum* can live at depths in fine substrata because of the ability to construct respiratory funnels (Figure 3.22) (Nichols 1959). This is important because it allows them to exploit a different food source as well as to obtain a greater degree of protection. Water that flows over the spatangoid body is discharged posteriorly into the interstices of the sediment or into posterior funnels.

Figure 3.20: Types of fascioles of spatangoids. A: Side view; B: posterior view; C: upper view; D: posterior view; E: lateral view. (From Melville and Durham 1966)

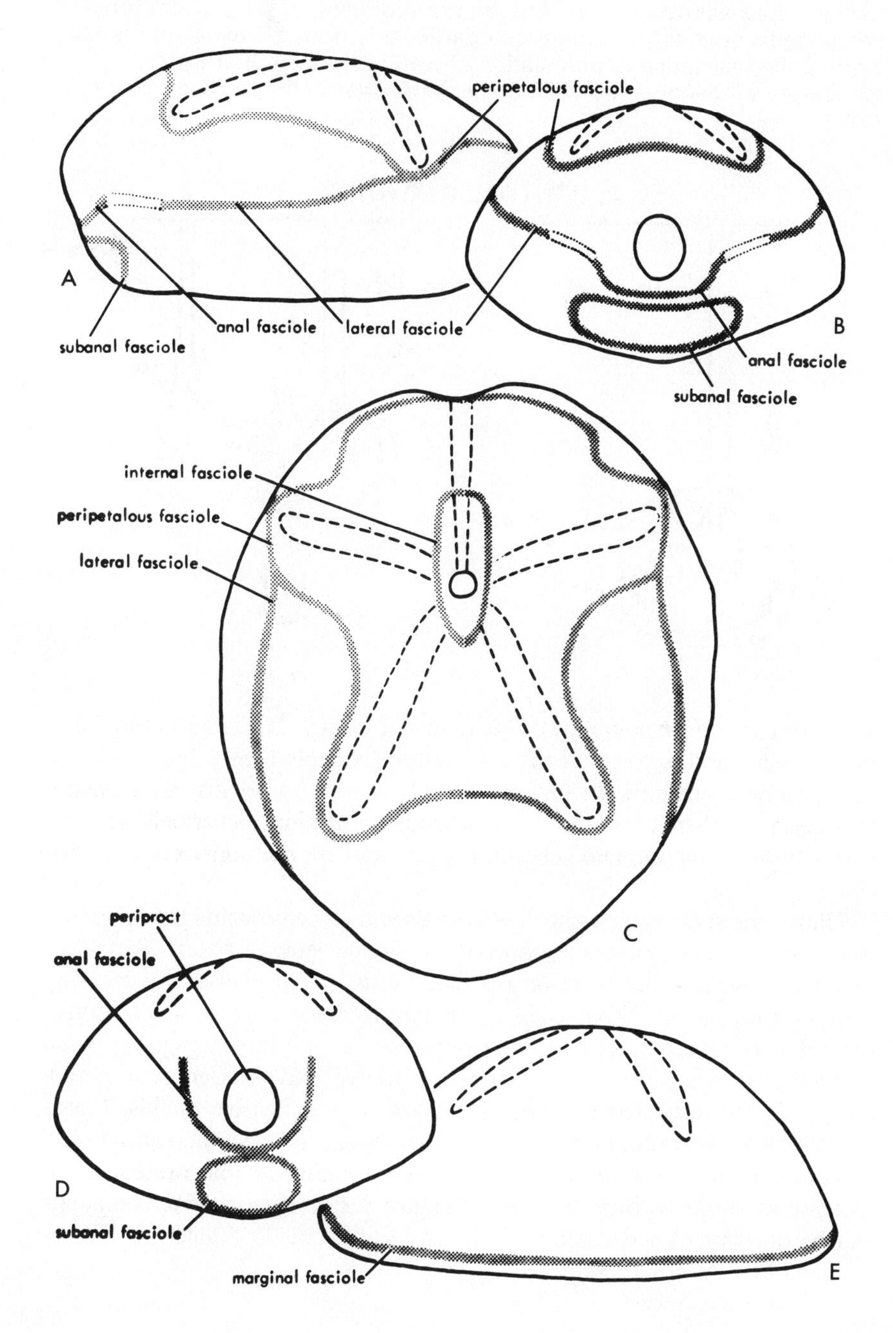

Figure 3.21: Arrangement of pumps in the tube system of *Echinocardium cordatum*. r_2: Frictional loss in the inhalant tube; r_5: frictional loss during back-flow of water in the tube from the high- to low-pressure sides; r_6: frictional loss as water seeps through the sediment. In sand, water may percolate through the sediment and particularly from the exhalant funnel. In mud, the resistance to percolation of water is so high that the old respiratory funnel must be used to discharge water. (From Foster-Smith 1978)

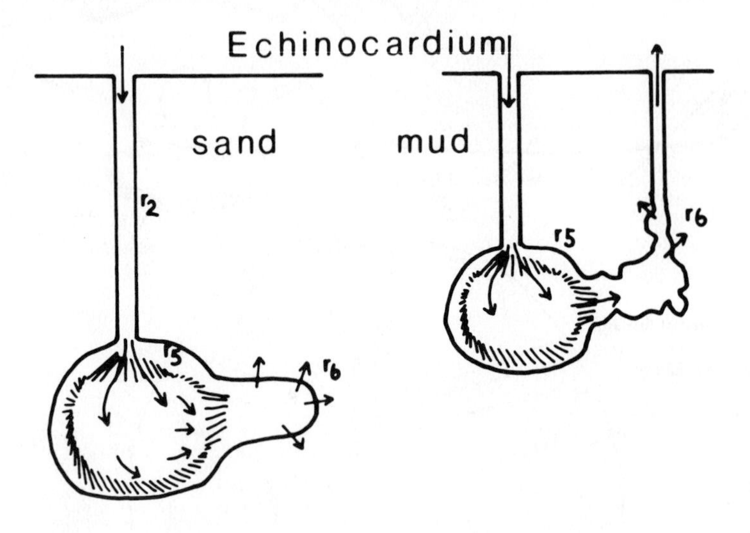

The number and permanency of the exhalent funnels is related to the depth of burrowing and porosity of the substratum (Nichols 1959). The discharge seems to be more difficult for juvenile *M. ventricosa* because they have a well defined subanal fasciole and construct two drainage channels whereas adults have an incomplete subanal fasciole and do not construct drainage channels.

Unlike the spatangoids, the cassiduloids and clypeasteroids are restricted to relatively coarse substrata associated with high-energy water. They have five petals because the anterior ambulacrum is not specialised for feeding. Because they occur in coarse substrata and do not burrow to great depths, the petals of clypeasteroids are not depressed as in the spatangoids. Even shallow burrowing can have an effect, however, and the degree to which the petals are restricted on the upper surface is indicative of this. Cassiduloids have evolved a higher test and the petals have become restricted to a limited region around the apical system, a change interpreted as an increase in ability to burrow deeper (Figure 3.23) (Kier 1966). The petals can be open or closed distally (Durham 1966), but the functional consequence of this is not known.

Figure 3.22: (a) *Spatangus purpureus* buried just below the surface of the substratum with the spines maintaining an opening to the surface and with an exhalant funnel shown. (b) *Echinocardium cordatum* buried below the surface of the substratum with the tube feet maintaining a respiratory, inhalant funnel and an exhalant funnel. (From Nichols 1959)

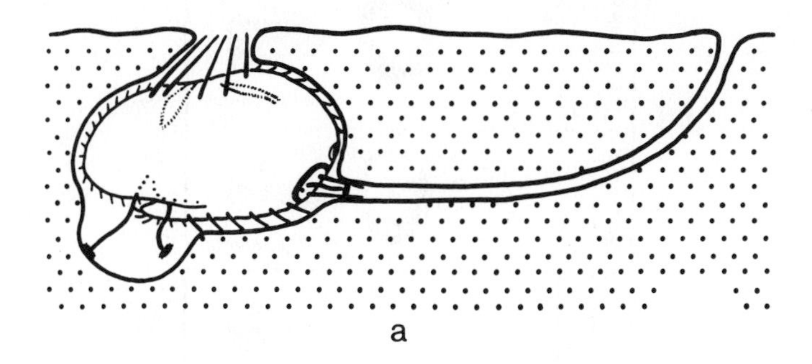

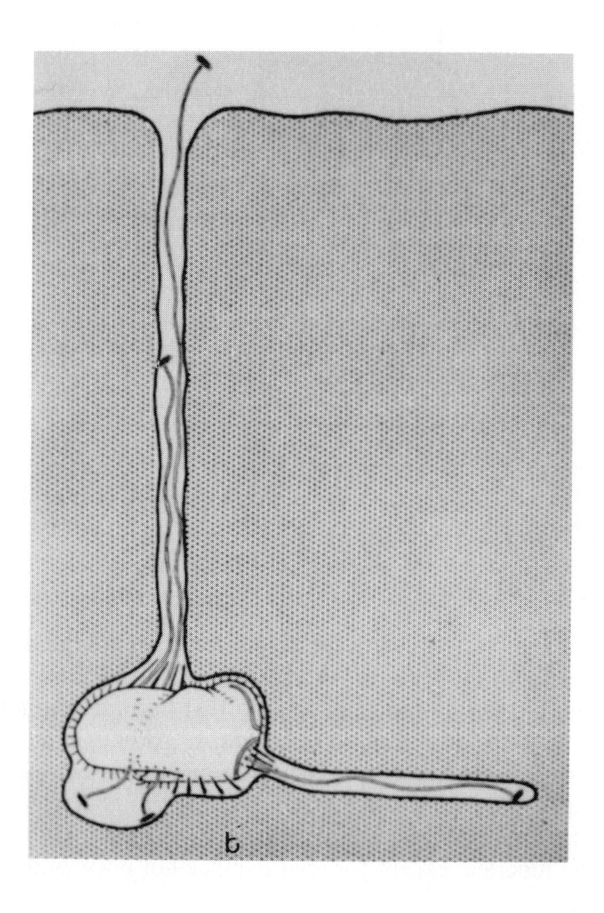

Figure 3.23: Upper and side views of cassiduloid genera showing evolutionary change in shape of test and length of petals. The petaloid area is shaded; the area below the petals is black. (From Kier 1966)

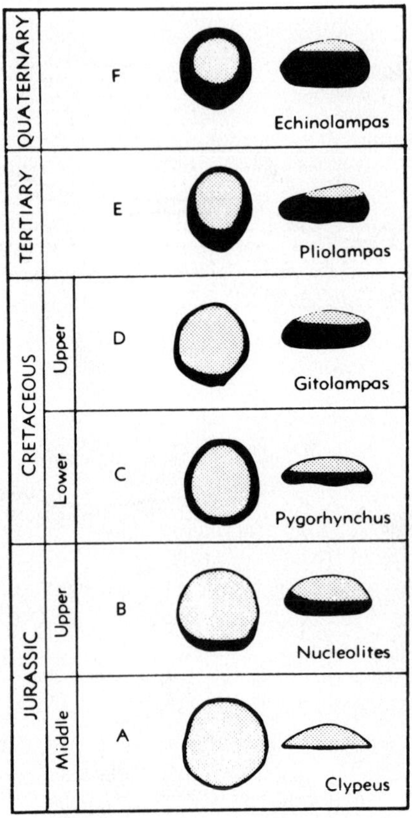

3.1.6. Holothuroidea

The tube feet, the body wall and the respiratory tree are the sites of oxygen uptake in holothuroids. The body wall lacks cilia, a phenomenon attributed to an original burrowing mode of life, so that currents cannot be moved across the body surface as in other groups. Although most groups of holothuroids use both the respiratory tree and tube feet, several groups have exploited a particular mechanism for oxygen uptake. Apodids lack a respiratory tree and tube feet, and take up oxygen across their thin body wall. Lacking tube feet except for anal papillae but possessing a thick body wall across which diffusion of oxygen must be limited, the burrowing molpadids have a respiratory tree for oxygen exchange. Elasipods lack a respiratory tree and have a thick or plated body wall, and presumably take up oxygen across modified tube feet.

The tube feet on the upper surface of epibenthic holothuroids are modified to form protuberances (papillate tube feet) which can be withdrawn. A long tube may even be extended from these protuberances in *Actinopyga* (Hyman 1955). Very few papillae occur in the epibenthic *Cucumaria frondosa* whereas the burrowing *Sclerodactyla briareus* has many (2300 in an average-sized individual) scattered hair-like papillae (Brown and Shick 1979). The radial water canals give off lateral podial water canals which can extend into the interradial areas and thus greatly increase the number and distribution of the tube feet or papillae. The tube feet of *S. briareus* do not exit the body wall through two pores as do the respiratory tube feet of echinoids. They consequently have no septum, and the ambulacral fluid of a tube foot–ampullar system, isolated from the radial water canal by a sphincter, flows up one side and down the other in a circular fashion as the result of ciliary action (Colacino 1973).

The greatest modification of the tube feet into papillae for oxygen uptake occurs in the deep-sea elasipods which have a thick or plated body wall and lack a respiratory tree (Hansen 1975). The papillae are found in the two dorsal ambulacra in most elasipods and may also be found along the lateral ambulacra along the lower body surface (Figure 3.24). Modifications of the water vascular system supply these papillae (Figure 3.25) The large water vascular dermal cavities may promote oxygen uptake by emptying ambulacral fluid into the papillae of the body wall by contraction of the body-wall muscles (Hansen 1975).

The branched pair of tubes known as the respiratory trees are found in the subclass Dendrochirotacea and the orders Aspidochirotida and Molpadida. Only the order Elasipoda, which has modifications of the body wall to facilitate oxygen uptake, and the order Apodida, which has extremely thin body walls, lack the respiratory trees. The latter arise from a cloaca and extend up around the gut (Figure 3.26). The cloaca is attached to the body wall by radial muscles. The trees are filled by means of a series of inhalations (Brown and Shick 1979; Crozier 1916; Robertson 1972). This stepwise inhalation may result from the high viscosity and inertia of water and the branched nature of the respiratory tree (Shick 1983). Inhalation involves contractions of the cloaca with the anal sphincter open. Exhalation involves relaxation of the cloacal muscles and collapse of the trees by the coelomic hydrostatic pressure with both the anal and tree sphincters open. Contraction by the respiratory trees themselves occurs (Crozier 1920). Even though respiratory trees are lacking in elasipods, radial muscles connect the cloaca to the body wall, which suggests that water may be taken into this portion of the gut for respiratory purposes.

Figure 3.24: A: Lateral views of two *Oneirophanta mutabilis affinis* showing the dorsal papillae which vary in number from 5 to 25. B. (a) View of the upper surface of the posterior unpaired dorsal appendage found on (b) *Psychropotes mirabilis*. (From Hansen 1975)

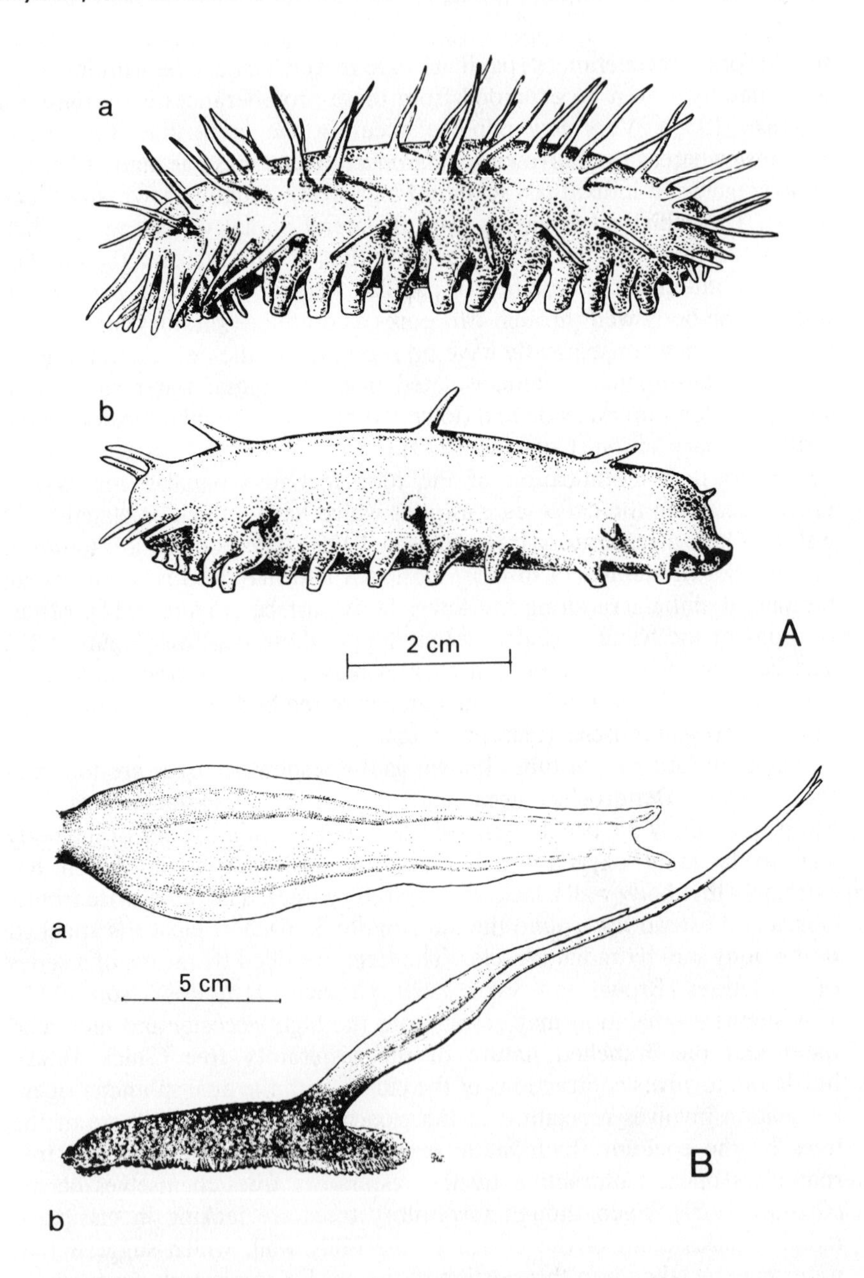

Figure 3.25: The water vascular dermal cavities of two tube feet of *Scotoplanes globosa* seen from the perivisceral coelom. (a) Water vascular cavity; (b) opening into a tube foot; (c) one of the ventro-lateral water vascular canals; (d) communication between the water vascular canal and the water vascular dermal cavity. (From Théel, in Hansen 1972)

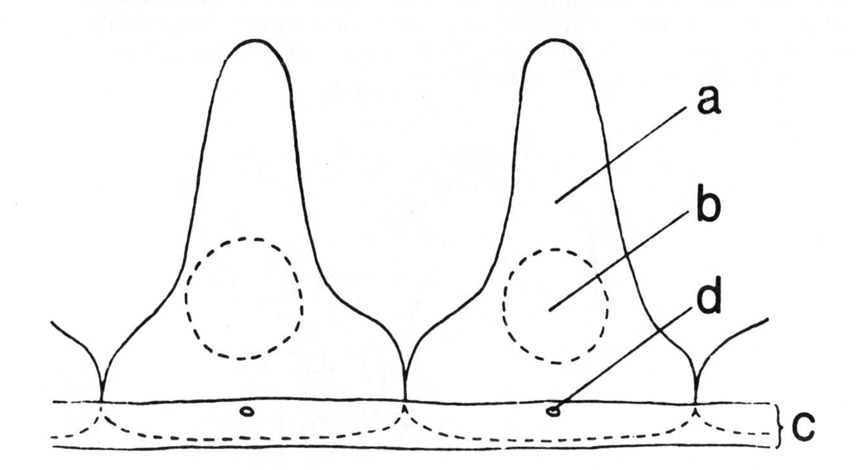

Inhalation usually involves rigid skeletal structures to withstand the negative pressures required. The body wall of holothuroids can be made rigid through the interaction of the collagen molecules present (Motokawa 1984; Wilkie 1984), but it is not known if this occurs during inhalation. However, a mechanism exists by which a holothuroid with a soft body wall could inhale despite the fact that an increase in coelomic pressure to antagonise the tension exerted by the contracting radial muscles would increase the hydrostatic coelomic pressure tending to collapse the cloaca (Wolcott 1981).

The total force supporting the body wall around the cloaca is equal to the difference between the coelomic and external pressures multiplied by the area of the body wall to which the radial muscles are attached. The total force countering cloacal expansion is the difference between the cloacal and coelomic pressures multiplied by the area of the cloaca. A net force causing expansion of the cloaca occurs because the attachment area of the radial muscles on the body wall exceeds the attachment area of the muscles on the cloacal wall (Figure 3.27). Very low cloacal suction would be required for filling because flow into the cloaca through the anus is not restricted. The increase in the coelomic pressure required for radial muscle antagonism is minimal. Relaxation of the body wall is not necessary for cloacal expansion. After the cloaca has filled, contraction of its circular muscles forces the water into the respiratory trees. Exhalation results from

Figure 3.26: Internal anatomy of (a) *Ocnus brevidentus* dissected from the left underside, with portions of the gonad removed, and (b) *Psolidiella nigra* dissected from the upper surface. an: Anus; c.r.: calcareous ring; g.d.: genital duct; g.tub.: genital tubule; int.: intestine; l.resp.: left respiratory tree; l.v.ir.: left ventral interradius; mad.: madreporite; mad.d.: madreporite duct (stone canal); oes.: oesophagus; p.v.: Polian vesicle; r.l.m.: radial longitudinal muscle; r.m.: retractor muscle; r. resp.: right respiratory tree; r.v. ir.: right ventral interradius; ten.: tentacle; tr.m.: transverse muscle. (From Pawson 1970)

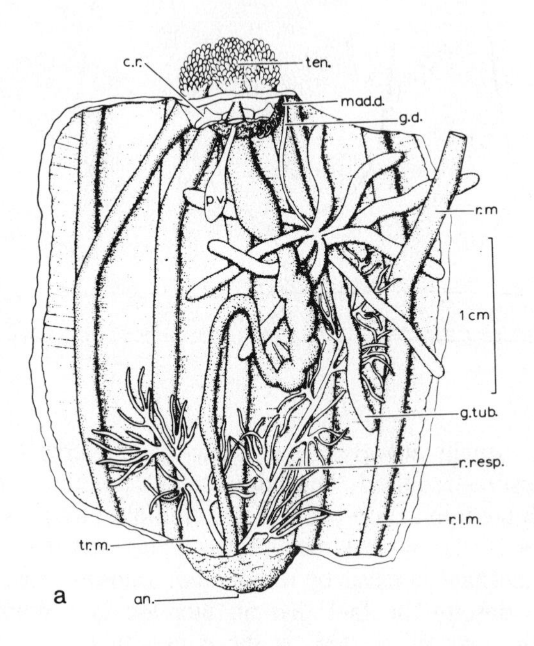

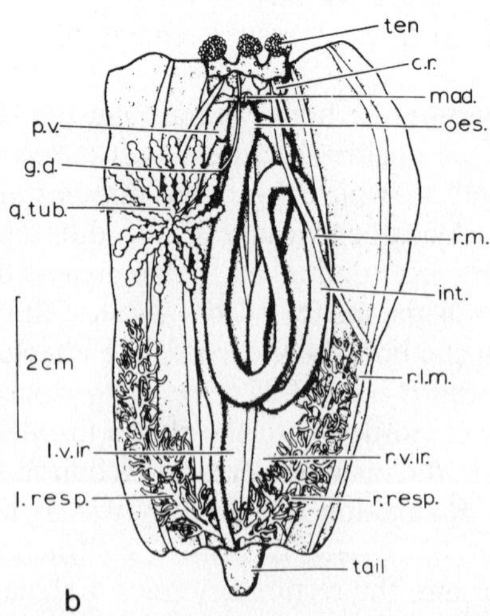

Figure 3.27: Forces acting on a hydrostatically supported body composed of coaxial cylinders. F_1 is the force perpendicular to the plane of section and tending to distend the body wall. R_b is the body-wall radius. F_2 is the force tending to collapse the cloaca towards the plane of section. R_{cl} is the cloacal radius. $P_{cl} = P_{co} [1 - (R_b/R_{cl})^2]$. P_{cl} and P_{co} are the cloacal and coelomic pressures, respectively. (From Wolcott 1981)

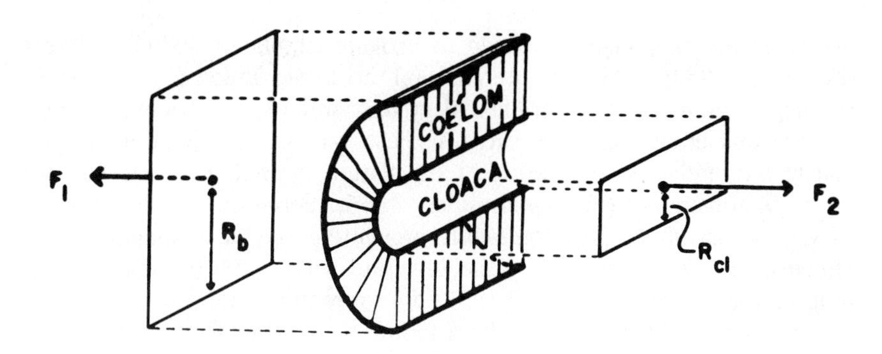

coelomic hydrostatic pressure which can have an active component causing rapid discharge.

The contribution of oxygen uptake via the respiratory trees to the total amount can be great (usually between 50 and 60%), depending on the contribution of other sites of uptake (Shick 1983). In *Cucumaria frondosa*, uptake via the tube feet and body wall is low, and the contribution of the respiratory-tree uptake of oxygen is directly related to body size and is important for oxyregulation (Brown and Shick 1979). The importance of size is also indicated by the fact that the respiratory tree develops late in some holothuroids (Théel 1882). The contribution of the respiratory tree is dependent upon the tidal volume and rate of pumping. The capacity of *Sclerodactyla briareus* to oxyregulate with declining oxygen levels involves an increase in tidal volume associated with little loss in pumping frequency (Brown and Shick 1979).

Because respiratory trees and tube feet are lacking in apodids, oxygen uptake is limited to diffusion across the body wall which is so thin as to be transparent. In the absence of external cilia, movement of water across the surface would require currents produced by movement of the body itself. It is possible that the characteristic puffed form of the apodid body wall is a mechanism to increase the surface to volume ratio. Unlike the situation with the respiratory trees or papillae, there is little possibility of regulating oxygen uptake. Consequently, it is not surprising that the rate of oxygen uptake by *Leptosynapta inhaerens* measured *in vitro* is more dependent upon external oxygen levels than species with respiratory trees (Shick

1983). However, *L. inhaerens* irrigates its burrow by peristalsis (Mangum and Van Winkle 1973), and it is possible that increased activity at low concentrations of oxygen could increase external convection.

3.2. CIRCULATION

Four systems have been proposed to provide circulation in echinoderms (Ferguson 1982b): (1) the perivisceral coelomic fluid, (2) the water vascular system, (3) the perihaemal system and (4) the haemal system. Conclusions about the involvement of these systems in circulation are primarily inferential because efficient circulation of their contents has not been demonstrated (Farmanfarmaian 1966; Ferguson 1982b; Walker 1982). Either there is a limitation preventing the development of an effective circulatory system or the low rates of respiration and nutrient utilisation (Lawrence and Lane 1982) do not require one.

The basic problem seems to be a failure to develop a closed or even a semi-closed systemic circulation. Greenberg (1985) pointed out that students of echinoderms are exceptional in seeing many fluid transport systems that lack characteristics typical of those in other phyla, and are consequently undistracted by hydrodynamic considerations. The perivisceral coelom and its extensions and the water vascular system are clearly inappropriate structurally for closed circulation. The structural arrangement of the perihaemal system as a coelomic tube around the haemal channels, and the absence of muscular elements would preclude its effective operation in circulation. The haemal system has the best potential for development into an effective circulatory system but, except for restricted circulation in some holothuroids, has not done so. Translocation of molecules within the echinoderm body seems to involve primarily the perivisceral coelom and the haemal system by some fluid circulation, by diffusion, or by coelomocytes.

3.2.1. Perivisceral coelomic fluid

The perivisceral coelomic fluid has been thought to be a route for translocation of molecules because it is in contact with the site of entry of oxygen across the body wall and nutrients from the gut. The concentration of oxygen in the perivisceral coelomic fluid varies with external levels (Farmanfarmaian 1966; Shick 1983), and the levels of nutrients vary with feeding (Bamford 1982; Ferguson 1982b; Walker 1982). Nutrients are also actively absorbed by the body wall (Bamford 1982) but actual movement of the nutrients into the perivisceral coelomic fluid has not been demonstrated. The perivisceral coelomic fluid seems to be the most probable route for translocation of oxygen. Although oxygen enters the body via the tube feet, the characteristics of the water vascular system make it unlikely that circulation of oxygen within the body occurs via the radial water canals.

The translocation of nutrients from one tissue or organ via the perivisceral coelomic fluid is not firmly established. The peritoneal surfaces of all tissues exposed to the perivisceral coelomic fluid can absorb nutrients (Ferguson 1982b), although it has been postulated that the outer coelomic sac of the gonads is selectively permeable (Walker 1980). The potential supply of nutrients to the developing ovary of *Asterias rubens* via the perivisceral coelomic fluid has been calculated to be inadequate for its needs (Beijnink, Van der Sluis and Voogt 1984).

The effectiveness of translocation of molecules via the perivisceral fluid would be inversely related to its volume, the distance between the site of origin of the molecules being translocated and their destination, the openness of the system, and mechanisms for producing fluid circulation.

The spaciousness of the perivisceral coelom varies among the classes. In crinoids much of the perivisceral coelom is filled with connective tissue, with the only non-occluded space being the axial sinus around the intestine (Figure 3.28) (Chadwick 1907). Extensions from the central body coelom

Figure 3.28: Vertical section through (a) the body of *Nemaster rubiginosa* with the interradius on the right, and (b) the axial organ. AN: anus, AO: axial organ, AS: axial sinus, CE: coelomic epithelium, CO: chambered organ, ES: oesophagus, FC: free cell, GT: glandular tubule with lumen visible, GT′: glandular tubule with lumen not visible, HV: haemal vessel, IN: intestine, MO: mouth, PC: perivisceral coelom. (From Holland 1970)

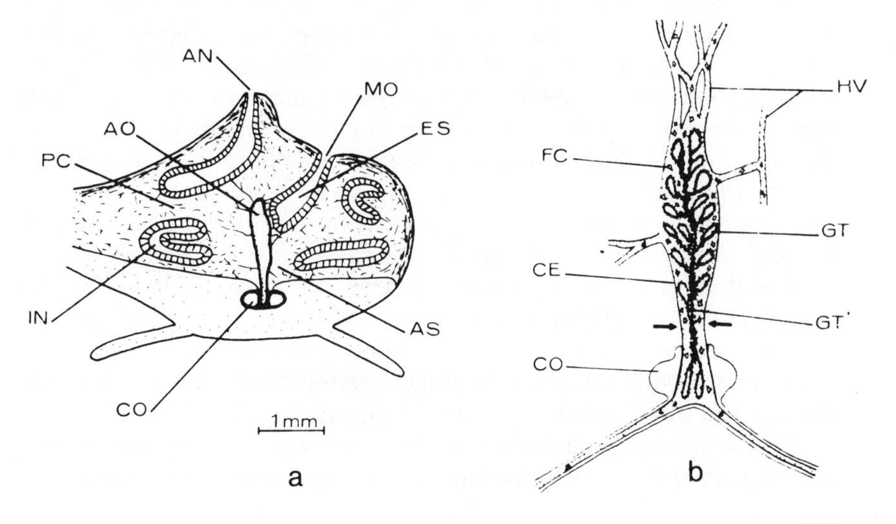

reach into the arms. Separate coelomic cavities in the chambered organ, distinct from the perivisceral coelom, occur in the lower part of the crinoid body (Chadwick 1907). Extensions of the chambers go into the cirri and into the stalk of stalked crinoids (Grimmer, Holland and Messing 1984). The diameter of the lumen of these tubes is small. The perivisceral coelom of all extant non-crinoid classes is not filled with connective tissue, but mesenteries occur and could provide barriers to circulation of perivisceral fluid except as surface currents. The extension of the perivisceral coelom of ophiuroids is restricted to a small tube because of the internal vertebral arrangement. The perivisceral coelom in the arms of asteroids is larger because the skeletal support is in the body wall, but may be small because of the small diameter of the arm as in spinulosids. The perivisceral coelom is largest in the echinoids and holothuroids, but can be very reduced in volume when the gonads are fully developed.

Circulation of the fluid in the perivisceral coelom and its extensions could result from flagellary currents or from changes in body form or gut form from muscular contractions. The epithelium covering the gonads and gut is flagellated (Ferguson 1982b; Walker 1982). However, the epithelial lining of the perivisceral cavity does not always have flagella as they are lacking in crinoids (Hyman 1955). They are also lacking in the ophiuroid *Ophiothrix spiculata* except on the dorsal surface near the arms, but occur extensively in *Ophiura luetkeni* and some other genera (Austin 1966). Flagellated pits occur in the median sagittal line of the dorsal wall of the aboral coelomic canal in the arms of crinoids (Chadwick 1907). Although they undoubtedly produce currents, their arrangement in pits rather than as a band indicates another function. Each epithelial cell of the coelomic tubes in the stalks of stalked crinoids has a flagellum (Grimmer *et al.* 1984).

The extensions of the perivisceral coelom into the arms of ophiuroids have flagellary currents which are distally directed on the upper side and downward directed on the lateral side (Figure 3.29) (Austin 1966). Flagellary currents in identical directions occur on the inner surface of the much more developed coelomic cavity of the arms of asteroids (Figure 3.30). They spiral in a distal direction over the surface of the gonads of asteroids (Walker 1979). Currents in the spatangoid *Meoma ventricosa* radiate outwards over the intestine and mesenteries to the test, and are pronounced over the intestine, tube-feet ampullae, and gonads (Chesher 1969). A directional translocation by ciliary currents has been reported for the molpadid holothuroid *Paracaudina chilensis* (Figure 3.31) (Yazaki 1930).

These currents may be effective in producing laminar currents along the surface and may be functional in distance translocation. Material injected into one arm of an asteroid appears subsequently in the others (Ferguson 1964). The currents produced by flagella do not extend away from the surface for a great distance. Thus flagellary currents could not be found at a

Figure 3.29: The perivisceral coelom in the arm of *Ophiothrix spiculata*. Dashed arrows indicate ciliary currents, solid arrows represent counter-currents. (From Austin 1966)

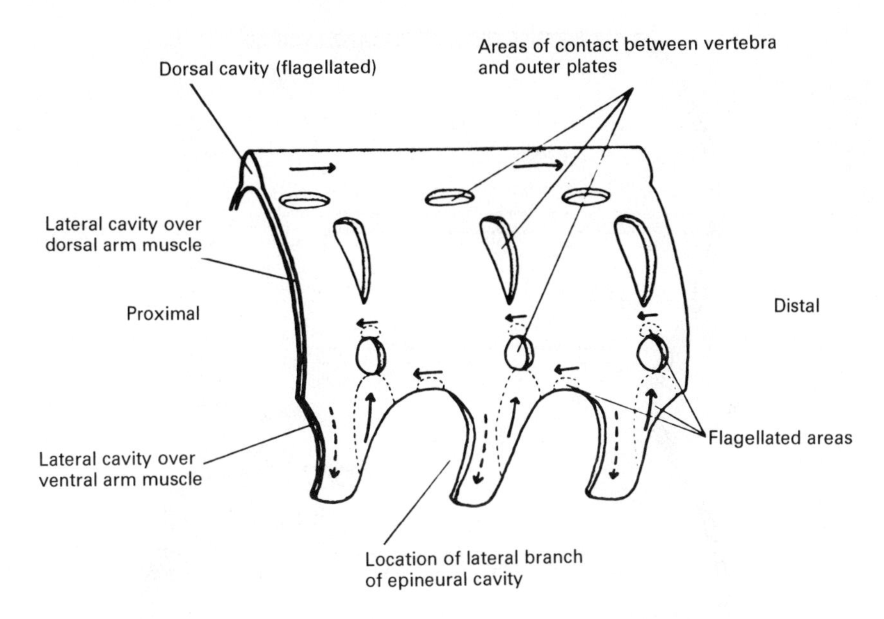

distance from the interior body wall of the regular echinoid *Lytechinus variegatus* (Hanson and Gust 1986). This lack of distance circulation would explain the hypoxic conditions that occur just outside the ovary wall and among the ovarian lobes of the echinoid *Strongylocentrotus droebachiensis* (Bookbinder and Shick 1986). Flagellary currents may be functional also to prevent boundary-layer effects and to promote transfer of molecules across the surface.

Currents in the perivisceral coelom of *L. variegatus* seem to result from movement of the Aristotle's lantern. The lantern is in continuous motion, even when not involved in feeding, and distinct currents are associated with its activity (Hanson and Gust 1986). Structural features of the lantern's musculature and attachment to the peristomial girdle corroborate this interpretation (Jensen 1985). The gnathophiurine jaws may similarly circulate coelomic fluid in some families of ophiuroids (Wilkie 1980).

Peristalsis of the gut occurs in all echinoderms and would produce some circulation of the perivisceral fluid. Because the level of activity of peristalsis depends upon the amount of food in the gut, circulation resulting from this mechanism would be variable. Changes in posture occur in some asteroids and holothuroids, which would result in displacement of the perivisceral fluid. The rate of change in posture is typically so slow that little

Figure 3.30: Currents (arrows) in the perivisceral fluid of *Asterias forbesi*. (From Budington 1942)

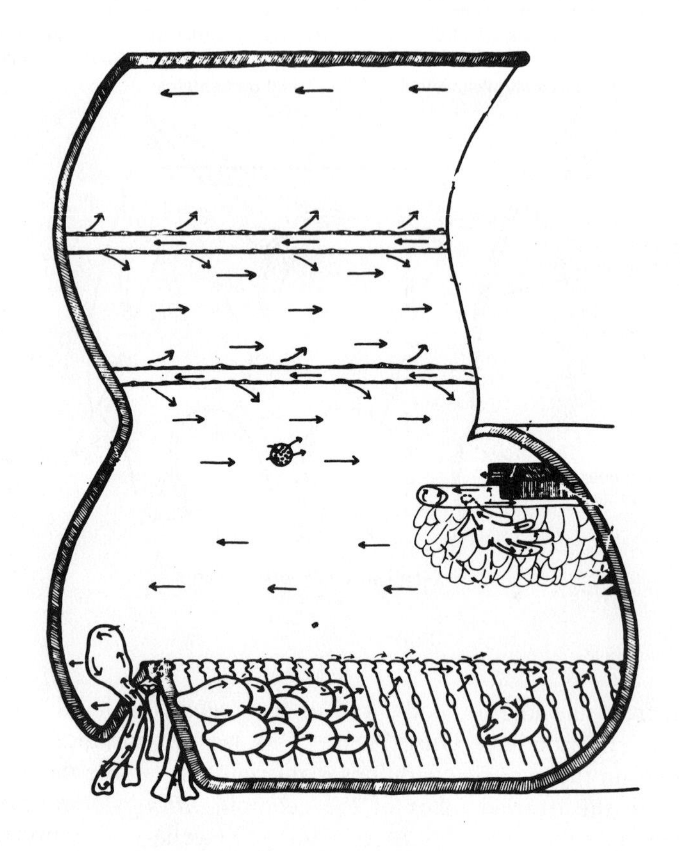

Figure 3.31: Course of currents of the perivisceral fluid in *Paracaudina chilensis*. The anterior end of the body is to the right. (From Yazaki 1930)

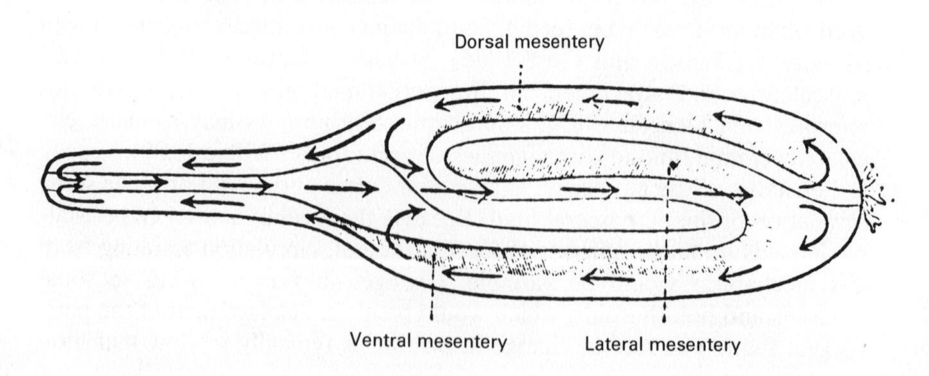

circulation would result except in such active forms as the synaptid holothuroids. Considerable circulation of the perivisceral fluid would result from the periodic filling and emptying of the respiratory trees of the dendrochirotid, aspidochirotid and molpadid holothuroids.

3.2.2. Water vascular system

The tube feet are the primary site of oxygen uptake, and the water vascular system might be expected to function in translocation of oxygen. However, the structure of the system makes circulation unlikely. The distribution is limited to the body wall and does not extend to internal sites. The structure and the internal circulation patterns in the tube feet and ampullae seem to be designed for transfer of oxygen to the perivisceral fluid, and not for circulation of the fluid in the water vascular system. For circulation of the fluid in the radial water canals to be functional in oxygen movement, it would be necessary for the fluid in one part of the canal to have a higher concentration of oxygen than the other. This might occur in those echinoids in which there is regional distribution of respiratory and suckered tube feet. Regional differentiation also occurs in tube feet in those holothuroids in which the bivium on the upper surface functions in oxygen uptake and the trivium on the lower surface functions in locomotion and attachment. It is extremely unlikely that the fluid in the radial water canals on the upper surface could circulate to those on the lower surface.

The use of the water vascular system for circulation also would be difficult because the radial water vessels dead-end and do not provide for Harveyan circulation. The mechanism for translocation of fluid in the radial water canals would have to be Galenic if muscular in origin, but could be circular if the cilia lining the canals differed positionally in the direction of their beat. Muscles in the radial water vessels function in control of fluid volumes associated with protraction of the tube feet, and there is no evidence that they are involved in translocation of the fluid. At this time, no evidence exists suggesting that the fluid in the water vascular system functions as a circulatory system.

3.2.3. Perihaemal system

The perihaemal system is the coelomic space underlying the oral nerve ring and the radial nerves, surrounding the bases of the gonadal ducts and the gonads, and the axial gland (Figure 3.33). The perihaemal system is closely associated with the haemal system throughout the body except in the gut (Cuénot 1948; Hyman 1955). Being of coelomic origin, the perihaemal system is lined with a flagellated epithelium. The flagellated cells

Figure 3.32: Sagittal section of *Neocrinus decorus*. aa: Water ring with radial water canals; ah: haemal ring; an: epithelial nerve ring; anp: deep nerve ring; as: spongy mass; bo: mouth; ca: arm ambulacral canal; cc: chambered organ; cnr: nerve commissure passing into the radials; ga: axial gland; i: section of intestine; lg: genital lacuna of the arm; n: aboral nerve centre; nb: brachial nerve; nl: lateral nerve of the arm; nta: nerve of the anal tube; oe: oesophagus; os: spongy organ; p: pores; pa: axillary primabrach; r: radial; re: rectum; rl: subepidermal lacunar network; sy: syzygy; tnb: brachial nerve trunk. (From Cuénot 1948)

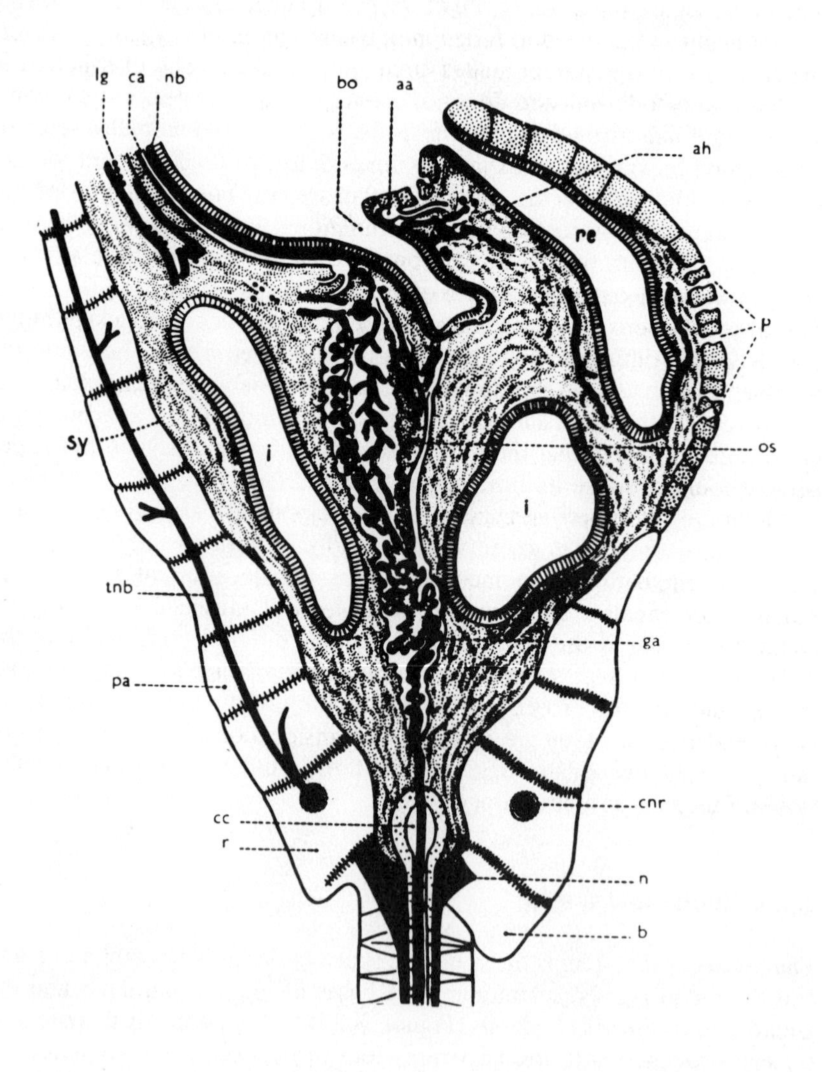

Figure 3.33: Diagram of the haemal and perihaemal systems of *Asterias*. aa: aboral perihaemal ring; cat: canal connecting the oral perihaemal ring and the intra-integumentary vessels of the disc; cmt: canals connecting the marginal sinus and the vessels of the intra-integumentary spaces of the arm; co: connection between the glandular sinus and the water canal (not shown); gb: axial organ; le: efferent lacunae of the intestine; lr: radial spaces; og: genital organ receiving a haemal vessel; ps: haemal pentagon on the stomach; st: terminal sinus. Arrows indicate the proposed direction of flow. (From Cuénot 1948.) According to Walker (1982), the branched intragonadal vessels are not present

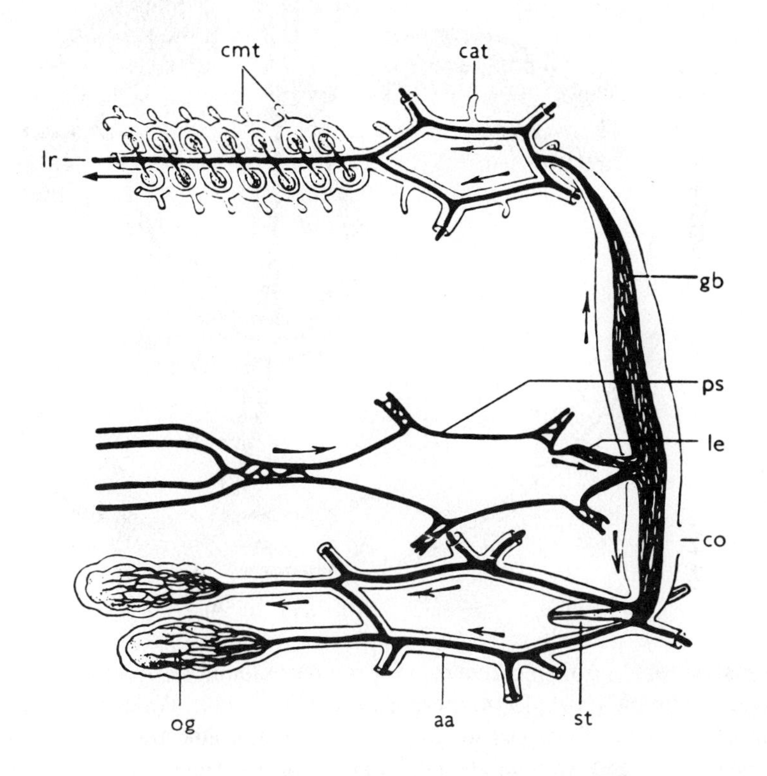

produce weak currents in the genital coelomic (perihaemal) sinus of asteroids and ophiuroids (Walker 1982). The composition of fluid contained in the perihaemal system is unknown, although that of ophiuroids is reported to lack cells and periodic acid Schiff-positive material except in the oral axial sinus (Austin 1966).

Hyman (1955) concluded that the function of the hyponeural vessel was to cushion and nourish the overlying nerve ring and radial nerves, but ascribed no function to the other coeloms of the system. With the difficulties noted regarding a role for the translocation of nutrients via the

Figure 3.34: Diagram of the haemal system of *Echinus* (inverted orientation). aab: Aboral haemal ring; c: collateral vessels; gb: axial organ; lb: vessel going to the axial organ; le: efferent vessel of the intestine; lg: genital vessel; lme: external marginal vessel; lmi: internal marginal vessel; lr: radial vessels; sg: glandular sinus; st: terminal sinus; v: spongy mass of the oral ring. Arrows indicate direction of proposed flow. (From Cuénot 1948)

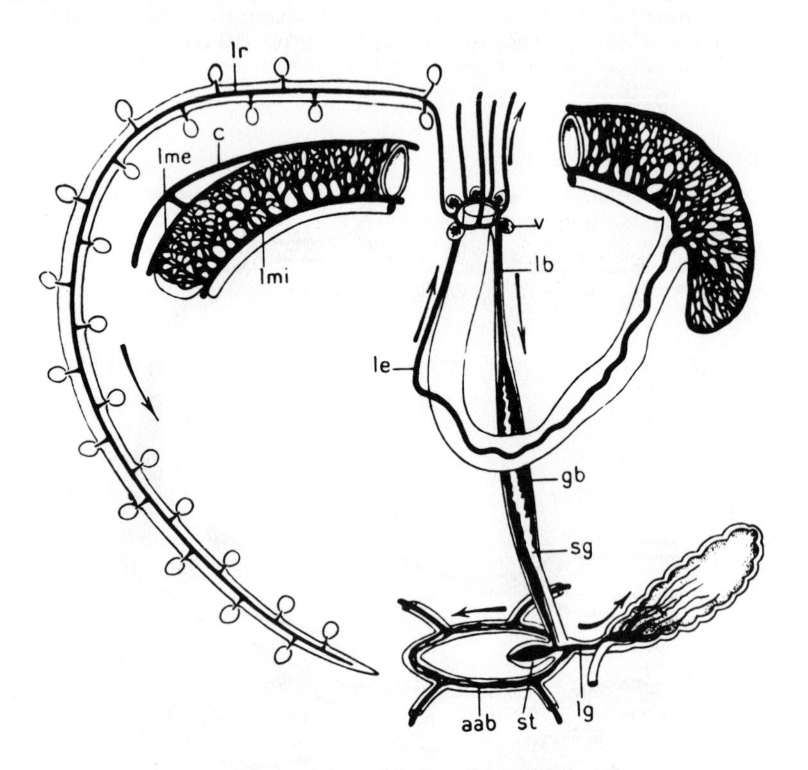

haemal system, a translocative role for the perihaemal system has been proposed (Beijnink *et al.* 1984; Ferguson 1982b, 1984; Walker 1982). The genital coelomic sinus occurs in asteroids, echinoids and ophiuroids. It becomes enlarged and nearly occluded prior to gametogenesis after the enlargement of the germinal epithelium in these groups (Walker 1982).

The axial sinus and other perihaemal coeloms of *Asterias rubens* (Broertjes, Posthuma, Den Breejen and Voogt 1980) and *Echinaster graminicola* (Ferguson 1970, 1984) receive nutrients from food, supposedly via the perivisceral coelomic fluid. Material in the axial sinus of *A. rubens* is translocated to the gonad (Broertjes and Posthuma 1978). Ophiuroids seem to have little potential for translocation of fluids by the perihaemal system (Austin 1966). No contraction of any perihaemal sinus occurs. Flagella are scarce or absent in most sinuses, and flagellary currents are weak or absent except at the ventral end of the oral axial sinus

Figure 3.35: Haemal system of (a) *Leptosynapta galliennei* and (b) *Holothuria tubulosa*. aa: Oral water ring; cg: gonaduct; ct: transverse canal connecting the internal vessel (li) of the descending intestine and the same vessel of the ascending intestine; e: muscular stomach; la: vessel anastomosing with the external vessel; le: external marginal vessel; lg: genital vessel; li: internal lacunar vessel; m: madreporite; md: dorsal mesentery; o: genital pore; oad and oag: right and left respiratory trees; oc: openings of connections between the coelom and the peripharyngeal space; og: gonad; pc: cloaca; r: beginning of rectum; u: area of sparse urns in the interradius AB; u′: area of dense urns in the interradius DE. v: Polian vesicle; vp: vesicles of the tentacles. (From Cuénot 1948)

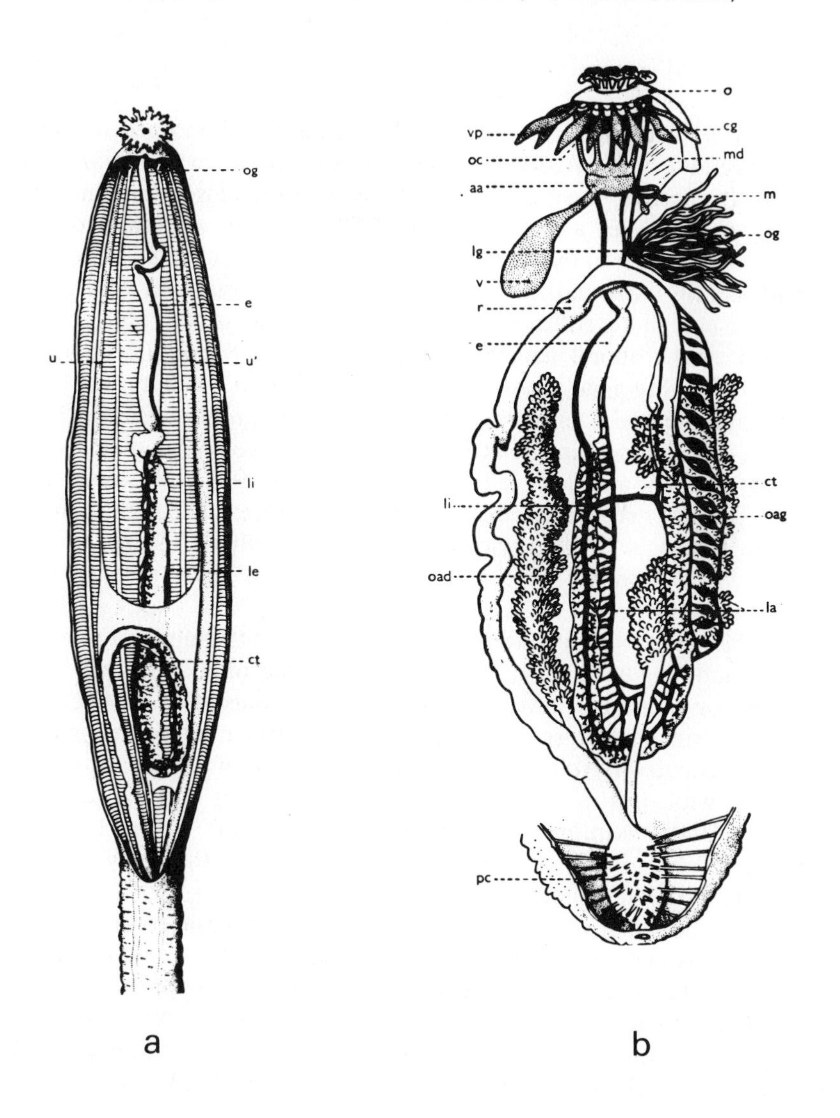

where flagellary currents from the oral axial sinus go into the ampulla.

The attribution of a translocative function to the perihaemal system is primarily due to its structural relationship with the gonad via the haemal genital sinus in asteroids. Its lack of a direct connection to the gut would require that nutrients be taken up via the perivisceral coelomic fluid. The area available for this is limited and a propulsive force to move the fluid from either the axial sinus gland or the hyponeural sinuses is not apparent. Thus, little evidence indicates that the perihaemal system functions in circulation.

3.2.4. Haemal system

Hyman (1955) defined the haemal system as a well developed but somewhat indefinite system of intercommunicating spaces in the strands and webs of connective tissue that fill the body interior. This definition is misleading as it implies a continuous fluid-filled lumen, and has led to an implicit assumption of the capacity of circulation. A fluid-filled lumen does not necessarily exist and the haemal system may consist only of a ground substance containing collagen, reticular fibres, and amoeboid phagocytes as in much of the aboral haemal system of *Asterias rubens* (Beijnink and Voogt 1984), the axial gland of *Diadema antillarum,* (Millott and Vevers 1964), and the oral haemal ring of *Psammechinus miliaris* (Burton 1964). The haemal channels in the stalk of *Democrinus conifer* contain densely packed granules and few cells (Grimmer *et al.* 1984). However, these studies did not necessarily consider seasonal changes. The genital haemal sinus of asteroids is filled with periodic acid Schiff (PAS) positive and mercuric bromophenol blue (MPB) positive fluid prior to and during gametogenesis (Walker 1982).

The haemal system is covered by a coelomic epithelium and may have circular and longitudinal muscles, and observations of either localised contractions or peristalsis have contributed to the concept of its circulatory role. The primary feature of the haemal system which led to the assigning to it of a circulatory role is that it connects the site of nutrient absorption (the gut) with sites of nutrient utilisation (the gonads and the tube feet) (Cuénot 1948). Nutrients are also actively absorbed by the body wall (Bamford 1982; Lawrence and Lane 1982), but whether they enter the perivisceral coelom is not known.

The development of the haemal system varies. Its basic components are (1) haemal strands, lacunae, or vessels associated with the gut, (2) an oral haemal plexus or haemal ring around the proximal part of the gut, (3) a radial haemal vessel which runs along the ambulacrum between the hyponeural vessel and gives off branches to the tube feet before ending blindly, (4) haemal vessels to the gonads, and (5) the axial gland which contains lacunae or vessels. All components except that of the gut are associated with the perihaemal coelomic system. The degree of development and specific arrangement of some features varies among groups.

In crinoids, a large number of haemal lacunae rather than vessels are associated with the intestinal wall, and a plexus occurs around the oesophagus (Figures 3.28, 3.32) (Cuénot 1948). Other branches from the oesophageal plexus beneath the tegmen lead to a genital haemal tube in each radius. The axial organ and the spongy bodies receive branches from the peripharyngeal plexus. The extension of the axial organ into the stalks of stalked crinoids is surrounded by the haemal channel which contains a mass of electron-dense granules (Grimmer *et al.* 1984). The abaxial side of this haemal channel is bounded only by a basal lamina, apparently without any cells with microfilaments.

In asteroids, an oral haemal ring gives off a radial haemal vessel that runs adjacent to the radial water canal and gives off branches to the tube feet (Figure 3.33) (Cuénot 1948; Ferguson 1982b; Hyman 1955; Walker 1982). An aboral haemal ring encircles the rectum. It is connected to the oral haemal ring by a vessel via the axial gland. A branch from the aboral haemal ring goes to each gonad and envelops it in a double-walled sac. Connections with the gut are variable and minimal (Beijnink and Voogt 1984; Ferguson 1982b). Contractions occur in the axial gland, gastric haemal tufts, and the aboral haemal ring (Hoffmann 1873; Gemmill 1914).

A very similar haemal system is found in ophiuroids except that haemal branches also go from the aboral haemal ring to the bursae (Austin 1966; Cuénot 1948; Hyman 1955). Peristaltic contractions occur in the genital axial gland, and localised contractions occur in the haemal vessels of the gut (Austin 1966). Nutrients are absorbed from the gut into the coelomic fluid and thence into the haemal fluid in *Ophioderma brevispina* (Ferguson 1985).

In echinoids, well developed haemal vessels are associated with the gut and lead either to the oral haemal ring or to the axial-gland plexus (Figure 3.34) (Cuénot 1948; Hyman 1955). The haemal vessels can be especially prominent on the stomach and diverticula of spatangoids (Koehler 1883). The oral haemal ring also gives off branches to the spongy body. Radial haemal vessels arise from the oral haemal ring, and genital branches arise from the aboral haemal ring as in the ophiuroids and asteroids. Contractile portions of the echinoid haemal system include the axial gland (Boolootian and Campbell 1964; Burton 1964; Millott and Vevers 1968); and the inner (Burton 1964), outer (Farmanfarmaian and Phillips 1962), and collateral (Burton 1964) vessels of the gut.

In holothuroids, the haemal vessels associated with the gut lead to an oral haemal ring around the pharynx, and the ring gives off radial haemal vessels as in the other classes (Figure 3.35) (Cuénot 1948; Hyman 1955). However, modifications in the system reflect basic differences in the body form and structure from those of the other classes. An axial gland is lacking so no haemal development occurs there. The single gonad is attached dorsally near the mouth and consequently receives the genital haemal vessel from the haemal vessel leading from the gut to the oral haemal ring. This structural difference is also associated with the lack of an aboral haemal ring.

The haemal vessels associated with the gut are little developed in the Dendrochirotacea, but can be extensively developed in two situations in the other subclasses (Herreid, LaRussa and DeFesi 1976). A *digestive rete mirabile* is associated with the descending small intestine in molpadids and aspidochirotids. A *respiratory rete mirabile* is associated with the ascending small intestine and intermingled with the left respiratory tree in aspidochirotids. This system is lacking in the molpadids and dendrochirotids even though they possess respiratory trees. Contractions of the dorsal vessel and its branches occur (Farmanfarmaian 1969; Herreid *et al.* 1976; Kawamoto 1927). Most of the large dorsal, ventral and transverse vessels are contractile in *Isostichopus badionotus* (Herreid *et al.* 1976).

These anatomical characteristics led to the conclusion that the function of the haemal system was to translocate nutrients from the gut to the gonads and tube feet. This interpretation is supported by the presence of coelomocytes and the high content of organic material within the haemal vessels, and the muscular contractions observed in various locations. The uptake of dyes, particles or nutrients from the gut into the haemal vessels associated with it has been demonstrated for only a few species of asteroids, echinoids and holothuroids (Beijnink and Voogt 1984; Broertjes *et al.* 1980; Ferguson 1982). The only quantitative data on the uptake of glucose into the haemal vessels of the gut of *Sclerodactyla briareus* indicate that little appears there in contrast to the amount that appears in the perivisceral fluid (Farmanfarmaian 1969). The gut is not the sole source of material in the haemal system, as nutrients are absorbed directly by the haemal system from the perivisceral coelomic fluid (Ferguson 1982b). The relative amount contributed to the material in the haemal system by the perivisceral coelomic fluid and directly from the gut is unknown.

The primary objection raised to assigning a circulatory function to the haemal system is the seemingly general ineffectiveness of the muscular contractions in moving the fluid (Perrier 1875). No observations on the movement of material in the haemal system of crinoids or asteroids have been made, although the possibility has been raised that there is a slow and gradual displacement of haemal ground substance resulting from contractions (Beijnink and Voogt 1984). In ophiuroids, peristaltic contractions of the genital axial organ cause movement of its contents towards the oral axial organ while contractions of the branches of the gut are weak and ineffective (Austin 1966). No movement of the fluid occurs within the outer vessel or collateral vessels of *Strongylocentrotus purpuratus*

(Farmanfarmaian and Phillips 1962), but an ebb-and-flow Galenic movement sometimes results in net unidirectional movement of material in the inner and collateral vessels towards the oral haemal ring of *Psammechinus miliaris* (Burton 1964).

Weak local contractions cause slight ebb-and-flow movements of the contents of the gut vessels of *Sclerodactyla briareus* (Farmanfarmaian 1969). In contrast, the vessels of the molpadid *Paracaudina chilensis* show rhythmic pulsations which move fluid from the rete mirabile to the posterior part of the small intestine, and supposedly result in a closed circulation of fluid (Kawamoto 1927).

A complex circulation via the haemal vessels occurs in *Isostichopus badionotus* (Figure 3.36) (Herreid *et al.* 1976). Peristaltic waves of the primary dorsal vessel move haemal fluid forwards to the digestive rete mirabile and the 150 to 200 hearts located between the dorsal vessel and

Figure 3.36: Haemal system associated with the gut of *Isostichopus badionotus*. The upper and lower parts of the diagram are the right and left sides respectively. 1: Oesophagus; 2: stomach; 3: accessory dorsal vessel; 4: edges of intestinal lamellae (plates) visible through the gut wall; 5: hearts; 6: primary dorsal vessel; 7: dorsal connective; 8: primary ventral vessel: 9: ventral transverse vessel; 10: ventral connective; 11: dorsal transverse vessel; 12: caudal end of posterior flexure of small intestine; 13: pulmonary vessel; 14: vascular follicle; 15: anastomotic plexus; 16: respiratory shunt vessels; 17: ventral collecting vessel; 18: infundibular vessel; 19: terminal tuft of pulmonary vessel; 20: dorsal connective from the secondary dorsal vessel; 21: secondary dorsal vessel; 22: constriction between the small and large intestines; 23: large intestine (avascular). Arrows indicate direction of flow. (From Herreid *et al.* 1976)

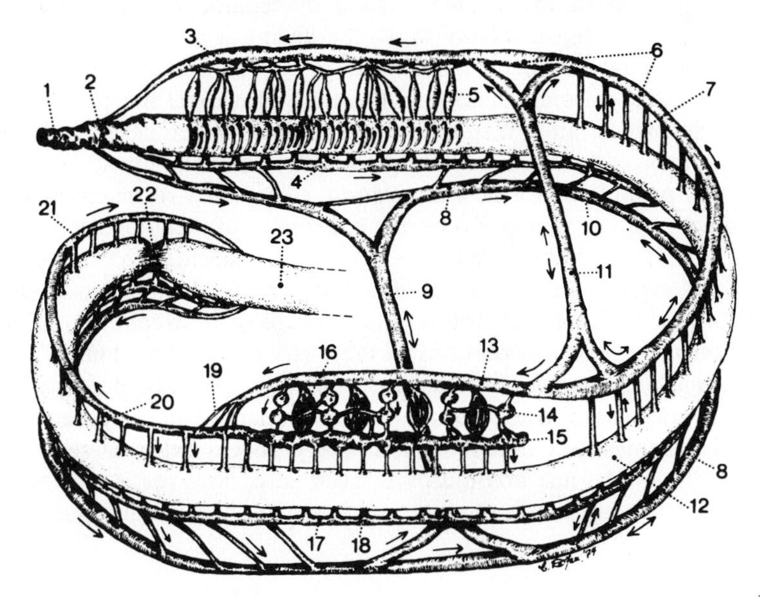

the descending small intestine. Peristaltic waves in the hearts force the fluid through the internal lamellae found only in this part of the intestine. The accessory hearts are probably necessary to produce sufficient pressure to force the haemal fluid through the increased surface area associated with the intestinal lamellae. The destination of the nutrients supposedly absorbed into the haemal fluid by these intestinal lamellae is unknown. The circulation of the gut vessels in *I. badionotus* is only within the gut itself and the respiratory rete mirabile. The gut haemal vessels do not extend into an oral haemal ring and do not appear to connect to the tube feet or gonad.

It is possible that the dissolved nutrients absorbed enter the perivisceral coelomic fluid through the respiratory rete mirabile. The circulation of the ascending small intestine is closely associated with the left respiratory tree. The pulmonary vessel branches off from the dorsal vessel and sends masses of small vessels to the small intestine. These small vessels intermesh with the tufts of the respiratory tree, and are either a vascular follicular network or a series of parallel vessels (the respiratory shunt complex). These small vessels merge and form a large anastomosing plexus just dorsal to the gut. The plexus becomes extensive towards the distal end of the small intestine and forms a second dorsal vessel. It is generally assumed that the respiratory rete mirabile functions in oxygen exchange, but the anatomical relationship may be excretory in function (Cuénot 1948; Hyman 1955). The vascular follicular system associated with the respiratory rete mirabile has characteristics indicating its involvement in coelomocyte destruction (Herreid, DeFesi and LaRussa 1977), and the products may be eliminated via the respiratory tree.

The function of the haemal system in echinoderms is not clear in general. Its anatomical relationships, the chemical nature of its contents, and the potential contractile nature of its myoepithelial covering lead to the suggestion that it is circulatory in function. However, the presence of a ground substance rather than a fluid and the lack of well developed musculature in most cases observed indicate that any circulation would be slow at best. It is possible that a slow circulation, particularly of macromolecules, would be adequate in a group having a low level of metabolic activity. It is unlikely that the haemal system is used to translocate small metabolites such as sugars, amino acids and fatty acids. The haemal system probably is not used in translocation of oxygen except in the aspidochirotes. The internal portions of the ovary (more than 1 mm deep) of the echinoid *Strongylocentrotus* are anaerobic (Bookbinder and Shick 1986; Webster and Giese 1975). The potential for development of the haemal system into a functional circulatory system exists as seen by the characteristics of some holothuroids. Even here, however, a true systemic circulation serving the body wall and gonad as well as the gut has not developed.

The presence of a ground substance and the lack of circulation has led to the suggestion that the haemal system is a repository in which materials are absorbed and re-released into the perivisceral or perihaemal coelom (Beijnink and Voogt 1985; Ferguson 1984; Grimmer and Holland 1979; Jackson and Fontaine 1984). This is possible, but does not explain the characteristics noted above. A repository function is not mutually exclusive of other functions.

3.3. LOCOMOTION

The extinct, asymmetrical carpoid echinoderms are thought to have been unattached, but with limited locomotion at best. All of the extant classes of echinoderms except the crinoids have been free-living throughout their fossil record, and even the contemporary crinoids are predominantly free-living. The basic body form of echinoderms is thus amenable to the requirements for this mode of life, but apparently either has not the potential for evolving rapid locomotion or the way of life does not have characteristics that would make rapid locomotion advantageous. Low levels of predation and/or the limitation of the kinds of feeding modes may be responsible for the failure to develop locomotory limbs.

3.3.1. Crinoidea

The suspension-feeding way of life does not require the free-living state, let alone the capacity for locomotion. However, the free-living state and the ability to locomote provide the capacity to change position to improve feeding potential and to escape from adverse physical conditions and predators. Breimer and Lane (1978) concluded that the stemless crinoids have seemingly evolved the capacity for locomotion to gain efficiency as sedentary feeding animals.

The free-living state and locomotion are not linked. The stalked isocrinids almost always lose their attachment. The cirri on the stalk provide temporary attachment to the substratum. They are thus free-living without the capacity to locomote. It would seem that the potential for passive movement alone is advantageous.

Although jointed, the single stalk lacks muscles. It would provide little potential for locomotory function and would actually be an impediment to locomotion. The comatulids have advanced further by losing their stalk completely at a small, early stage. The loss of the stalk and its role in raising the crown by passive lift produced by currents, as well as the difficulties of locomotion with a large crown, may have been factors in the evolutionary decrease in size of crinoids. The ability to move to exposed, elevated positions allows comatulids to be 'functional stalked crinoids' without the possession of a stalk (Meyer and Macurda 1977).

The arms are responsible for locomotion in crinoids. Breimer (1978) described the basic anatomy of the arms of crinoids. They are supported by a longitudinal series of skeletal elements, the brachials. All recent crinoids are grouped in the subclass Articulata because of the elaborate articular faces of their brachials. The articulations are either muscular or ligamentary.

The muscular articulation (Figure 3.37) is more common. Paired flexor muscles occur on the inner (adoral) parts of the articular facets and an extensor ligament on the outer (aboral) parts of the facets. The muscles and ligaments are on opposite sides of a transverse ridge that serves as a fulcrum. In many recent crinoids, the facets are not straight, but oblique to the arm's longitudinal axis. Usually the socket for the attachment of the pinnules is on the higher side of the brachials, at least in the middle and distal portions of the arms. In many recent crinoids, however, the pinnules

Figure 3.37: Muscular brachial articulations in crinoids. (1) Straight muscular articulation in radial of *Endoxocrinus alternicirrus*. (2) Oblique muscular articulation on brachial of *Neocrinus decorus*. (From Breimer 1978)

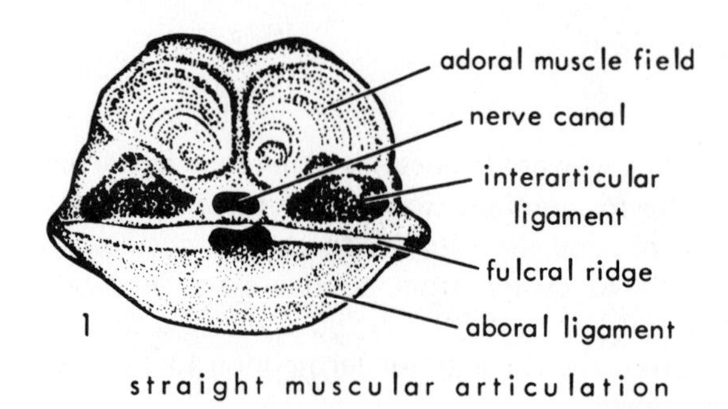

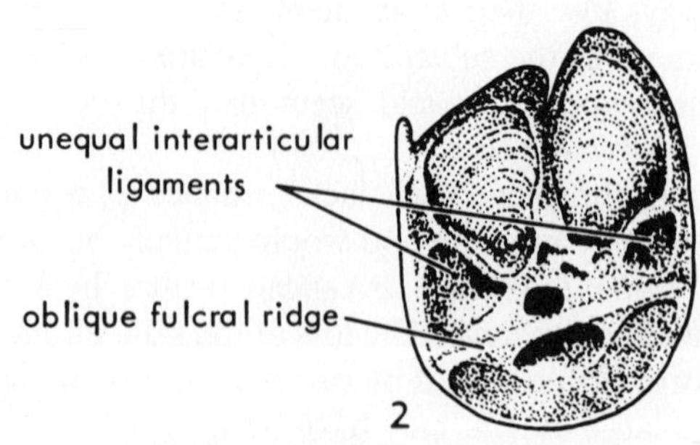

are attached to the lower side of each brachial (Figure 3.38). This reversal of the articulations superimposes at least two different longer axes of articulation and tends to increase the flexibility of the arm with lateral pressure. This reversal is most clearly seen in swimming antedonids, and is less developed in some creeping comasterids.

The role of ligamentary articulations in arm locomotion in crinoids has been little analysed. The presence of ligaments in the arms of crinoids could provide the mechanism for altering the degree of rigidity of the arms and pinnules so that they can function in rapid and complex movements (Meyer 1971). Synostial articulation is simplest and phylogenetically oldest, with the brachials having nearly flat interfaces and short ligaments over most of the joint face (Breimer 1978). In syzygial articulation (Figures 3.38, 3.39), the ligament fibres are very short and located mainly in depressions. These depressions are opposite each other and do not interlock with the ridges that separate the depressions. This type of articulation is confined to the arms and is specially developed to increase the mobility of the arms of vagile crinoids. The syzygial contact allows slight mobility in all directions, and also may prevent torsion from breaking extended arms.

In synarthrial ligamentary articulation, two larger ligament bundles occur on either side of a fulcral ridge (Figure 3.39). These are weakly developed in creeping crinoids (Comasteridae) with a resulting minimal degree of flexibility of the arms. Differential flexibility is possible in two directions as a result of this articulation.

Figure 3.38: (a) Proximal part of free arms of *Florometra serratissima* with pinnules arising from lower side of the brachial. (2) Distal part of free arms with pinnules arising from the upper side of the brachial. (From Breimer 1978)

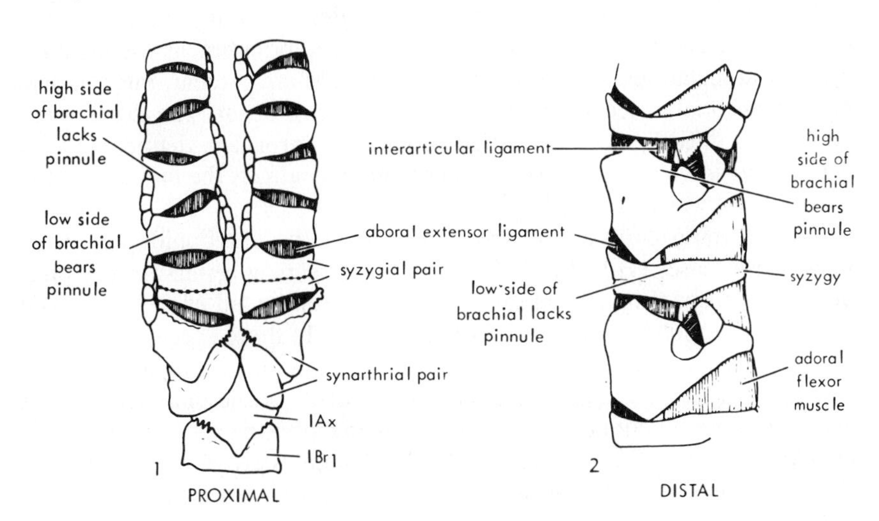

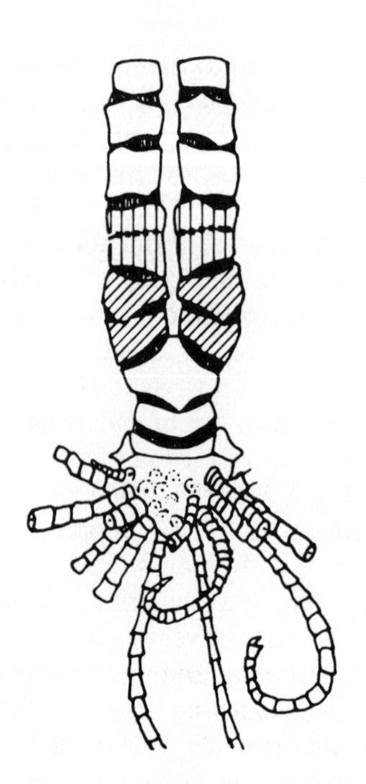

Figure 3.39: Proximal arm portions of *Eumorphometra hirsuta,* showing synarthrial articulations (oblique ruled) between first and second secundibrachs and syzygial articulations (vertical ruled) between the third and fourth secundibrachs (from Breimer 1978)

The most regular distribution of synarthrial and syzygial articulation is in comatulids (Breimer 1978). Comatulids have only two primibrachs (the initial brachials arising from the body and supporting a branch) in each ray. These primibrachs have synarthrial articulations (Figure 3.40) except in the zygometrids which have syzygial articulations. Syzygies may be repeated regularly in the free arms of comatulids, but synarthries are not. This means that the free arms may be twisted to some degree in any direction. Isocrinids can have more than two primibrachs. The primibrachs usually have syzygial articulations when only two primibrachs are present, but can also have synarthrial articulations when more than two occur. The distribution patterns of these ligamentary articulations in the arms are characteristic of genera and species. The distribution undoubtedly affects the mobility of the individual as well as its feeding.

The creeping comasterids are bulkier than the antedonids and have thicker arm bases (H.L. Clark 1915, 1921). Creeping results from a pull-and-push mechanism carried out by the cirri with the aid of the arms. The anterior arms pull and the posterior arms push. In those species with arms of unequal length, the longer ones are always anterior during locomotion. The sharp, recurved hooks on the distal segments of the pinnules of most comatulids are functional in providing attachment to the substratum during crawling (Figure 3.73) (Gislén 1924). The pinnules are closely applied to the substratum where the arms touch down (Meyer and Macurda 1977). It

Figure 3.40: Distribution of synarthrial (oblique ruled) and syzygial (vertical ruled) brachial pairs in (a) *Cenocrinus asterius* and (b) *Glyptometra investigatoris* (from Breimer 1978)

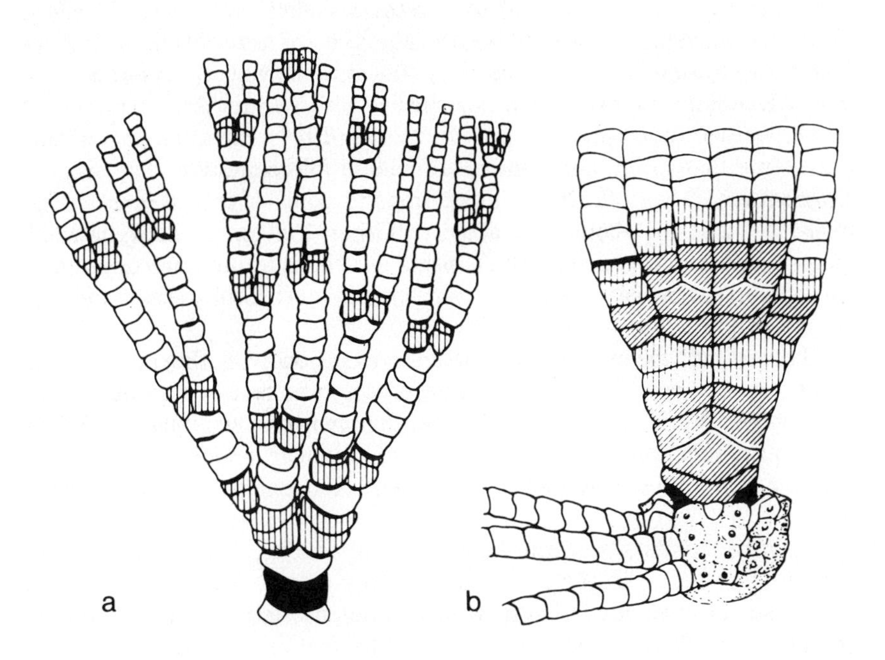

is possible that a duo-adhesion system (adhesion, de-adhesion) occurs (Hermans 1983).

All comatulids except the Comasteridae can swim. Their swimming has been described by W.B. Carpenter (1876), Chadwick (1907), and H.L. Clark (1915). Swimming begins by a release from the substratum with all arms spread horizontally. The arms are then alternately raised upwards with the pinnules flexed inwards and then moved downwards with the pinnules extended. Species with bifurcated arms alternately raise and lower the corresponding half-arm of each set. Multiarmed species seem to behave similarly, with successive use of each arm of a set. Swimming is limited to a distance of several metres.

3.3.2. Asteroidea

Most asteroids do not obtain their food from their water column and they must locomote to obtain food. However, the type of food and the mode of capture do not require rapid locomotion. Consequently, the body form has retained radial symmetry which is usually pentamerous but may be multiple. Locomotion is radially directed and any arm can lead (Romanes 1885; J.E. Smith 1965).

Use of the entire arm for locomotion in asteroids would require the arm to be slender and flexible, and to possess a skeletal system for localised controlled movement along its length. The arms of asteroids have muscles that allow flexion and orientation of the arm. However, except for the family Benthopectinidae which has species with slender arms with a well developed pair of longitudinal muscles (Fisher 1911), the arms do not have a structure that would allow them to function in locomotion. The arms of *Cuenotaster* are very flexible with ventral and dorsal sheets of muscle, and might be used in locomotory activity (Fisher 1940). The lack of axial, longitudinally arranged skeletal elements like those in the arms of crinoids and ophiuroids probably precludes a predominant role for the asteroid arm in locomotion.

The assumption by asteroids of an orientation with the radial water canals on the lower body surface allowed the tube feet to assume a locomotory role. Variations in the tube feet that can have significance in their locomotory function include the number of ampullae associated with them, whether or not they have suckers, and their arrangement in the ambulacral groove.

The tube feet pass between the ambulacral ossicles. The arrangement of the tube feet can be in either two rows or four (Figure 3.77). For example, Asteriidae have quadriserial tube feet, whereas Astropectinidae, Pentacerotidae and Asterinidae have biserial tube feet. Two rows of tube feet may occur the entire length of the ambulacral groove or four rows may occur in certain regions. Thus in *Porania* in the family Gymnasteridae, the tube feet are quadriserial only in the median portion of the arm, and in *Neomorphaster* in the family Stichasteridae, the tube feet are quadriserial until near the end of the arm where they are biserial (Sladen 1889). The quadriserial arrangement of the tube feet results in a considerable widening of the arm. The functional significance of the increase in number of tube feet per unit length of the arm probably has more importance in the strength of attachment than in locomotion.

Most of the expansion of the tube foot in asteroids is due to the contraction of the radial canal, with complete protraction involving the contraction of the ampullae (Nichols 1972). The ampullae would provide for greater and more forceful extension of the tube feet than possible in ophiuroids. The ampullae associated with the tube feet may have one or two lobes. Although the ampullae ordinarily are either completely one- or two-lobed in a species, *Hymenaster quadrispinosus* has single ampullae proximally, but double ampullae distally. The functional significance of the number of lobes has not been ascertained, but it probably involves the volume of the fluid that the reservoirs can hold and the surface to volume ratio. The two-lobed ampullae may produce greater pressure in the tube feet (Heddle 1967). Greater control over the extent of elongation of the tube foot may be possible with the two lobes because each is separately innervated (Nichols 1969).

The presence of two-lobed ampullae can be related to the suckerless tube feet in paxillosids that burrow (Luidiidae, Astropectinidae, Goniopectenidae) (Nichols 1969), but the paxillosid Porcellanasteridae have single ampullae associated with suckerless tube-feet (Fisher 1911). Some forcipulatids (such as *Pisaster brevispinus, Asterias forbesi* and *Pycnopodia helianthoides*) have suckered tube feet that are extended and used to excavate the substratum for feeding but not for burrowing (Figure 2.20).

Locomotion by asteroids on horizontal surfaces seems to be a stepping process whether or not the tube feet have suckers (Jennings 1907; Kerkut 1953; J.E. Smith 1947, 1950) (Figures 3.41, 3.42). The retracted tube foot is oriented in the direction of movement, protracted and, when fully extended, swung back through an angle of about 90° with the sucker pressed against, but not necessarily firmly attached to, the substratum. The

Figure 3.41: (A, 1 to 4) The successive stages of the ambulatory step in *Asterias rubens.* (B, 1 to 4) The conditions of contraction and relaxation of the muscles of the foot during the successive stages of static posture of the ambulatory step. The orienting (postural) fibres are stippled. (From J.E. Smith 1947)

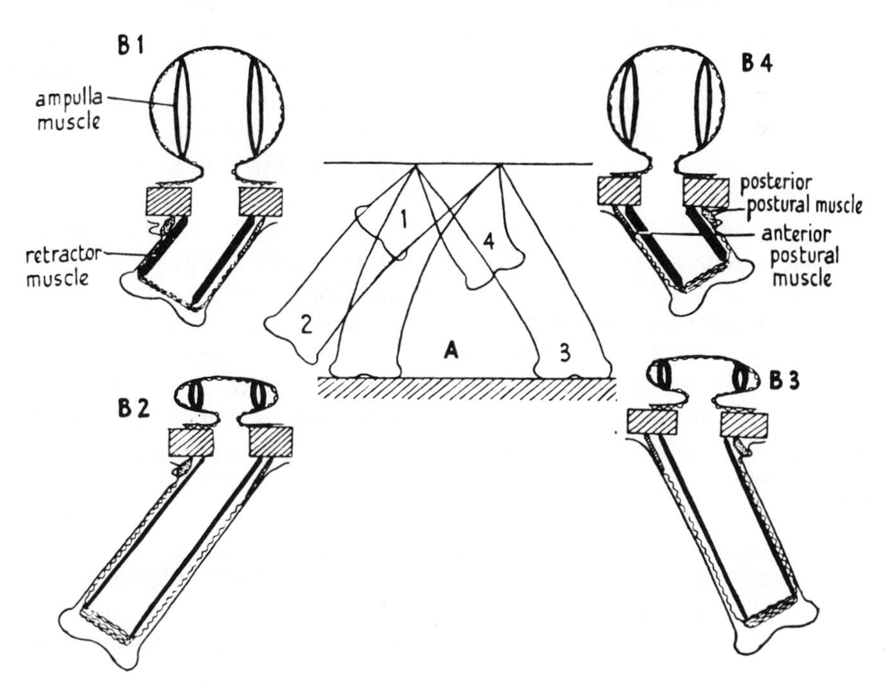

Figure 3.42: Successive positions of a tube foot during the ambulatory step. The foot is protracted and attached to the ground; the disc then remains fixed in position while the base and body are moved over it; the foot is then released, retracted and repointed. Note that the foot is bent during the attached part of the step. (From Kerkut 1953)

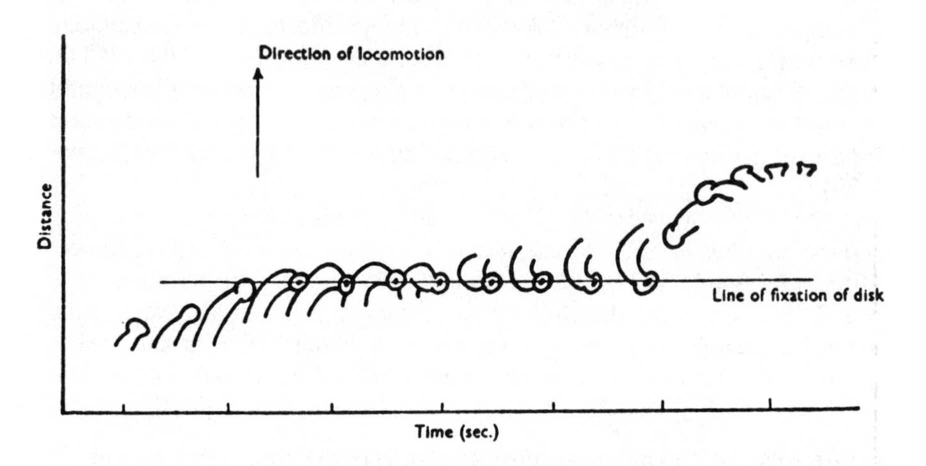

pendulum-like motion thrusts the base of the foot forwards relative to the sucker. The foot is retracted and reoriented at the end of the backswing. The tube foot thus acts as a strut that is used as a lever. Tube feet at any given point on both sides of the ambulacrum act in synchrony, and successive waves of activity occur along the ambulacrum in metachronal fashion. The tube feet of all of the arms function simultaneously in locomotion, those of each of the rays necessarily having a different direction of movement in relation to the arm axis (Jennings 1907). This ability results from the orienting (postural) muscles of the tube feet (J.E. Smith 1947).

The suckers of the tube feet provide an attachment to solid substrata and allow asteroids that have them to function in high-energy water or on angled surfaces. It should be noted that many apparently suckerless tube feet can produce a miniature sucker through localised contraction of the tip of the tube foot (Romanes 1885). In locomotion on an angled surface or against a current, the tube feet do not function as a lever but by traction, the suckers attaching and the tube feet retracting to pull the individual (Kerkut 1953). Many asteroids with suckered tube feet live on muddy substrata where suckers would be non-functional in locomotion (Fisher 1911).

The attachment of the asteroid tube foot to a hard surface is a combination of adhesion by mucus and of suction. Each contributes about half of the total attachment capacity for a tube foot of *Asterias vulgaris* (Paine 1926, 1929). The average suction force of a single tube foot is about 29 *g*. The development of suction by the tube foot of *Asterias rubens* involves longitudinal muscles that extend from the stem of the tube foot to a connective tissue plate in the centre of the disc (Figure 3.43) (J.E. Smith 1947). Elevation of the longitudinal muscles provides the suction. Radial muscles on the disc itself pull down the centre of the disc to release the suction. Connective tissue in the disc prevents a decrease in diameter during the retraction of the tube foot. The adhesive aspects of attachment of the tube foot in asteroids (as well as in echinoids and holothuroids in which the tube feet function in locomotion) may require a duo-gland adhesive system (Hermans 1983). Two cell types would be required to produce adhesive and de-adhesive mucopolysaccharides.

Vertical burrowing of the entire individual to bury itself occurs in the paxillosids. Although they ingest substratum and feed on infauna, burrowing is probably also related to predator avoidance. Members of other orders (e.g. forcipulatids) excavate the substratum for feeding but do not burrow. *Luidia ciliaris* and *Astropecten irregularis* use lateral movements of the tube feet along with a widening of the ambulacral groove and a narrowing of the aboral surface (Figure 3.44) (Heddle 1967). The arms do

Figure 3.43: Conditions of contraction and relaxation of the ampullar muscles and retractor fibres of the foot during (A) protraction, (B) retraction, and (E) localised bending of the tube foot. Contrasting muscles are represented by thick lines, relaxing muscles by thin lines. The positions of the longitudinal and circular fibres of the connective-tissue sheath of the tube foot when it is (C) protracted and (D) retracted. (From J.E. Smith 1947)

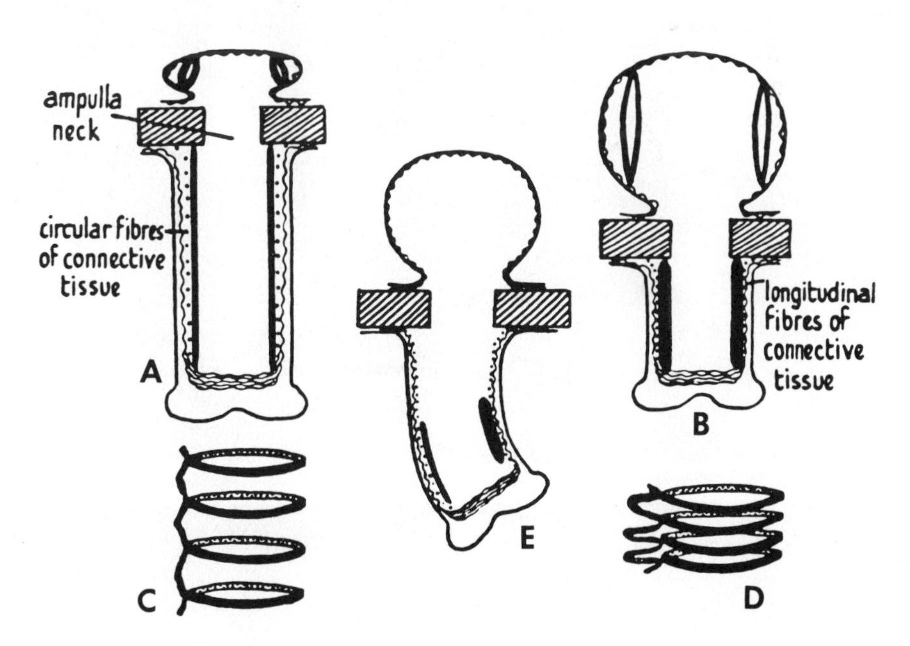

Figure 3.44: Transverse section of the arm of *Astropecten irregularis* in the (a) walking and (b) digging postures. ad.o.: adambulacral ossicle; a.g.: ambulacral groove; a.o.: ambulacral ossicle; i.p.: inframarginal plate; p.: paxilla; s.o.: superambulacral ossicle; s.p.: supramarginal plate; t.f.: tube foot. (From Heddle 1967)

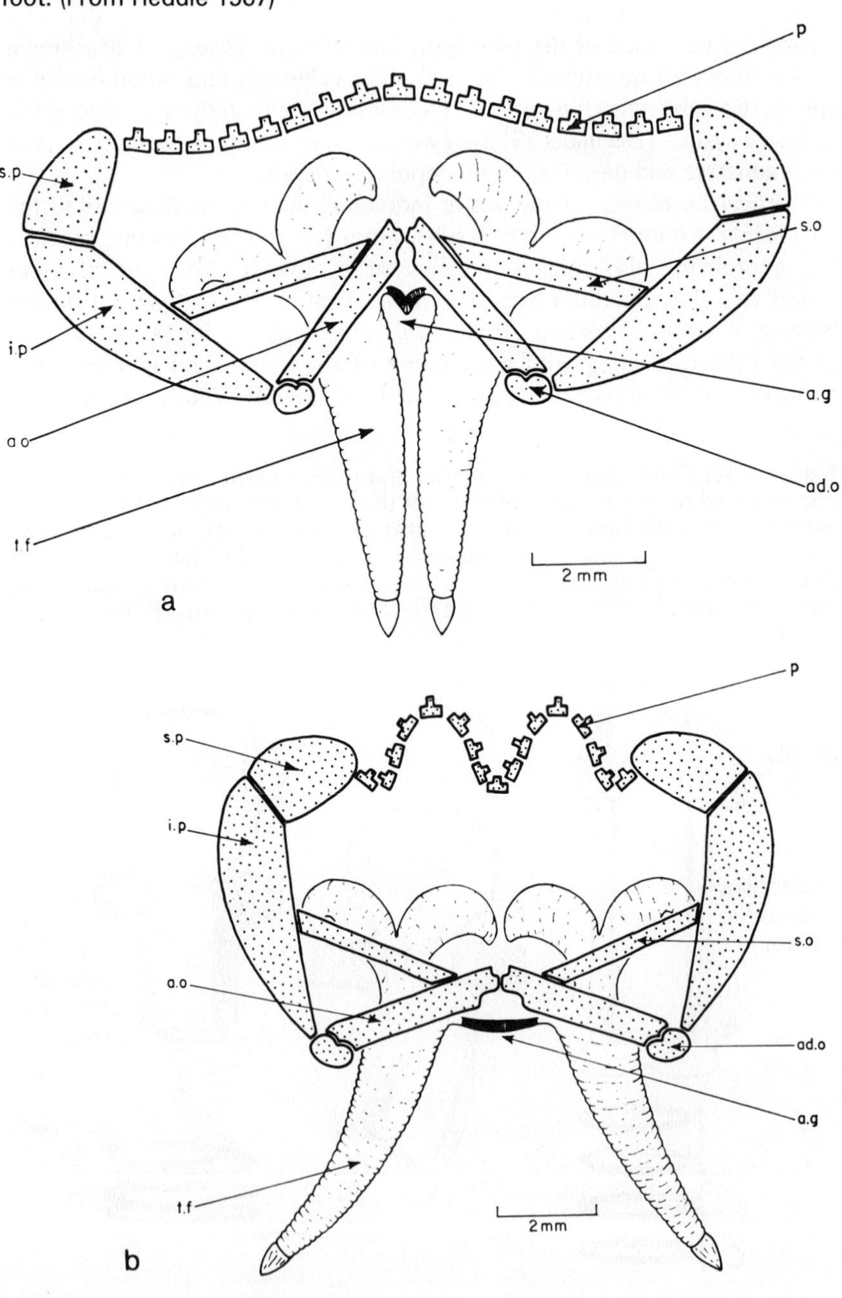

not have the necessary anatomical features to provide undulations that could function in burrowing. Burrowing in these asteroids is vertically only. Horizontal, directional burrowing would be difficult because of the arms. The spines themselves have no constraints that would prevent them from developing the necessary radial motion. Only the tube feet are used in burrowing by the mudstar *Ctenodiscus crispatus* (Shick *et al.* 1981). Individuals use the tube feet as levers or anchors and can plough forwards into the sediment with a leading arm or sink vertically. Horizontal locomotion by buried individuals has not been observed.

Passive locomotion in the water column occurs in juvenile *Asterias* and *Henricia* several centimetres in diameter (A. Agassiz 1869).

3.3.3. Ophiuroidea

Locomotion in ophiuroids is primarily a function of the entire arm. The critical development in the arm of ophiuroids was fusion of the opposite ambulacral ossicles to form single pieces (vertebrae) that fill most of the interior of the arm (Figure 3.87) (Nichols 1969). This makes increased mobility possible, for it permits the arm to twist and turn. Each arm segment is an independent unit because the tube foot is no longer set in two ossicles. The degree of skeletal development can be related to activity, even in congeneric species. *Ophioderma appressum* is delicate and agile whereas *O. brevicaudum* and *O. devaneyi* are heavily calcified and sluggish (Hendler and Miller 1984).

Two major types of articulation occur between vertebrae. Members of the order Ophiurida, which includes most Recent genera, have subcylindrical vertebrae generally with zygospondylous articulation in which ball-and-socket joints allow twisting (Figure 3.75). It is generally thought that these arms move laterally and function for active locomotion, but not vertically. However, extreme vertical flexion occurs in the arm-coiling response of *Axiognathus* (= *Amphipholis*) *squamata* (Emson and Wilkie 1982). This capacity is attributed to the great development of the upper proximal condyle which allows it to function in both lateral and vertical flexion. The euryalinid ophiuroids (the basketstars) have streptospondylous articulation, the vertebrae typically articulating by broad hourglass-shaped surfaces that allow the arms to coil vertically (Figure 3.75). This facilitates locomotion on irregular surfaces, and euryalinids are well known climbers.

Locomotion of ophiuroids over surfaces is by sinusoidal movement of the arms based on their vertebral arrangement (J.E. Smith 1965). The central skeletal articulating vertebrae have opposed pairs of synergistically contracting and relaxing muscles that cause the arms to flex. The individuals move either with one or two arms leading, the next pair moving synchronously in rowing motion, and the remaining arms or arm trailing. The co-ordinated movement of the arms is probably best developed in those active species that have relatively short arms. Those species that have long arms are not active carnivores but are particulate feeders.

The disc is elevated during locomotion. Although they lack suckers, the tube feet function to some extent in locomotion by helping to grip the substratum either by friction or adhesion and by providing the fulcrum on which the arm is pivoted (J.E. Smith 1937). The tube feet of active *Ophiothrix fragilis* are maintained in a state of extension and seem to provide the only means of contact with the substratum. The tube feet of active *Ophiocomina nigra* are not only extended but also in constant motion. The tube foot of *O. nigra* has a terminal knob well provided with glandular elements, indicating an adhesive nature.

The behaviour of the arms in climbing by euryalids has not been described, but specialised arms are present in some species. The primary branches of each arm of *Astrophyton muricatum* divide to form secondary branches. The middle secondary branches form the longer feeding branches while the outer branches form the shorter and thicker locomotory branches (Figure 3.45) (Wolfe 1978). Spines and hooks facilitate grasping.

The tiny commensal *Nannophiura* uses the fine hooks on the broad, flattened ends of the arms to crawl about the spines of its host, the echinoid *Laganum* (Mortensen 1933).

Figure 3.45: Feeding (F) and locomotory (L) arms of *Astrophyton muricatum* (from Wolfe 1978)

The tube feet of *Ophionereis reticulata* have a direct locomotory function (May 1925). The tube feet are in complete extension along the entire length of the arm. When the weight of the arm rests on a particular set of tube feet, the distal part of the tube foot bends and forms a base on which the arm is levered forward. The tube feet of *Ophiocomina nigra* can also have a similar direct locomotory function that can be used on both horizontal and vertical surfaces (J.E. Smith 1937). In sequence, the tube feet are extended and the side of the terminal knob and distal part of the tube foot are applied to the surface. The terminal knob turns so that only its tip is in contact with the surface, the distal part of the tube foot is bent at an angle to the rest of the tube foot, the tube foot straightens and becomes vertical, the arm is levered forward, and the terminal knob is detached with a jerk (Figure 3.46). *Ophiocomina nigra* uses the tube feet for slow locomotion to move food, springing forward by a rowing motion of the four trailing arms after the leading arm touches the food (Fontaine 1965). The minimal involvement of the tube feet in ophiuroid locomotion may be due in part to the minimal provision of reservoirs for the tube feet.

Amphiurid ophiuroids use the tube feet to burrow vertically (Figure 3.47) (Woodley 1975). Burrowing is initiated by lateral movements of the tube feet outwards from beneath the disc and proximal part of the arms. Alternate pairs of the tube feet are synchronised in moving across and up between the spines so that a ridge of particles is thrown up along the sides of the arms and disc. The ophiuroid begins to sink under its own weight into the substratum. The arms make a lateral fold near the disc in two or three adjacent arms. The folds move sinusoidally along the arm distally,

Figure 3.46: The successive phases of movement (from a to f) of the tube foot of *Ophiocomina nigra* climbing a vertical surface (from J.E. Smith 1937)

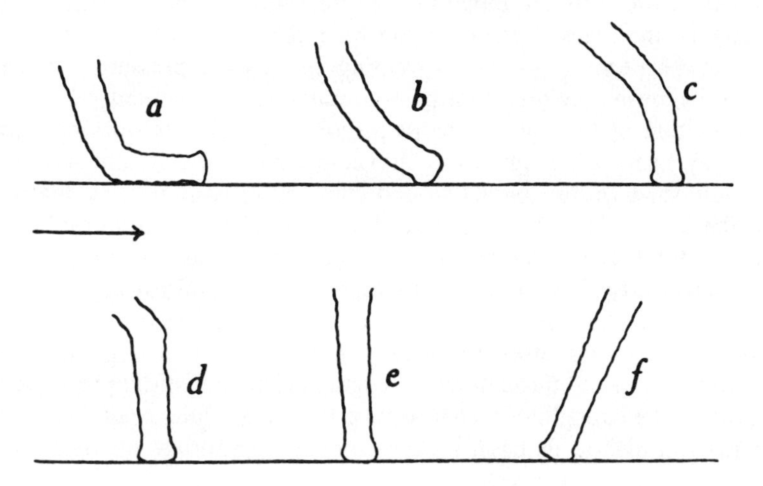

Figure 3.47: View of the arm of *Amphiura filiformis* from below showing the tube feet in various phases of burrowing activity: the tube feet strike outwards from the arms, and swing laterally across and up between the spines. Approximately alternate pairs of tube feet are in phase. (From Woodley 1975)

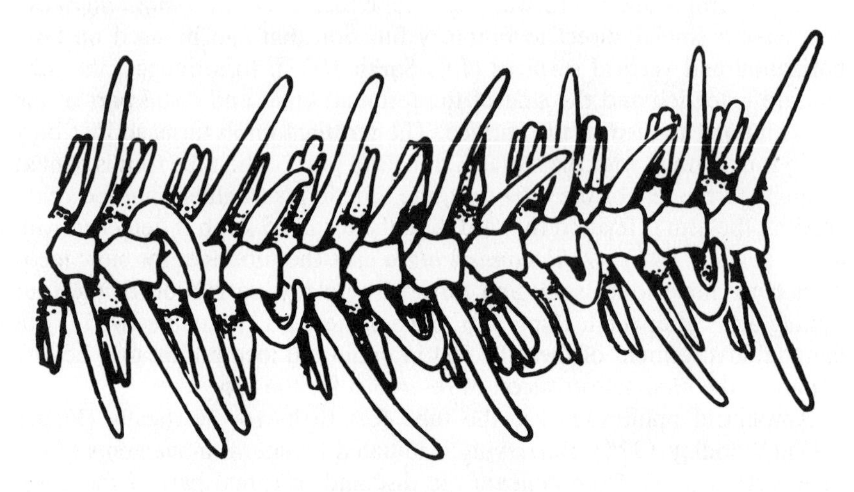

bringing the arms down into the substratum (Figure 3.48). When most of the arms are buried, they fold and serve as an anchor and pull the rest of the disc at an angle into the substratum.

In amphiurids, an accessory mid-dorsal vesicle with sphincters occurs in the radial water canal between adjacent pairs of tube feet (Figure 3.49) (Woodley 1967, 1980). The tube foot itself has a proximal bulb, longitudinal muscles, and a double-spiral lattice of connective tissue. A valve separates the tube foot from the radial water canal. The tube feet are extended by pressure developed in the radial water canal primarily by the elasticity of the walls. Sphincters between the tube-foot pairs close the canal between each segment, preventing the loss of pressure down the canal and allowing the tube-foot pairs to function independently.

Contraction of the bulb develops turgor and performs working movements; any changes in length from bulb contraction are transient and small. The significance of the spirally wound connective tissue fibres is that they allow the tube foot to bend without buckling. The system allows flexing when the distal stem muscles contract. These properties allow rapid alternation between rigidity and flexibility that is useful in burrowing.

The strength of the ophiuroid tube foot is affected by its shape (Woodley 1980). The strength of the parabolic tube foot (e.g. *Amphiura filiformis*) is constant throughout its length, a characteristic appropriate for burrowing. The strength of a conical tube foot (e.g. *Ophiura albida*) is low at its tip and high at its base, a characteristic appropriate for functioning

Figure 3.48: Sequence of *Amphiura chiajei* burrowing into mud (a, 15 s after initiating; b, 30 s; c, 60 s; d, 105 s). a and b: Mud particles are thrown up by proximal tube feet. a, b, c: Arm (arrow) ploughs into substratum. (From Woodley 1975)

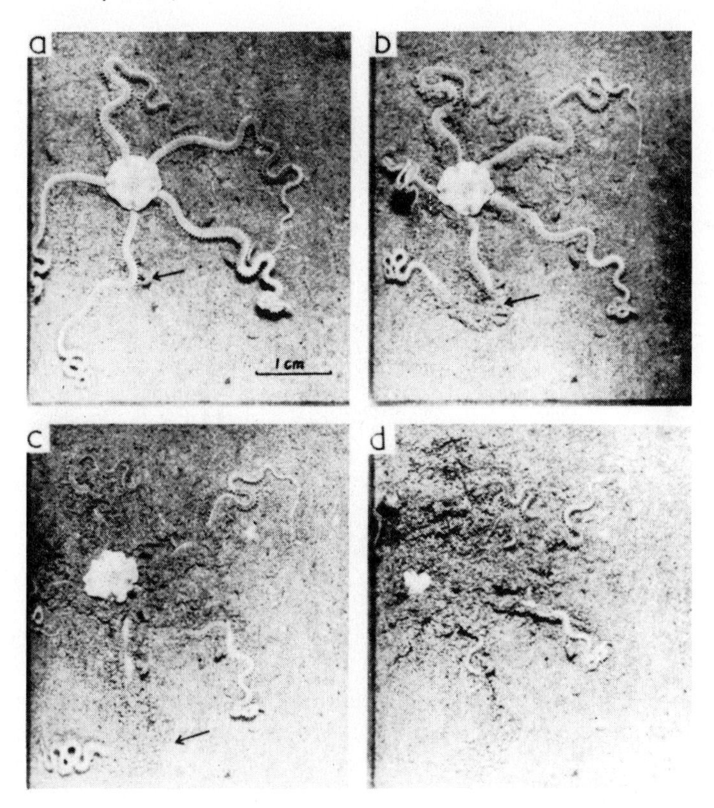

during arm locomotion when the tube feet are thrust down to provide frictional contact with the substratum.

A deep-sea ophiuroid *Bathypectinura heros* swims by vigorous thrashing of its arms (Pawson 1982a). Shallow-water ophiuroids move their arms in swimming motion when released into the water column, but have not been observed to swim in the field. Small juvenile ophiuroids (*Ophiopholis* with five arm-joints) have passive locomotion in the water column (A. Agassiz 1869).

3.3.4. Echinoidea

The variety of foods eaten by regular echinoids does not require rapid locomotion. Apparently no modification of pentamerous symmetry is

Figure 3.49: Part of the radial water vascular system of *Amphiura filiformis* (longitudinal collagen fibres external to the double-spiral lattice of the stem have been omitted). (From Woodley 1980)

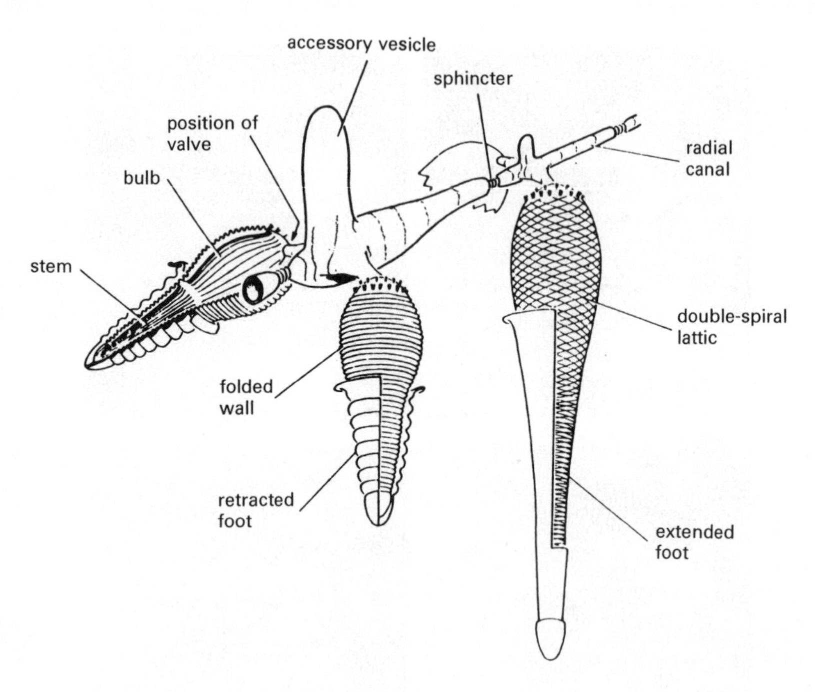

necessary for adequate rates of locomotion to occur, and the ability to locomote in any direction should actually be advantageous. Similar consideration should apply to the irregular echinoids, and their development of bilateral symmetry should not be related to increasing the rate of locomotion, but to attaining unidirectional locomotion within a particulate substratum and to ingestion (Kier 1974). Bilateral symmetry allows asymmetrical development of the spines and musculature for increased efficiency of operation.

Unidirectional locomotion by irregular echinoids does not necessarily involve a change in body form. In the clypeasteroids, the test profile can be ovoid (e.g. *Fibularia*), campanulate (e.g. *Clypeaster*), or discoid (e.g. *Dendraster*); and the ambitus rounded (e.g. *Laganum*) or acute (e.g. *Echinodiscus*) (Durham 1966). In the atelostomatids, cassiduloids have a rounded profile and an ambitus which can be nearly perfectly circular, but a trend towards elongation probably reflects the increased depth of burrowing (Kier 1966). Spatangoids may be quite rounded in profile, but do have an oval to a more or less elongate shape (Mortensen 1950). Irregular echinoids can also be quite large. *Plagiobrissus grandis* has horizontal dimensions of 220 by 160 mm and a height of 55 mm (Kier 1974).

These characteristics also indicate the lack of a need for rapid locomotion and the ability of non-streamlined forms to locomote on or within particulate substrata.

Locomotion in regular echinoids results from activity of the tube feet and spines, the role depending upon the group. Young of the cidaroid *Goniocidaris canaliculata* creep upon the spines of the parent by the aid of five large peristomial tube feet (Thomson 1876). Adult cidaroids have very small suckers on their tube feet and use only the spines for locomotion but with considerable dexterity. Cidaroids from which the spines have been removed cannot right themselves (Lawrence 1976a). Cidaroids can even climb vertical glass rods by use of the spines (Prouho 1888). The diadematids have only weakly suckered tube feet and use their spines as levers for rapid locomotion over horizontal surfaces. The tube feet are used for movement on angled surfaces. *Arbacia punctulata* uses its spines more than its tube feet for locomotion (A. Agassiz 1873). The regular echinoid *Lytechinus variegatus* uses its spines and tube feet in similar ways (G.H. Parker 1936). The spines can function as levers in any direction due to the ball-and-socket attachment to the test (Figure 3.111). In contrast to cidaroids, regular echinoids such as *L. variegatus* which have short spines use the tube feet primarily for righting, and can right themselves if the spines are removed (Lawrence 1976a).

The spines on the lower surface may be modified in regular echinoids living on soft substrata (A. Agassiz 1881). Echinothuriids have club-shaped fleshy spines or hoof-shaped spines (Figure 3.50). The diadematid *Plesiodiadema indicum* has extremely long spines, those on the upper surface ending in a hoof. Micropygid diadematids have club-shaped spines on their lower surface.

Regular echinoids show an evolutionary trend of an increase in the number of ambulacral plates and a compounding of the plates that has resulted in an increase in the number of tube feet (Figure 3.107) (Kier 1974). As with the asteroids, this is probably more related to increasing the capacity to anchor to the substratum than to locomotion. The increase in the width of the ambulacra would decrease the area available for spines and decrease the ability to use the spines for locomotion and protection (Kier 1974). Consequently, smaller ambulacral spines evolved along with the increase in ambulacral width. Ampullae associated with the locomotory tube feet of echinoids are large and the tube feet are capable of great extension.

In contrast to the ophiuroids and asteroids, no regular echinoid uses suckerless tube feet for locomotion. However, because regular echinoids such as *Lytechinus variegatus* can locomote on firm sand, one can presume that the tube feet function in the same manner as those of asteroids, and that the suckers simply provide greater capacity for attachment to the substratum and for locomotion on angled hard substrata or against a current.

Figure 3.50: View of the lower surface of *Pleurechinus bothryoides* showing spines modified into hooves (from A. Agassiz 1881)

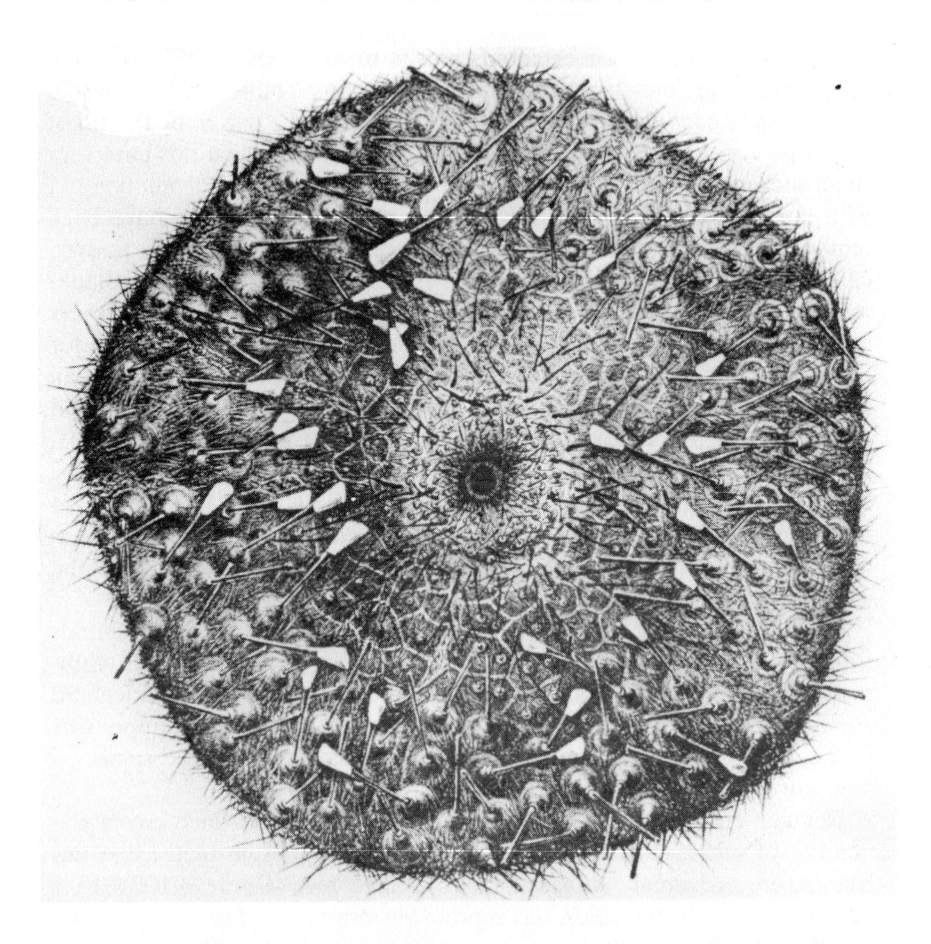

The suction disc of echinoids differs from that of asteroids in being a ring formed of ossicles rather than a connective-tissue plate (Figure 3.51).

The tube feet of echinoids do differ from those of asterozoans in being very close together. As a consequence, the radial water canals do not contribute to the protraction of the tube feet, and extension is solely due to the ampullae (Nichols 1972). The tube feet associated with attachment and which can function in locomotion have isopores with large periporal areas, relatively small pores, and a thick connective tissue and muscle–fibre layer as well as suckers (Figure 3.13) (A.B. Smith 1978, 1980a). The ratio of surface area to volume is unimportant to such a tube foot, and the area of attachment is more critical. For mechanical reasons, it is advantageous for the stem to spread out slightly towards its base so that the contracted tube

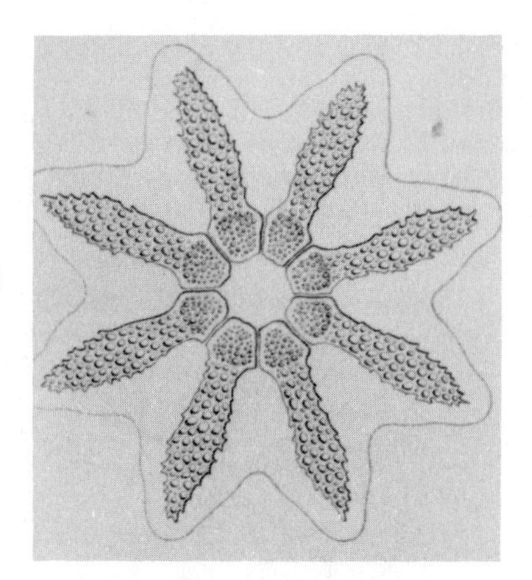

Figure 3.51: Calcareous ring in the sucker of *Strongylocentrotus droebachiensis* (from Lovén 1883)

foot is more capable of compensating for lateral forces. The great development of these features seems far beyond that which would be necessary for locomotion, and the tube feet probably function primarily for attachment in high-energy environments.

Locomotion in the sand dollars such as *Echinarachnius parma* is primarily by spine movements (G.H. Parker 1927; G.H. Parker and Van Alstyne 1932), although juvenile *Dendraster excentricus* use tube feet in locomotion (Chia 1969). The spines are best developed over the anterior portion of the lower surface where their distribution is bilaterally symmetrical in relation to the axis of movement. The area occupied by the locomotory spines varies with species (Figures 2.42 and 3.52), but it is not known whether this is adaptive for locomotion or rather the consequence of an inverse relation with the degree of development of the food grooves. Waves of co-ordinated spine movement pass from the anterior edge of the test posteriorly. The spines move backwards in a vertical motion and recover more laterally. The tube feet are involved minimally, primarily in piling up particles on the anterior upper surface, a behaviour that may be more related to burrowing than to locomotion *per se.* In rotation of the individual, the spines of the anterior half on both sides of the axis move together towards either the left or the right. The marginal spines of *E. parma* may promote rotation by providing drag (Ghiold 1983).

The sand dollar *Mellita quinquiesperforata* also moves primarily by spine movement (Hyman 1955; Salsman and Tolbert 1965). The spines are much more able to effect locomotion on sand than on mud or coarse gravel (Bell and Frey 1969; Weihe and Grey 1968). *M. quinquiesperforata* has elongate (1200 to 1300 μm) spines on the lower surface that are responsible for locomotion, and shorter (440 μm) club-shaped spines on the upper surface to move particles (Ghiold 1979). The inability of *M. quinquiesperforata* to burrow in mud may be due to the inability of the club-

Figure 3.52: Views of the lower surfaces of clypeasteroid and laganoid sand dollars showing the distribution of locomotory and geniculate spines. The foot grooves are indicated by the solid lines. A: *Scaphechinus mirabilis*; B: *Astericlypeus manni*; C: *Rotula augusti*; D: *Arachnoides placenta*; E: *Mellita quinquiesperforata*; F: *Dendraster excentricus.* Bar = 2 cm. (From Ghiold 1984)

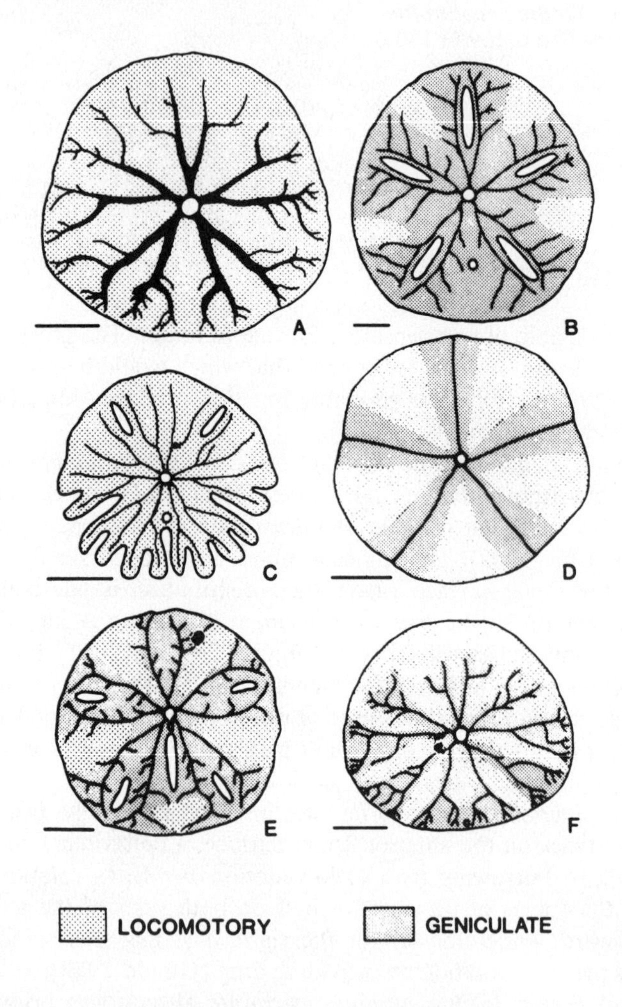

shaped spines on the upper surface to manipulate the very fine particles that fall between the spines and accumulate on the surface. The test margin spines occur around that ambitus and function in burrowing by shovelling particles (Ghiold and Seilacher 1982).

The tiny clypeasteroid *Echinocyamus pusillus* does not burrow in the sense of excavation, but does use the accessory tube feet to move particles that may be larger than the individual itself (Ghiold 1982; Nichols 1959;

Telford 1983). The spines cannot transport particles along the test because their arc of rotation is insufficient to reach adjacent spines. The spines thus merely support the individual particles on the test surface. It is assumed that the spines and tube feet on the lower surface are responsible for forward locomotion.

The cassiduloids have large locomotory spines on their lower surface and a naked sternal area which is thought to increase burrowing efficiency (Kier 1966). The locomotory spines of *Cassidulus caribaearum* rotate in a flattened ellipse, swinging posteriorly and slightly away from the body surface during the power stroke and closer to the surface during the return (Figure 3.53) (Gladfelter 1978). The peripheral locomotory spines are more perpendicular to the body surface than the more axial ones. The movement of the spines passes as a continuous wave posteriorly, with the

Figure 3.53: View of the lower surface of *Cassidulus caribaearum* with most of the lateral spines removed. Their position is indicated by the large tubercles. The remaining spines show position and direction of movement of the spines at one point in time. The large ventro-lateral spines move in flattened ellipses, with the power stroke posterior and away from the ventral surface, and the recovery closer to the ventral surface and of longer duration. A: Anterior phyllode; B: locomotory spine, showing pattern of movement; C: antero-lateral phyllode, showing perforations for large buccal podia; D: oral spines, forming screen behind the mouth; E: lateral area of locomotory spines, showing tubercles; F: sand grain being passed between two tube feet of the anterior phyllode. (From Gladfelter 1978)

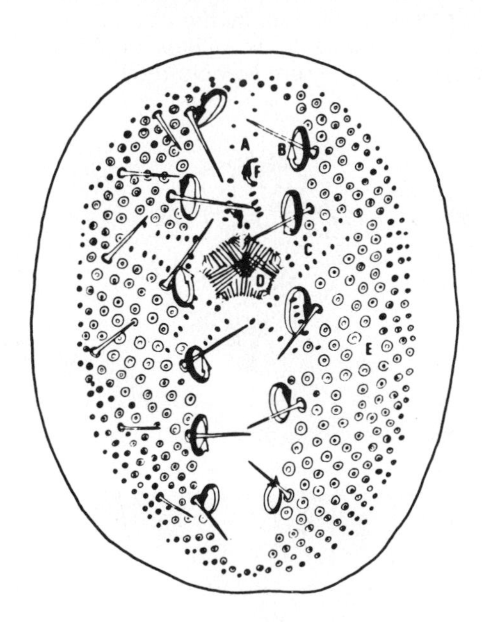

wave front parallel to the transverse rows of spine bases. The timing of the passage of a locomotory wave is such that the most anterior and most posterior spines are in nearly the same phase. This results in continuous action during locomotion. The movement of the spines in the two sides of the axis are 180° out of phase so that the power phase alternates from the anterior of one side to the other. This should increase the efficiency of burrowing.

The spatangoids bury and locomote through the substratum by moving particles with their spines on the plastron and lateral interambulacra (Figure 3.54) (Nichols 1959). In burying, *Spatangus purpureus* and *Echinocardium cordatum* move the lateral spines in such a way as to shift material from below and cast it up in mounds on either side while some forward movement results from the action of the plastron spines. Forward movement is minimal and burying is almost vertical. Although the presence of plastron spines is not necessary for burying in these two species, they are necessary for burying in *Lovenia elongata* (Ferber and Lawrence 1976). The locomotory spines vary in number, location on the plastron, and shape. *Echinocardium cordatum* is found in fine sand or sandy mud and has spatulate spines on the plastron. *Spatangus purpureus* is found in substrata of larger particles and has almost pointed spines on the plastron. Although the plastron of many spatangoids is flat, that of others including

Figure 3.54: A: Views of the (a) upper, (b) under, and (c) lateral surfaces of *Echinocardium cordatum* showing the large locomotory spines. B: Successive stages of the initial burrowing process showing the side and front mounds of substratum piled up by the locomotory spines of the undersurface. The dotted line is drawn to indicate the amount of forward movement during burrowing. (From Nichols 1959)

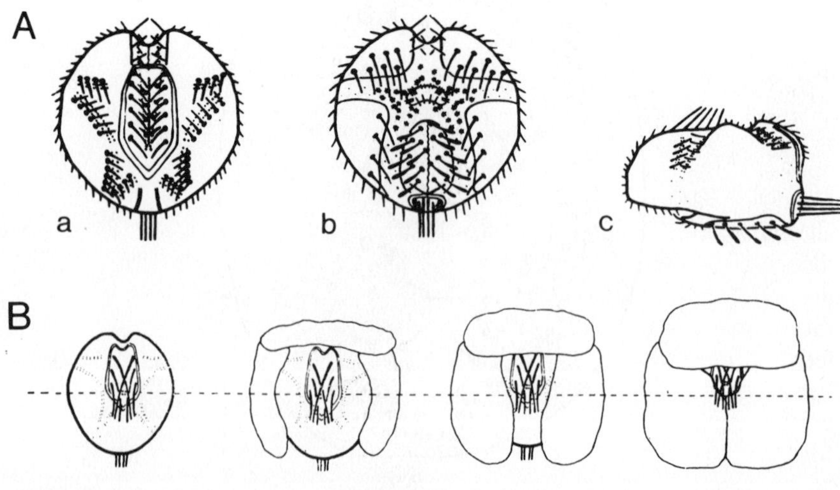

E. cordatum is keeled where the locomotory spines occur. This increase in area of locomotory spines without an increase in body width should increase their effectiveness.

These locomotory spines of spatangoids have deeply sunken areoles (scrobicules or camellae) that form large bulges on the inside of the test (Figure 3.55) and provide attachment for the strong muscles (Ferber 1976). The highly excentric, scale-like spine bases and mamelons produce an efficient motion in restricted directions (Fischer 1966). The spines are arranged in bilaterally symmetrical clusters around the otherwise naked plastron and peristome and show metachronal activity (Ferber and Lawrence 1976). As with the clypeasteroids, the sizes of particles that the spines can manipulate have distinct limits.

Body construction may constrain the evolution of burrowing echinoids (Seilacher 1979). If burrowing is a means of microphagous feeding, the morphological features must meet the requirements for that function as well as for burrowing *per se.* The outline of sand dollars (commonly broader than long) and of many heart urchins would be adaptations primarily for feeding. Modifications to promote locomotion of sand dollars include complete flattening of the lower body surface to provide an effective locomotory sole, flattening of the upper body surface and production of mucus to avoid deviation from horizontal burrowing by drag on the non-locomotive surface, and a sharp ambitus with steering marginal spines.

3.3.5. Holothuroidea

Suspension-feeding holothuroids show little movement and many have effectively resumed a functionally pelmatozoan existence. This seems contradictory to the trend towards the eleutherozoan existence of crinoids. The difference is probably due to the ability of holothuroids to attach firmly to the substratum and a body which is little susceptible to damage in high-energy environments where suspended food is continuously available. Holothuroids seem less susceptible to predation which would reduce their need for locomotion for diurnal protective behaviour.

Surface-feeding holothuroids, in contrast, must locomote to obtain food, although water currents do carry food to some reef-dwelling forms. Rapid locomotion is unnecessary for such feeding types. Bilateral symmetry (the presence of three ambulacra, the trivium, on the lower surface) results from the epibenthic existence in which the application of a large number of tube feet to the surface is advantageous for locomotion or attachment. Pentamerous symmetry is maintained in burrowing holothuroids that do not use the tube feet for locomotion and in apodids that use their tentacles for locomotion.

Figure 3.55: Views of the undersurface of *Lovenia elongata* showing (a) the large locomotory spines, and (b) the large, asymmetrically buttressed tubercles with deeply sunken areoles (from Mortensen 1951)

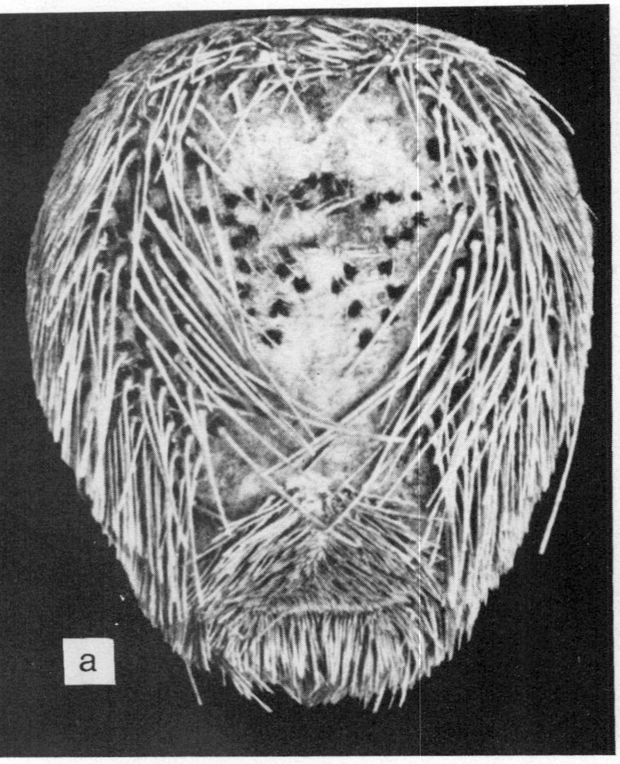

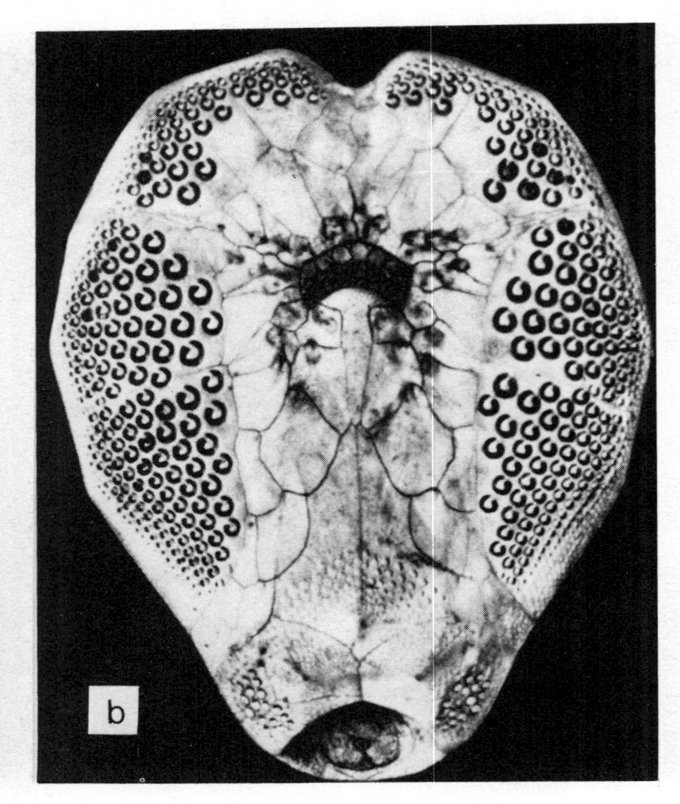

As in the echinoids, only suckered tube feet seem to be used for locomotion in holothuroids except in elasipods. Tube feet with suckers were probably lacking in primitive holothuroids, and locomotion in the echinoid way would not have been possible until they developed (H.B. Fell and Moore 1966). Tube feet are used for locomotion by three extant groups of holothuroids. Dendrochirotids and aspidochirotids have tube feet with suckers for locomotion on firm substrata. The deep-sea elasipods have enormous tube feet which lack suckers for locomotion on soft substrata (Figure 3.56). The exact mode of action of the tube feet on horizontal surfaces is unknown. Because the holothuroids can move over particulate substrata, their tube feet probably function in the same way as those of asteroids in such situations. The suckers on the tube feet function in

Figure 3.56: The elasipod *Scotoplanes* in the process of walking. Note the extended tube feet and the constrictions in the body (from Hansen 1972)

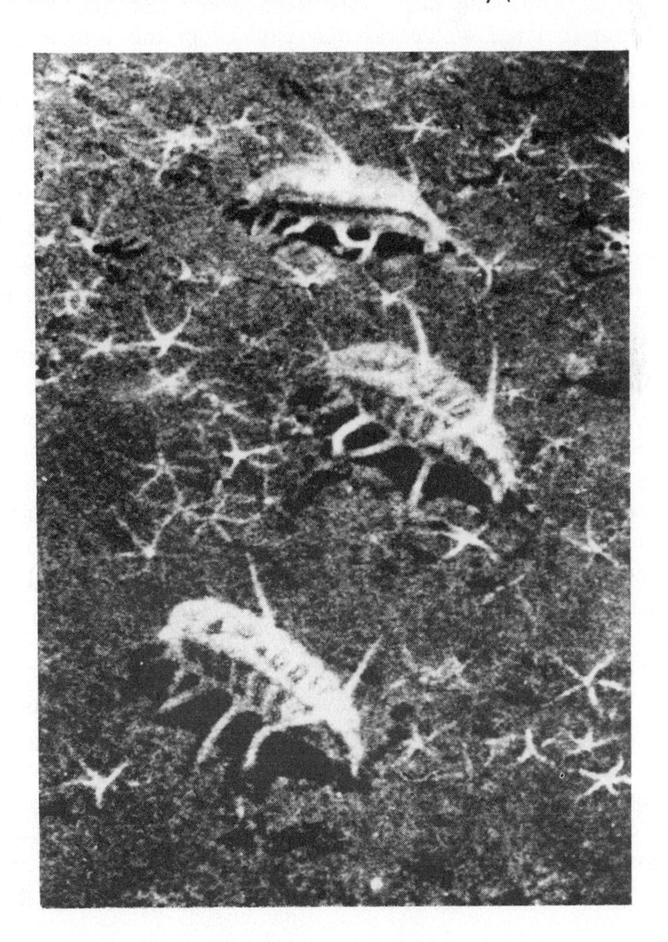

locomotion on angled hard surfaces and in high-energy environments. The suckers have a ring of calcareous ossicles in their discs like those of echinoids. As in asteroids and echinoids, the number of tube feet on the ventral surface of aspidochirotids and dendrochirotids can be increased by irregular coursing of the radial water canal. As with the asteroids and echinoids, the concentration of tube feet on the ventral surface is probably for secure attachment as well as for locomotion.

The protraction of the ambulacral tube feet has not been well described. Species such as the dendrochirotid *Pawsonia* have accessory ampullae, whereas the aspidochirotid *Holothuria* has none (Nichols 1972). In *Holothuria*, extension of one tube foot involves contraction of the neighbouring ones. The contractile radial water canals are not used and, in the absence of ampullae, extension of a tube foot is caused solely by contraction of adjacent tube feet.

The elasipods are deep-sea holothuroids found on soft substrata. They have few and unusually large tube feet which are not connected to ampullae but to large water vascular cavities in the ventral body wall (Figures 3.24, 3.25, 3.56). These tube feet raise the individual from the bottom. Hansen (1972) suggested that contraction of the body wall forces fluid into the tube feet to extend them, and that retraction and orientation movements might be the same as in asteroids.

Small synaptid apodids such as *Synaptula hydriformis* use their tentacles to move about on solid substrata, probably through the adhesion–dehesion mucus process. The pentactula larva of *Leptopentacta elongata* also moves on the surface and burrows by use of its tentacles before the tube feet appear (Chia and Buchanan 1969).

The flexible body wall and its musculature (a layer of circular muscle and five longitudinal muscle bands in the radii) allow locomotion based on peristalsis (Heffernan and Wainwright 1974). Peristalsis is defined as any muscular contraction moving along a radially flexible tube in such a way that each component wave of circular, longitudinal or oblique muscular contraction is preceded or followed by a period of relative relaxation of all similarly oriented muscle within a given tubular segment. Peristaltic movement based on a hydrostatic skeleton is effective for burrowing, but has the disadvantage that contraction at one point affects the entire body and influences all muscles (Barrington 1967). In addition, contractions must be done slowly to be efficient.

'Direct arching peristalsis' involves a wave of contraction passing anteriorly along the dorsal longitudinal musculature, followed by a wave of contraction of the dorso-ventral musculature. As the dorso-ventral musculature subsequently relaxes, a wave of contraction of the ventral musculature passes along the body tube. Although the component muscular contractions have not been described, the muscular waves occur in *Isostichopus badionotus* (Figure 3.57) (G.H. Parker 1921) and *Astichopus*

Figure 3.57: Lateral view of *Isostichopus badionotus* showing the passage of a single peristaltic wave over the body. A: Resting position, B: initiation of wave at posterior end, C: wave at middle of body, posterior end of the body returned to the surface of the substratum, D: wave leaving the anterior end. (From G.H. Parker 1921)

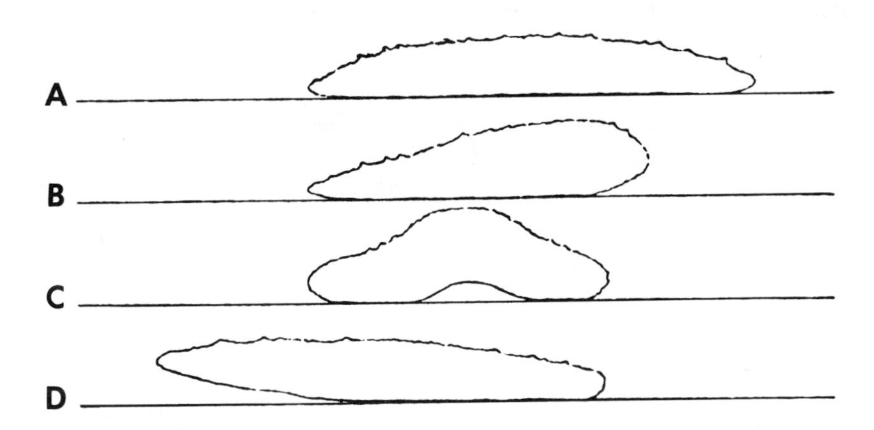

multifidis (Glynn 1965). The peristaltic waves move at speeds of *c.* 0.40 and 0.60 cm s^{-1}, respectively. One peristaltic wave at a time occurs in *P. parvimensis*, and a second begins before the preceding finishes in *A. multifidis.*

'Direct overlapping peristalsis' is the most frequent locomotory mechanism in *Euapta lappa* (Figure 3.58) (Heffernan and Wainwright 1974) and *Opheodesoma spectabilis* (Berrill 1966). Here, peristalsis moves in the same direction as the individual. A wave of contraction of the longitudinal musculature passes along the body, and during the latter part of its duration occurs simultaneously with a wave of contraction of the circular muscles. Relaxation of the circular and longitudinal muscles occurs simultaneously. This system also differs from the previous one in that the body cylinder's major axis is not elevated. An effective arch is formed by the decrease in diameter of the body cylinder as contraction of the circular muscles releases the body wall from the substratum. The necessary anchoring of the body during the process is thought to be due to the tentacles and the microscopic anchors in the body wall (Figures 3.83, 3.84). The anchors catch the substratum when pulled posteriorly and release when pulled anteriorly. Catching would also be facilitated by inflation of the body with stretching of the epidermis. This type of peristalsis seems to fit the descriptions of locomotion in *Sclerodactyla briareus* and *Cucumaria curata* on a solid surface, but with anchoring being provided by the tube feet (A.S. Pearse 1908 and Brumbaugh 1965, respectively). Locomotion on particulate substrata would not be possible because of the

Figure 3.58: (a) The body wall of *Euapta lappa* showing seven warts and the muscles that delineate them; solid lines indicate muscles in a state of tonus; broken lines indicate relaxed circular muscles. (b) Anterior part of body with a direct overlapping peristaltic wave passing from right to left. B: Initial contraction of longitudinal muscle; C: contraction of circular muscle and further longitudinal muscle contractions; D: relaxation of all muscles, body wall distended by coelomic pressure; E: state of tonus. (From Heffernan and Wainwright 1974)

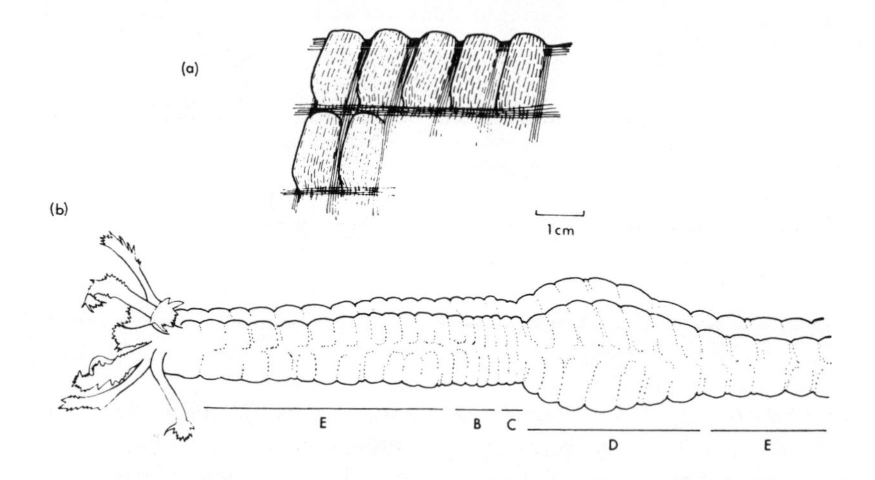

inability of the tube feet to attach.

'Direct longitudinal peristalsis' has longitudinal muscular contractions but no circular muscular contractions. This occurs as *Euapta lappa* moves across surfaces with minimal friction and also on vertical surfaces. The tentacles provide the only attachment. Each wave of contraction of the longitudinal muscles begins by pulling the posterior end forwards (or upwards). The region of longitudinal compression is propagated anteriorly and causes a displacement of the anterior end. At this point, the tentacles do not need to continue their full degree of attachment. While some of the tentacles maintain their attachment, others reach forwards and secure a new attachment.

The apodid *Leptosynapta tenuis* burrows by scraping sand laterally with the oral tentacles and advancing the body by antiperistaltic locomotor waves (Figure 3.59) (H.L. Clark 1899; Hunter and Elder 1967). Alternate thrusts push the body against the sides and head of the burrow, perhaps to compact the sides with mucus and to loosen the material ahead. Slippage of the individual is prevented by the elevation of the anchors in the body wall by the contraction of the circular muscles (H.L. Clark 1899). The anchors are much denser and longer in the posterior part of the body. Because the contraction of the body begins next to the head and moves forwards, the

Figure 3.59: Burrowing activity of *Leptosynapta tenuis* (the widths of the burrow of the synaptid are exaggerated). A: Funnel-shaped feeding depression, B: faecal mound, C: site of burrow extension, D: site at which new feeding funnel will open, E: tentacles, F: bottom of U-shaped burrow. (From Powell 1977)

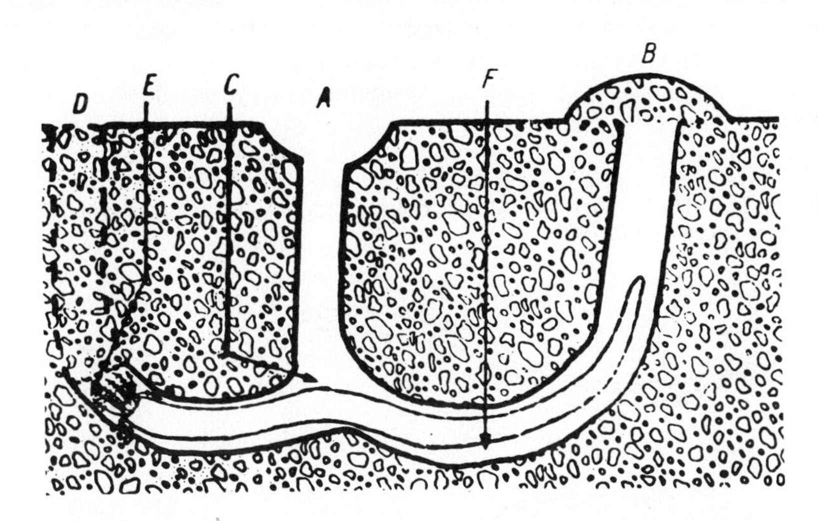

head end is pushed forwards, the anchors there lying flat in the skin. Particles are ingested in the burrowing process and the burrow behind the individual is filled with defaecated material (Myers 1977). Anchors are similarly used for locomotion in *Synapta* (de Quatrefages 1842).

The molpadid *Paracaudina chilensis* initiates burrowing by contraction of the circular muscles of the anterior end resulting in a sudden thrust, relaxation with a slight retraction, and then a shortening of the anterior portion of the body by contraction of the longitudinal muscles (Yamanouchi 1926). *Sclerodactyla briareus* burrows by peristaltic action of the body wall with the tube feet attached to a solid object (A.S. Pearse 1908). In the absence of a solid object for the attachment of the tube feet, burrowing is more difficult and slower. Peristaltic burrowing by *Pseudocucumis mixta* is similar, but the mid-ventral tube feet also move particles.

The lack of calcification, the flexibility of the body wall, and its potential for elaboration have permitted the evolution of swimming in many holothuroids. Most swimming holothuroids are deep-sea elasipods, with others being deep-sea synallactid aspidochirotids, and several apodids (Pawson 1976). In each of these cases, the holothuroids have considerable reduction in the thickness of the body wall. Undulations of the body or of body extensions are the basis for the swimming action.

Bathyplotes natans swims by powerful and quick sinuous bends of the body in an up-and-down direction (Sars 1867). *Leptosynapta albicans* extends the body and describes a sinusoidal path during swimming (Figure 3.60) (Glynn 1965). The complete wave of contraction over a 20 cm long individual was estimated to require *c.* 2 s, with swimming speeds of 1 m min^{-1}. This anterior movement is in contrast to the posterior direction of movement reported for *Labidoplax dubia* (Hoshiai 1963). Young *Leptosynapta inhaerens* swim more aimlessly by scissor-like flexions of the body (Figure 3.61) (Costello 1946). Basic behaviour aspects in swimming occur, because the similarly constructed *Synaptula hydriformis* does not swim (Glynn 1965).

Figure 3.60: Undulatory swimming movements (a to b) of *Leptosynapta albicans* (from Glynn 1965)

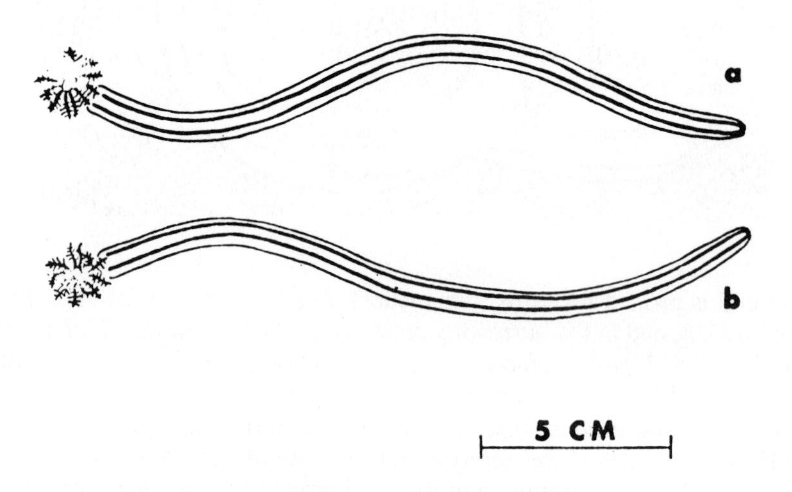

Figure 3.61: Swimming movements of *Leptosynapta inhaerens.* The effective stroke (a to b, c to d) is directed towards the base of the U. A rotation of the entire body through 180°, on the axis of the U, occurs between b to c. (From Costello 1946)

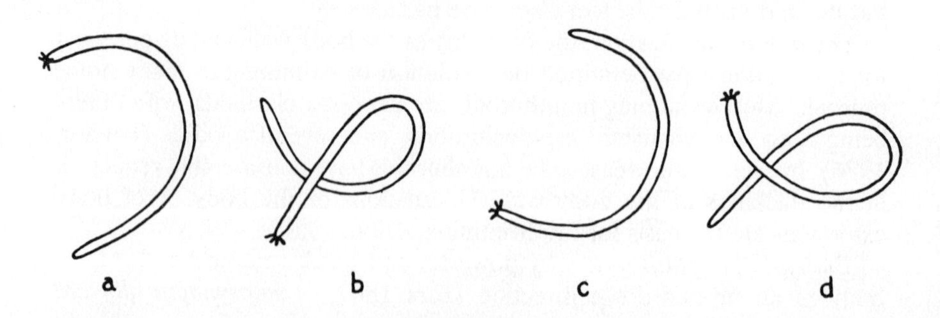

Swimming movements are either by undulations of the entire body as in the aspidochirotid *Paelopatides* (Figure 3.62) (Billett, Hansen and Huggett 1985; Pawson 1976) or by movement of the flattened brim or veil around the body as in the elasipod *Enypniastes* (Figures 3.63 to 3.65) (Ohta 1985; Pawson 1982a). A most unusual swimming behaviour is that of *Peniagone diaphana*, which swims with the long axis held vertically with slow simultaneous stroking by the tentacles and the postanal fan (Barnes *et al.* 1976). The movement of the tentacles provides the thrust while the movement of the postanal fan counteracts the shift in body axis. With the extremely thin, transparent body wall and a fluid-filled coelom, individuals must be near neutral buoyancy and little effort should be required to maintain position.

Figure 3.62: The aspidochirotid *Paelopatides grisea* swimming by undulation of the lateral brim (from Billett *et al.* 1985)

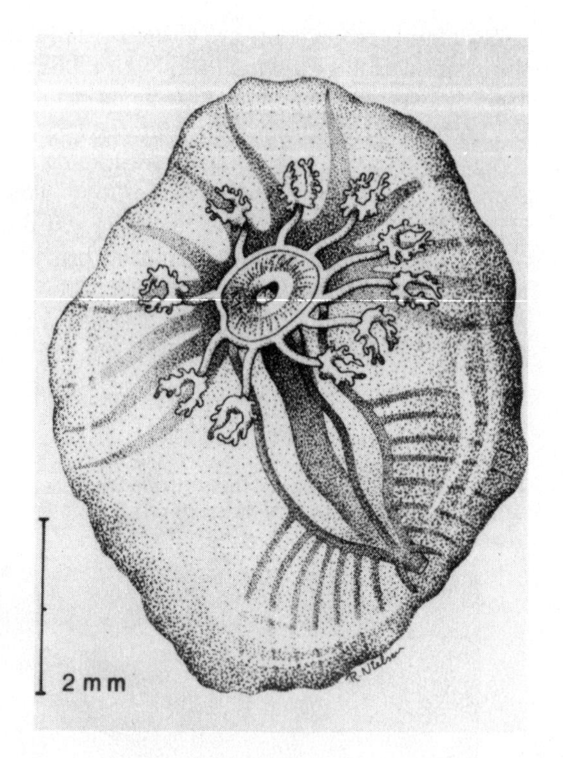

Figure 3.63: View from below of juvenile *Enypniastes diaphana* showing bifid tentacles connected directly to the water ring. The anterior brim has twelve large papillae. Eight pairs of ventro-lateral tube feet form a continuous series around the posterior third of the body. (From Billet *et al.* 1985)

Figure 3.64: Swimming behaviour of *Enypniastes eximia* (from Ohta 1985)

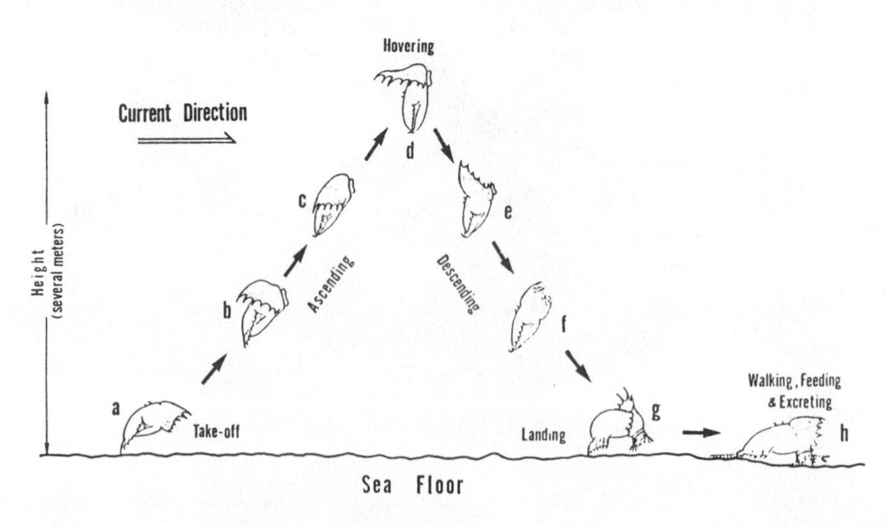

Figure 3.65: Postures of *Enypniastes eximia* in the (a, b) ascending phase, and (c) descending phase (from Ohta 1985)

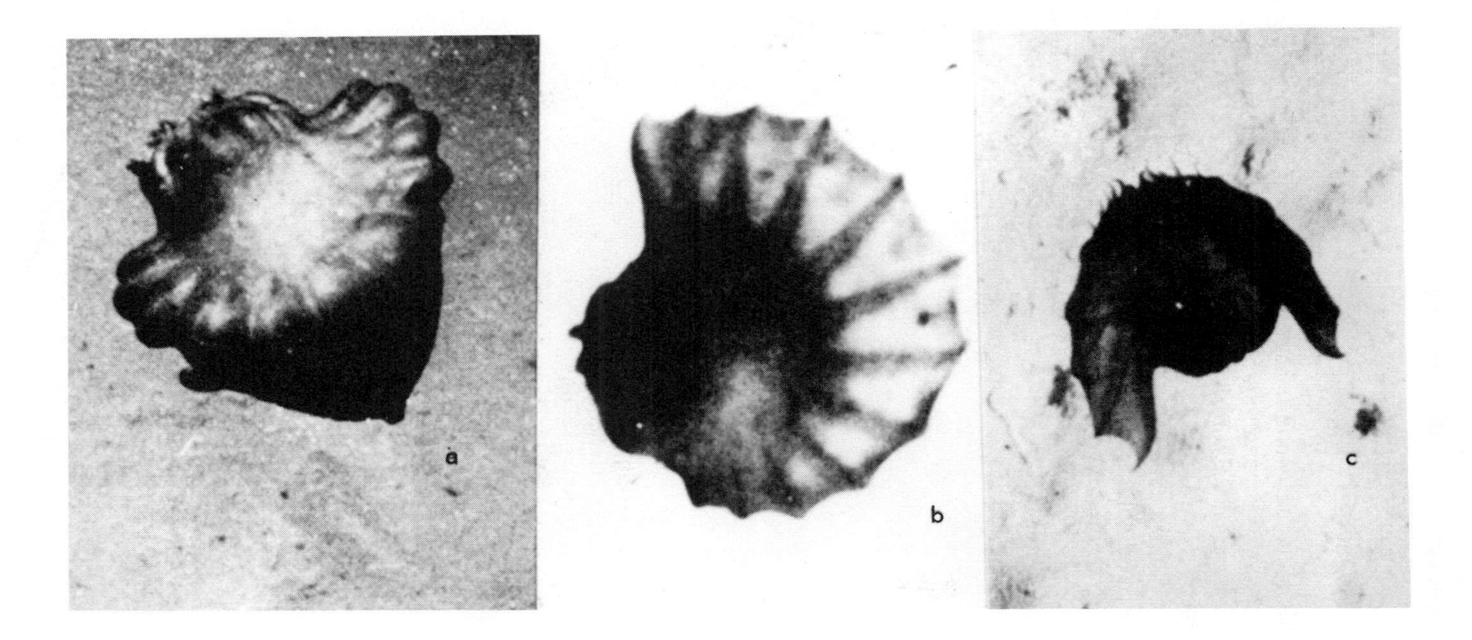

3.4. MAINTENANCE OF POSITION

3.4.1. Crinoidea

Crinoids show an evolutionary trend towards relinquishing a permanent attached way of life and reducing the stem (Bather 1900). This can be seen in the extant classes of the Articulata. However, loss of a permanent attachment requires the development of alternative methods of attachment to the substratum. Whether permanent or not, attachment to the substratum by crinoids is a feature of their skeletal system and this demonstrates its versatility. Attachment in crinoids is by an expanded terminal disc, a radix, or by cirri (H.W. Rasmussen and Sieverts-Doreck 1978). A tapered column was used as a prehensile tail by some extinct classes (Figure 3.66), but is not found in any extant group.

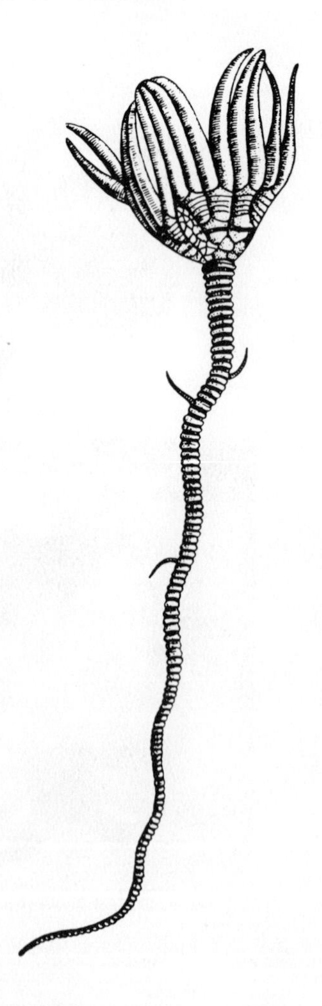

Figure 3.66: Tapering stem of *Woodocrinus macrodactylus* (from de Koninck, in Ubaghs 1978)

Permanent attachment involves using the stereom nature of the skeleton to cement the end of the column to a solid substratum. The Recent stalked millericrinid *Calamocrinus* has a discoid type of holdfast which cements to rocky bottoms or shells as well as to stems of other crinoids. The 'stem' of the recent cyrtocrinid *Holopus* is a mass of stereom cementing the cup to a solid substratum (Figure 3.67). The stalked bourgueticrinids are attached by either irregular terminal plates on solid substrata or by branched roots which may have a terminal plate on soft substrata (Figure 3.68). Great plasticity occurs, because even congeneric species can have either one or the other form (A.M. Clark 1977). The branches from the two main roots of *Bathycrinus aldrichianus* penetrate up to 10 mm into the sediment and increase the surface area for anchorage (Macurda and Meyer 1976).

Although the isocrinids are stalked throughout life, true cirri are found in whorls on the internodal columnal plates. These cirri move rapidly and are used as hooks to grasp (A. Agassiz 1878). At some stage in the life of

Figure 3.67: Attachment plate of *Holopus rangi* (from Carpenter 1884)

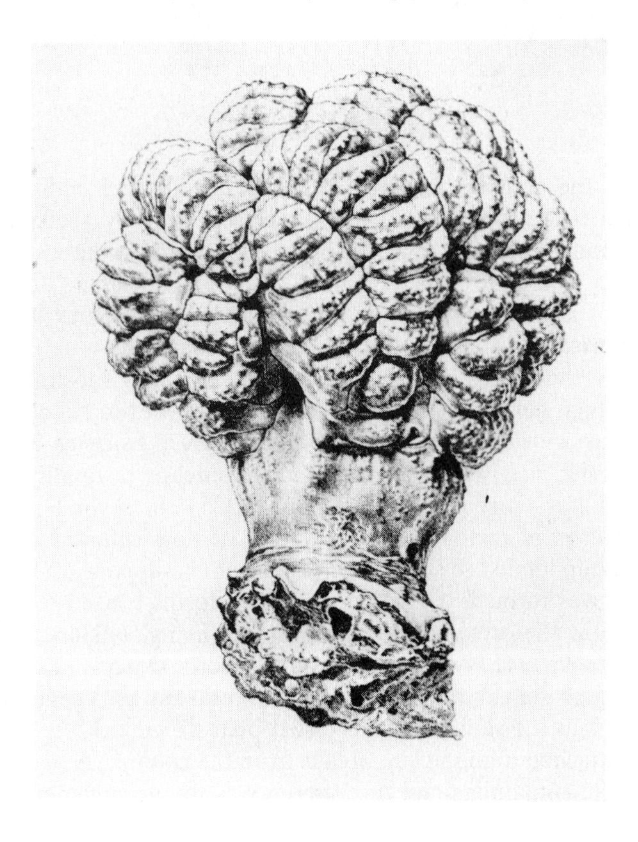

Figure 3.68: Distal end of the stalk of (a) *Democrinus brevis* showing irregular attachment plate, and (b) *Democrinus parfaiti* showing 'roots' (from A.M. Clark 1977)

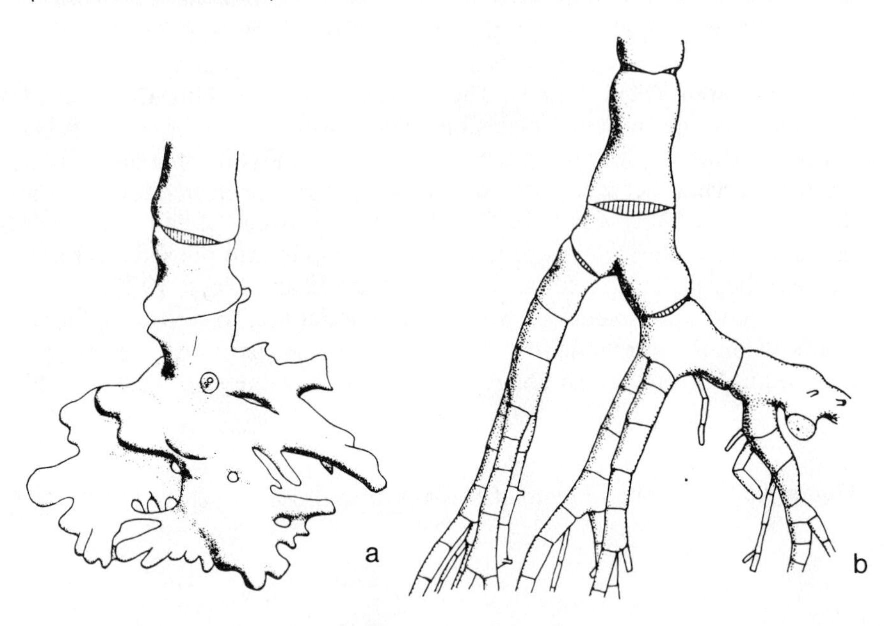

isocrinids, the column becomes detached distally and ends at an articulation below a node which becomes closed over by secondary skeletal material. Individuals are attached by or supported on the distal verticils of cirri found along the stalk (Figure 3.69). Differences in the cirri of the columns of isocrinids seem adaptive (Macurda and Roux 1981). *Diplocrinus macleananus* has a short stem with numerous long cirri by which individuals attach to alcyonarians, sponges, or rocky substrata in areas of relatively high current velocities (Macurda and Meyer 1976). In contrast, *Annacrinus wyvillethomsoni* has a column over twice as long, with far fewer, shorter cirri with which individuals anchor in muddy substrata in areas with less current velocity (Roux 1975). In several groups of the stalked Mesozoic crinoids, the formation of new columnals immediately below the cup tended to stop so that a permanent uppermost columnal (the proximal) was formed. The column in such forms tended to break immediately below the proximal producing a free-living stalkless individual but one without cirri (H.W. Rasmussen and Sieverts-Doreck 1978).

The end of the column of the pentacrinoid larva of recent comatulids is expanded into a disc with a fenestrated plate (Figure 3.70), and the individual is attached until the cup breaks from the column. A major difference between the comatulids and the isocrinids is the arrangement of the cirri.

Instead of being encumbered with a stalk with cirri at intervals along it, which restricts movement, the stalk is completely eliminated and the cirri are concentrated on the ossicle at the bottom of the cup, the centrodorsal plate. The time at which the centrodorsal separates from the pentacrinoid larval stalk is variable. It occurs when 20 or 30 cirri are present in *Antedon tenella* but only 10 in *Antedon bifida* (P.H. Carpenter 1884). Chadwick (1907) stated that *Antedon bifida* does not detach from the stalk until the cirri are sufficiently developed for attachment.

The absolute size of the centrodorsal at the time of separation is also quite variable. After formation of the first two whorls of cirri, cirral

Figure 3.69: Stalked isocrinid *Cenocrinus asterius* supported on distal verticils of cirri (from Messing 1985)

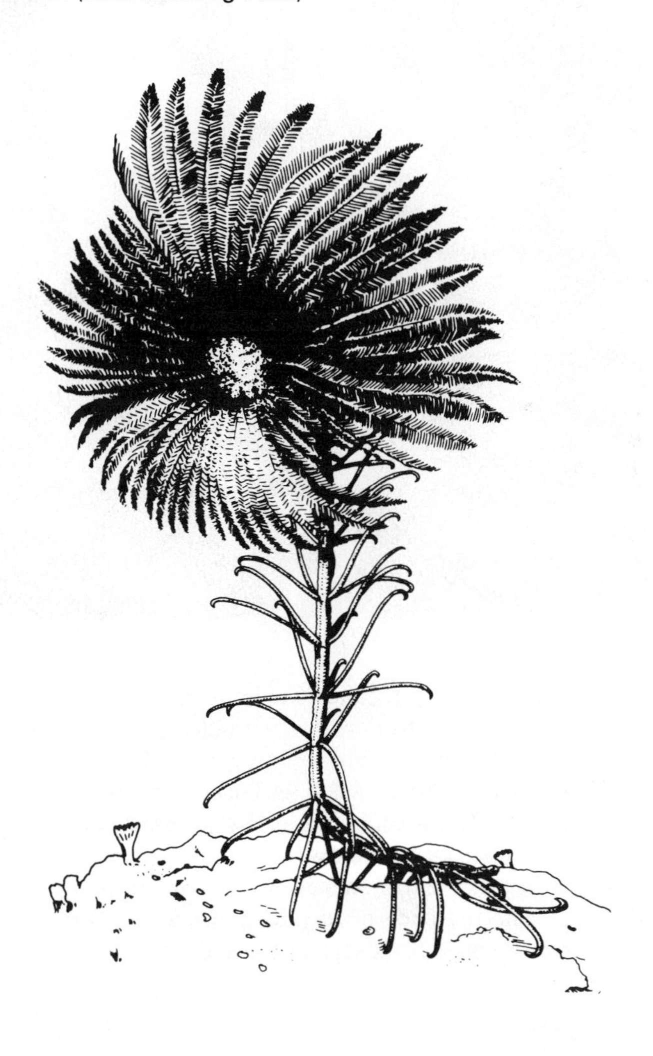

Figure 3.70: Attachment plate of the pentacrinoid of *Florometra serratissima*. Remnants of the cement are indicated by the arrowhead (from Mladenov and Chia 1983)

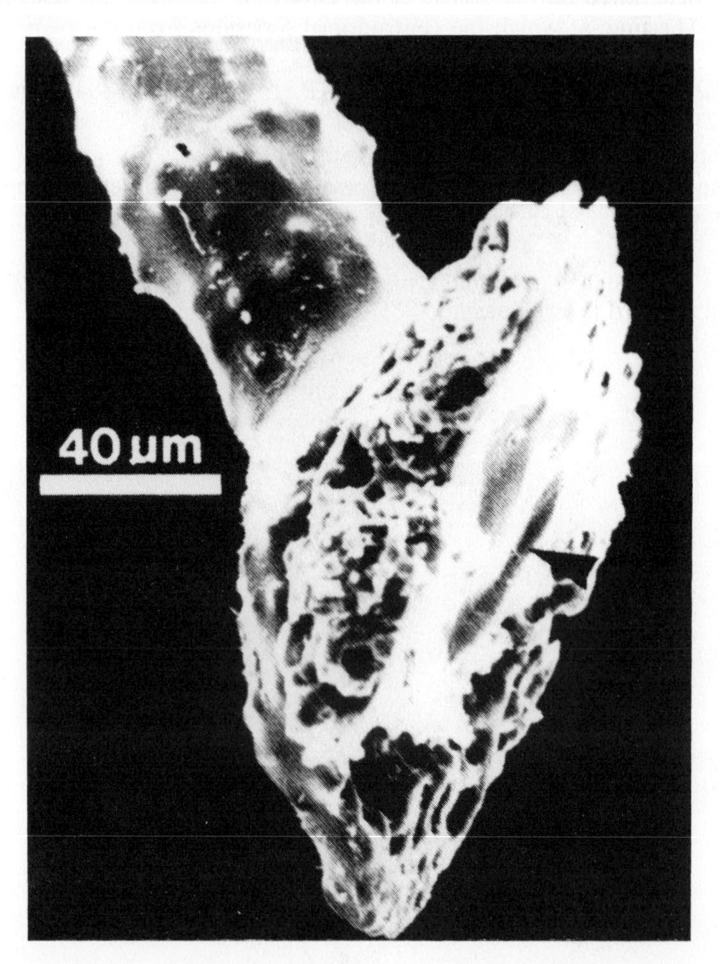

development often shows no regularity. The length, number and arrangement of the cirri on the centrodorsal is variable (Figures 3.71 and 3.72). The centrodorsal of comasteraceans is never conical or columnar and has large cirral sockets which are crowded and usually formed into one or two irregular circles (H.W. Rasmussen and Sieverts-Doreck 1978). In contrast, the centrodorsal of antedonaceans can be discoidal, conical or columnar, and has numerous small cirral sockets which can form up to 10 to 20 distinct columns. The presence of serrate spines and of tubercles and the number and dimensions of cirral segments are considered adaptive (Meyer and Macurda 1977).

Figure 3.71: Centrodorsals of (A) *Mariametra,* (B) *Lamprometra,* and (C) *Zygometra* showing different numbers and arrangements of cirrus sockets (from Rasmussen and Sieverts-Doreck 1978)

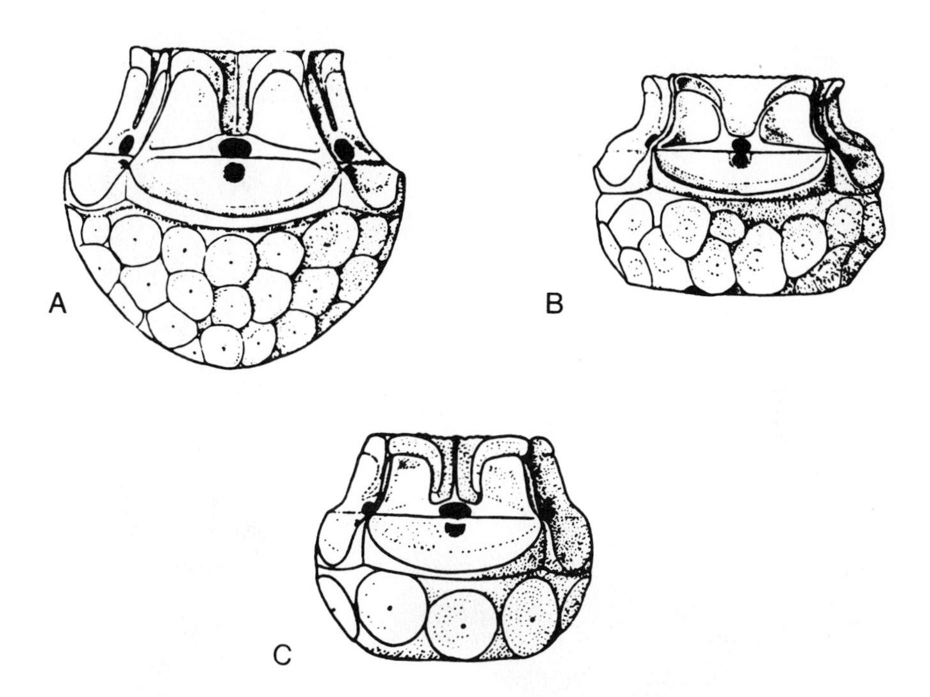

The characteristics of the centrodorsal and the cirri can be related to the species' way of life (Meyer 1973). *Nemaster grandis* has a convex centrodorsal which is covered with cirri except for a small polar area. Its cirri are long (about 35 mm), stout and numerous (about 50) and composed of about 35 short segments which can attach to gorgonian stalks and sponges. *Tropiometra carinata carinata* has a narrow, discoidal centrodorsal with lateral cirri only. The cirri are short (15–20 mm) and seem to be functional for attachment within crevices. The cirri can become extremely long. Those of *Antedon longicirra* are 80 mm or more (P.H. Carpenter 1884). This development of the cirri also occurs in *Pontiometra andersoni* and may represent a functional reversion to the stalked condition (Meyer and Macurda 1980).

Many deep-sea comatulids live on mud or even on globigerina ooze. P.H. Carpenter (1888) proposed that the comatulids occurring on mud use the cirri as stilts to elevate the body.

A final development in attachment by crinoids is the use of the arms. The recurved hooks (grapnels) on the distal pinnules of *Comanthus bennetti* and *Pontiometra andersoni* are used as tethers to assist in main-

Figure 3.72: Cirri and proximal portions of arms of (A) *Antedon elegans* and (B) *Antedon discoidea* (from P.H. Carpenter 1888)

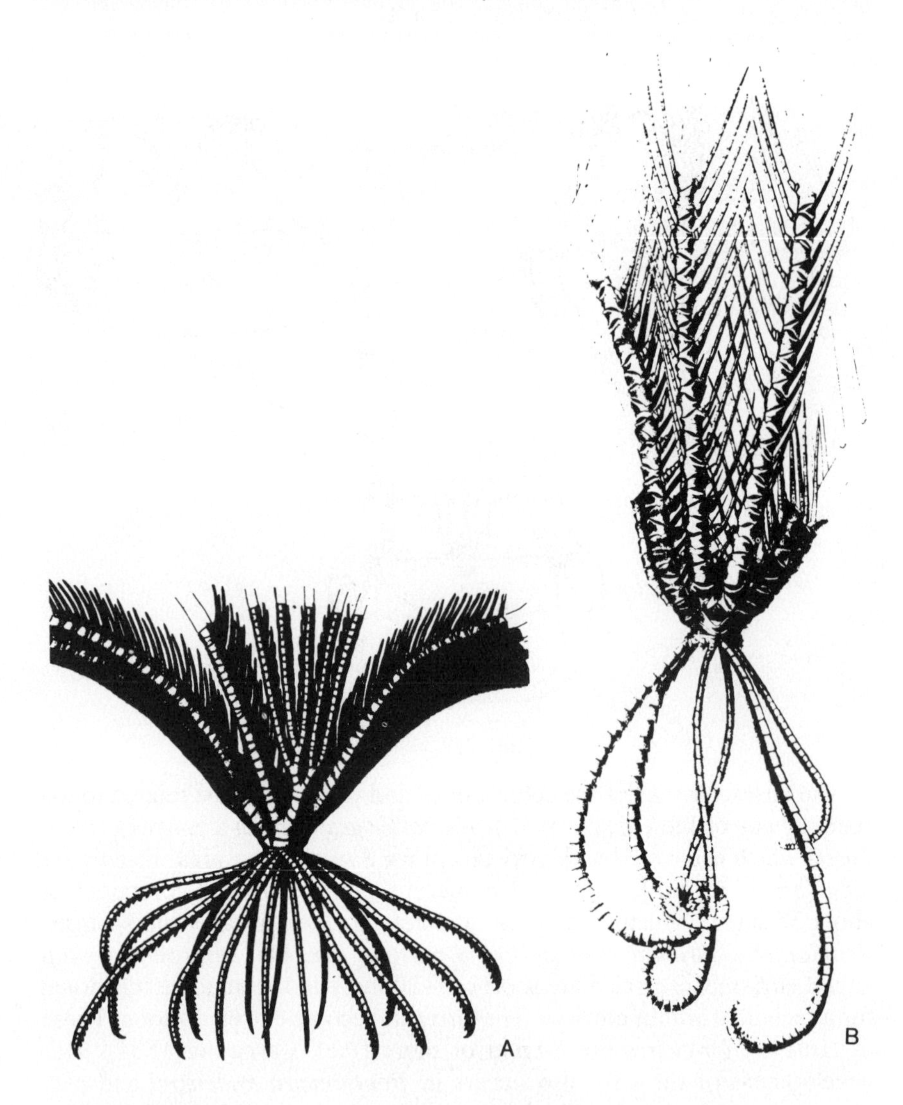

taining position (Figure 3.73) (Meyer and Macurda 1980). Species such as *Comanthus parvicirrus* which have no or only weakly developed cirri attach by the arms and pinnules, while others which have few or no cirri such as *Comanthina schlegeli* live within the reef infrastructure where the means of attachment are less essential.

Figure 3.73: Recurved hooks (grapnels) of the distal pinnules of *Comatella nigra* (after Clark 1921, from Breimer 1978)

3.4.2. Ophiuroidea

It is not known whether the depth of occurrence of burrowing ophiuroids is related to maintenance of position or to avoidance of predation. The length of the arms would determine the depth of burrowing. Non-adjacent arms of juvenile *Ophiophragmus filograneus* grow faster than the other three (Figure 3.74). This concentration of material into only two arms may permit earlier and deeper burrowing (Turner 1974).

Exposed ophiuroids use their arms and appendages as prehensile organs to attach to substrata or to otherwise maintain position. The use of the arms as prehensile organs is related to their vertebral structure (Figure 3.75). The ball-and-socket articulation (zygospondylous) found in most ophiuroids has restraints that allow primarily lateral movements. The hourglass articulation (streptospondylous) found in euryalid ophiuroids has an orientation that allows vertical movements. In this group, the arms and their branches can roll upon themselves and can coil around objects.

The prehensile unbranched arms of *Asteroporpa annulata* coil around

Figure 3.74: Juvenile *Ophiophragmus filograneus* with the two long and three short arm pattern (from Turner 1974)

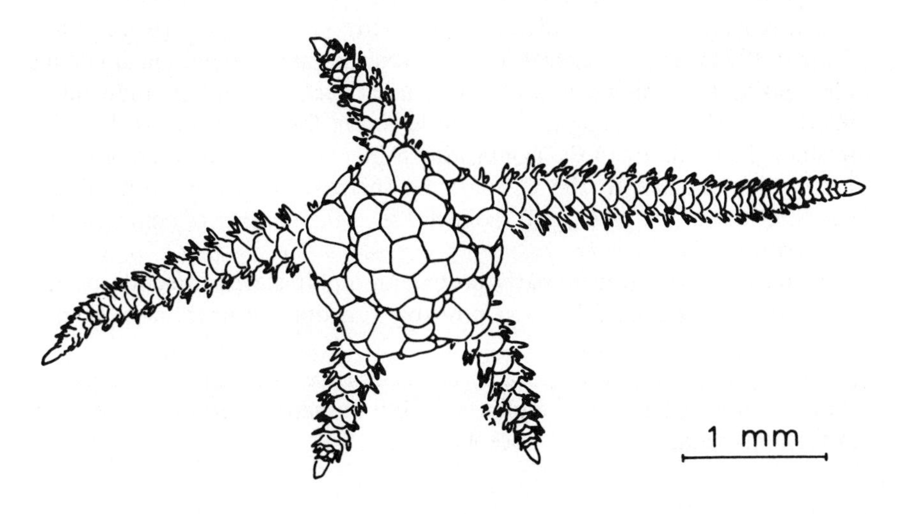

Figure 3.75: (A) Zygospondylous articulation in *Ophiura* and (B) streptospondylous articulation in *Gorgonocephalus*. 1: Proximal face, 2: distal face. abh: Aboral hinge, adh: adoral hinge, dg: dorsal groove, dm: dorsal muscle attachment, dn: dorsal nose, hd: horizontal dumbbell, rc: radial canal, vd: vertical dumbbell, vm: ventral muscle, vn: ventral nose. (From Spencer and Wright 1966)

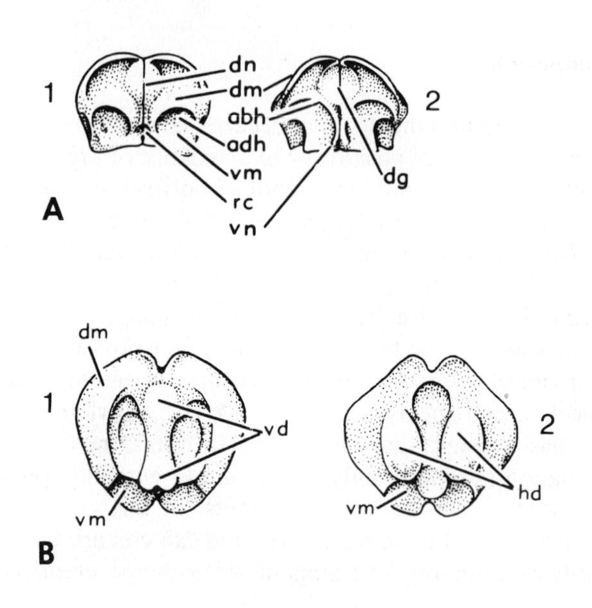

coral branches (Hendler and Miller 1984). The locomotory arms of the basketstar *Astrophyton muricatum* grasp the substratum and are used as prop roots (Hendler 1982). The ophiurid *Ophiothrix fragilis*, which has zygophiuroid joints, is epibenthic on a variety of substrata in dense aggregation reaching *c.* 2000 individuals per metre (Figure 2.14) (Brun 1969; Warner 1971). These aggregations are important in individuals maintaining their position in high current flow. Under such conditions, individuals interlock their supporting arms while lowering the disc towards the substratum. The stability of the aggregation is increased by the decrease in the frictional resistance to the current, the increase in the inter-individual cohesion within the patch, and the increase in the number of contacts with the substratum.

Hooked spines occur in early postmetamorphic stages of some species of ophiuroids (Figure 3.76) and probably function for anchorage. Spines are modified into anchorage hooks in adults of various ophiuroid families (Gislén 1924). They are sporadic in occurrence and little developed in some Ophiothricidae and Amphiuridae, but numerous and well developed in Ophiomyxidae and Gorgonocephalidae.

Figure 3.76: (A) Upper surface and (B) undersurface of a newly metamorphosed individual of the ophiuroid *Ophiothrix savigny* showing hooked spines (from Mortensen 1938)

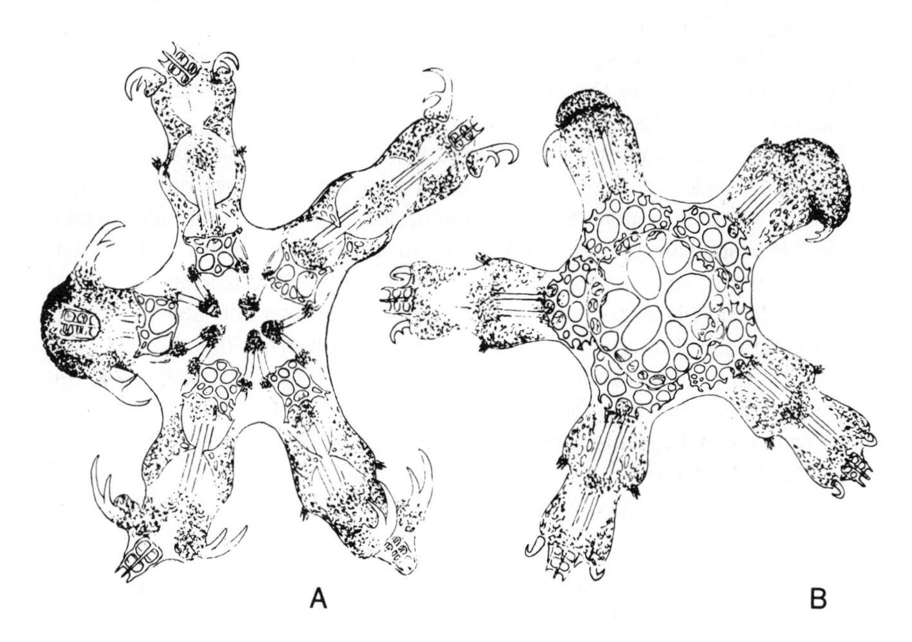

3.4.3. Asteroidea

Anchorage in asteroids, echinoids and holothuroids is by suckered tube feet, which probably evolved independently in the three groups (Nichols 1969), in contrast to the modes in crinoids and ophiuroids. The use of the tube feet for this function may be more effective than cement, roots or prehensile organs. Asteroids, echinoids and holothuroids occur in extremely high-energy locations where crinoids or exposed epibenthic ophiuroids are not found. Differences in feeding may be involved, however. The evolution of adhesive suckered tube feet was probably associated with the invasion of solid substrata in high-water-energy environments which required the capacity to attach. However, the possession of suckered tube feet cannot be taken to mean that a species lives on a solid substratum, let alone that there is high-energy water. Representatives of asteroids, echinoids and holothuroids with suckered tube feet also live on particulate substrata. Most species of asteroids that live on globigerina ooze have suckered tube feet (Fisher 1911)

In addition to the strength of their adhesive properties (Hermans 1983), the effectiveness of the tube feet in attaching the individual to the substratum is a function of the number of tube feet, the strength of the stem, and the area of the sucker and its characteristics. An increase in the number of tube feet for attachment could result from an increase in the total length of the ambulacra by lengthening the arm or increasing the number of arms. However, these changes would result in consequences for other characteristics of the body. Many asteroids show an increase in the density of the tube feet.

The paxillosids do not have suckered tube feet. The valvatids and spinulosids have suckered tube feet found in two parallel rows on each arm (biserial). The polyphyletic forcipulatids usually have tube feet which alternate (especially proximally) to form four or six rows along each ambulacral groove (Figure 3.77). This results from a decrease in the size of the ambulacral plates so that they are more numerous and develop a zigzag appearance. Forcipulatids which have multiple rows of tube feet pass through the biserial condition ontogenetically. The restriction of the quadriserial condition to the proximal part of the arm in zoroasterids may be associated with slenderness of the distal part of the arm. The greater breadth of the proximal part of the arm may be related to the presence of the pyloric caeca and gonad in this region only which, in turn, gave lateral

Figure 3.77: (a) Quadriserial tube feet of *Distolasterias stichtantha* and (b) biserial tube feet of *Pythonaster murrayi* (from Sladen 1889)

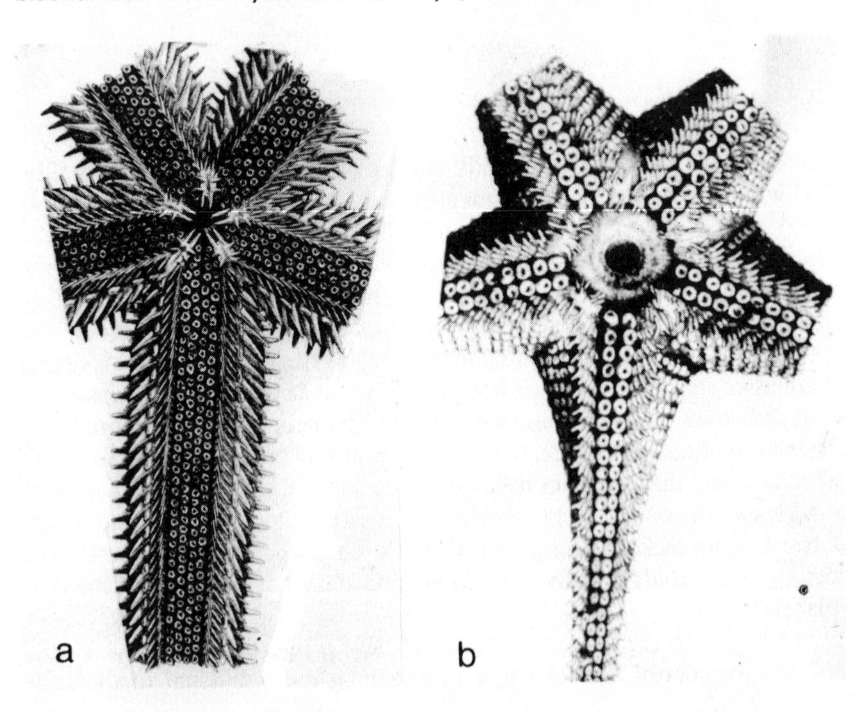

space for the quadriserial condition. Zoroasterids are deep-sea asteroids which live primarily on soft bottoms (Fisher 1928; Sladen 1889) and feed on the sediment and small organisms (Jangoux 1982). The use of the suckered tube feet there would not be for attachment nor for forcing open bivalves, but may function in the capture of prey.

3.4.4. Echinoidea

A general correlation exists between the degree of development of suckered tube feet and the occurrence of regular echinoids. Cidaroids have tube feet that are poorly developed suckers and are not found exposed in high-energy water.

The tube feet of the euechinoids may be suckered along the entire length of the ambulacrum or primarily from the ambitus to the mouth (Nichols 1969). These latter function in attachment. This is reminiscent of the restriction of the ambulacra of the extinct *Tiarechinus* to the portion of the test in contact with the substratum and the lower part of the exposed position (Figure 3.78).

Figure 3.78: (a) Upper and (b) side views of the Triassic echinoid *Tiarechinus princeps* showing extension of the ambulacral to only just above the ambitus (from Lovén 1883)

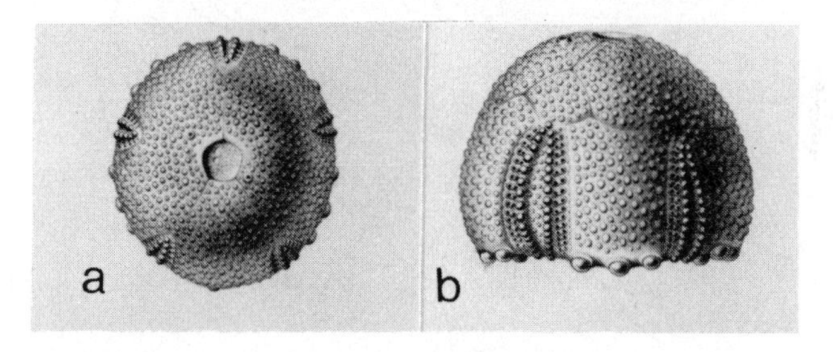

The width of the ambulacrum of cidaroids is one-fifth to one-quarter that of the interambulacra (Melville and Durham 1966). A pronounced change occurred between the Triassic and the Jurassic with the width of the ambulacrum increasing from a mean of 24% to one of 42% of that of the interambulacra (Kier 1974). The increase in area resulted from the arrangement of the pore pairs in arcs associated with compounding the ambulacral plates (Kier 1974).

Areas of enlarged or increased number of pores adorally (phyllodes) also increase the attachment ability (Figure 2.31). Genera such as *Heterocentrotus* and *Colobocentrotus* occupy extremely exposed areas with high water energy and are noted for this characteristic (Figure 3.79). However,

Figure 3.79: Views of *Podophora podifera* from (A) the side, (B) the undersurface showing the numerous tube feet of the phyllodes bordering the peristome, and (C) a portion of the test from the ambulacral region near the apical system showing the modification of the primary spines. (From A. Agassiz 1908)

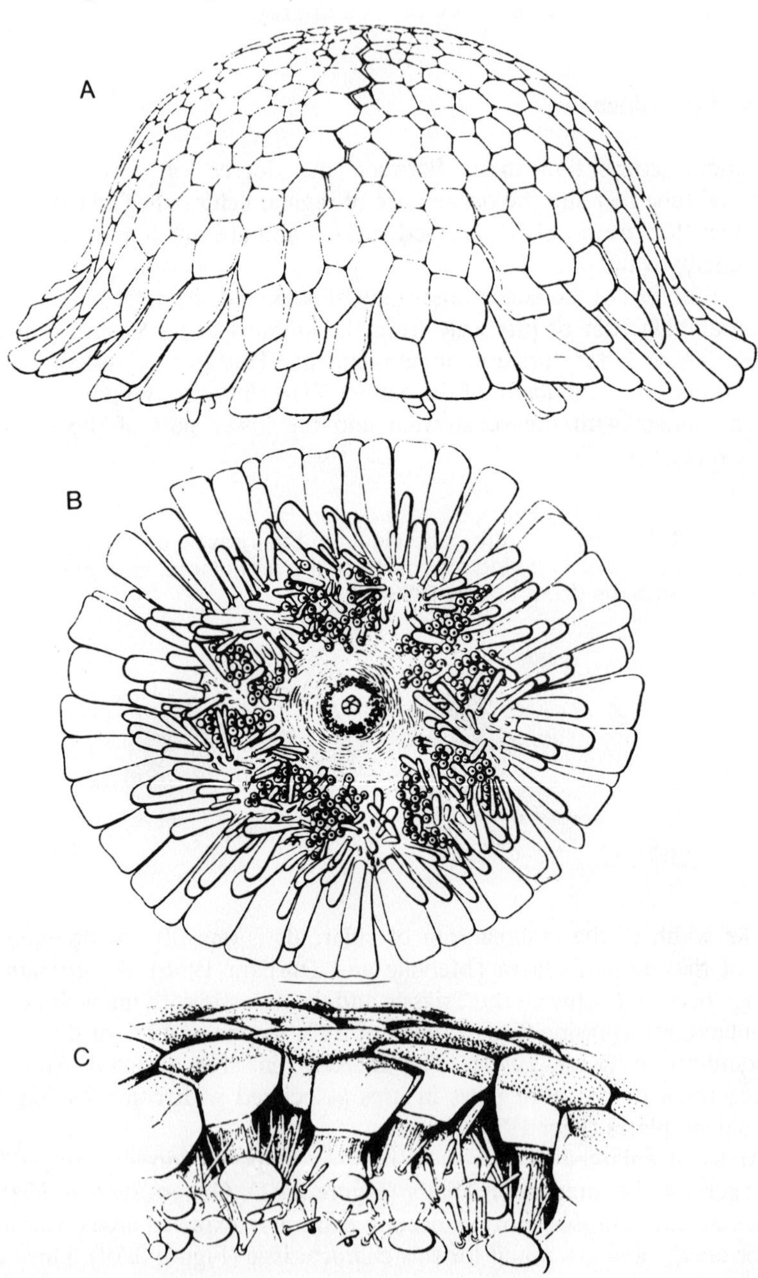

as with asteroids, species of echinoids with suckered tube feet such as *Lytechinus variegatus* can be found on particulate substrata.

In addition to the number of tube feet, the strength of attachment is determined by the strength of the suckered tube foot itself. This is determined by the tensile strength of the stem and the adhesive and suction power of the terminal sucking disc (A.B. Smith 1978). The suckered tube feet which have the greatest strength of attachment have small pores and a large attachment area, a large and well developed sucking disc, a well developed connective tissue layer, and retractor muscles developed to such an extent that the lumen of the stem can almost be occluded when the tube foot contracts (Figure 3.13). The strength that can be developed by the sucker would depend upon the strength of the retractor muscles within the disc and the development of the supporting elements there, as well as the strength of the adhesive material secreted by the disc. Several genera of regular echinoids differ in their ability to withstand prolonged pull (Märkel and Titschack 1969; Sharp and Gray 1962), but this ability has not been correlated with all of the structural variables present.

Although they are not attached to the substrata, clypeasteroids which do not burrow deeply may be subject to dislocation and injury by currents and turbulence. Juvenile *Dendraster excentricus* 5 to 10 mm in length contain sand in an intestinal diverticulum which ranges from nearly 0 to almost 25% of the total body dry weight (*c.* 50 mg) whereas individuals larger than 10 mm contain no sand in the diverticulum (Figure 3.80) (Chia 1985). Whether this results in increased stability for the small individuals as hypothesised is not known.

The flattened form of sand dollars is functional for shallow burrowing and for minimising drag and facilitating position maintenance (Figure 3.81) (Telford 1981, 1983). The low, domed profile acts as a hydrofoil and generates lift. The lift generated in a function of both body shape and the presence/absence of lunules. The critical velocity increases with the cross-sectional area to width ratio (Figure 3.82).

The lift generated by *Echinarachnius parma* is greater than that of *Mellita quinquiesperforata.* Compared to *E. parma*, *M. quinquiesperforata* has a thinner edge, a shallower dome, and a greater camber anteriorly and lesser posteriorly than *E. parma.* The lower surface of *E. parma* is nearly flat whereas that of *M. quinquiesperforata* is slightly concave. These geometrical differences contribute to the differences in lift. Occluding the lunules increased the lift in *M. quinquiesperforata*, but it was still less than that of *E. parma.* Small *M. quinquiesperforata* are lifted at lower current velocities (26 cm s^{-1}) than large individuals (43 cm s^{-1}) as functions of surface area and weight (Telford 1983). The lunules in *M. quinquiesperforata* provide a route for water flow from below the individual and reduce lift. Ambital notches provide the same function. Depressions on the lower surface of the sand dollars facilitate movement of water to the lunules and

Figure 3.80: The digestive tract and diverticula of a juvenile *Dendraster excentricus* (upper test removed). A: Aristotle's lantern, E: oesophagus, S: stomach, I: intestine, R: rectum, D: diverticulum with pouches. (From Chia 1985)

Figure 3.81: Forces acting on the sand dollar in a current (from Telford 1983)

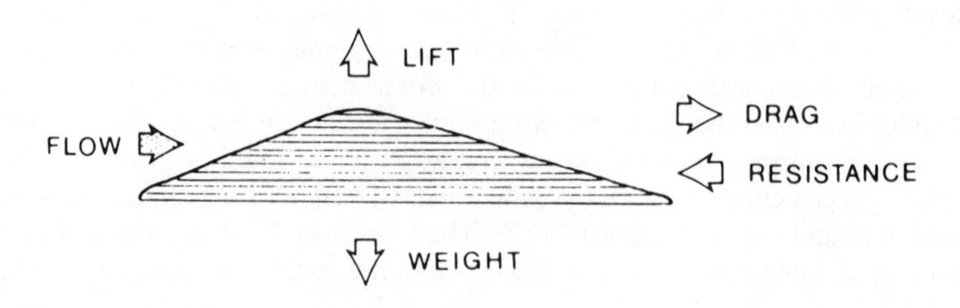

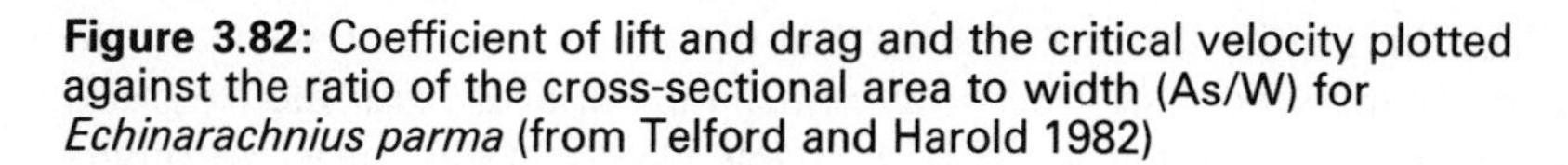

Figure 3.82: Coefficient of lift and drag and the critical velocity plotted against the ratio of the cross-sectional area to width (As/W) for *Echinarachnius parma* (from Telford and Harold 1982)

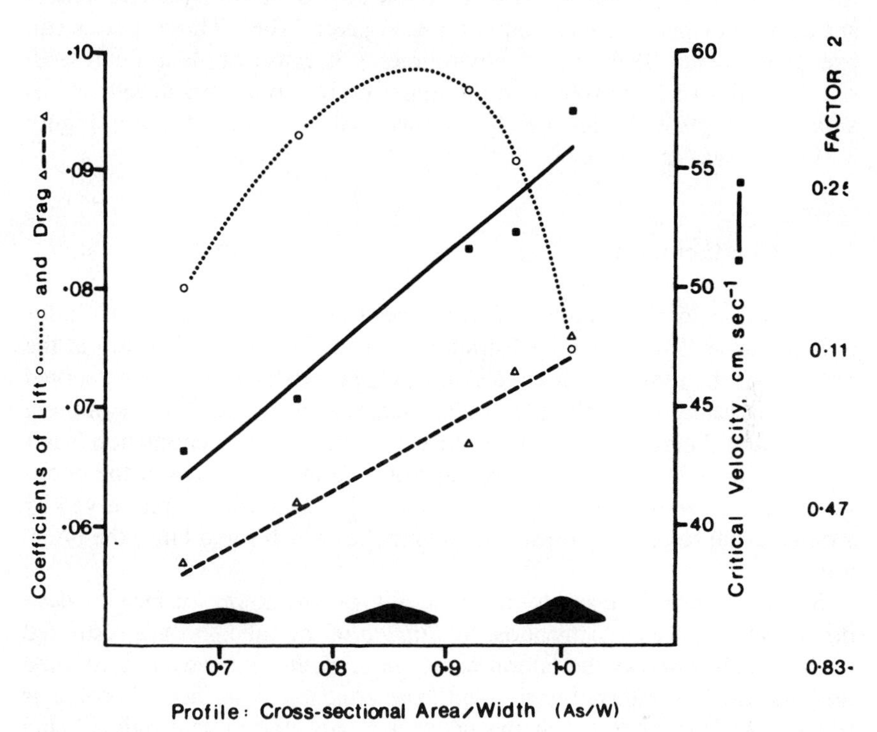

notches. This provision for decreasing lift as well as the thinner edge and flatter dome may be the reason that lunulate sand dollars can be larger in diameter than non-lunulate ones.

Modification of the body form to resist turbulence is found in several tropical echinometrids that live in the high-energy intertidal. Genera in this family have a non-circular ambitus. This can be pronounced in *Echinometra* (with the long transverse axis being from ocular I to genital 3) and *Heterocentrotus* (with the long transverse axis from ocular II to genital 4) (H.B. Fell and Pawson 1966). The elliptical form of the test is exaggerated by the greater length of the primary spines in the long axis in *Heterocentrotus*, which wedges its elliptical form into the interstices of coral on the reef ridge. Individuals on the reef flat have thicker spines than those in calm waters (Dotan and Fischelson 1985). *Echinometra* uses its smaller spines similarly in coral or in burrows that it excavates (Lawrence and Sammarco 1982). The presence of only small individuals in burrows while large ones live in protected tide pools suggests a size-related phenomenon (Lawrence 1983).

The elliptical form is not restricted in the family to such semi-protected genera. Some species of *Colobocentrotus* are elliptical (with the long axis from ocular II to genital 4) (H.B. Fell and Pawson 1966), whereas others are circular or pentagonal in outline (A. Agassiz 1908). These species can live in extremely high-energy environments because of their depressed, streamlined form. The spines on the upper body surface are all reduced to a uniform length and thickened to produce a close smooth covering (Figure 3.79) (A. Agassiz 1908).

3.4.5. Holothuroidea

The increase in the number of attachment suckered tube feet in holothuroids which live on solid substrata has come about both by a shift in the position of the radial water canals and by a broadening of the area served by a water canal. The shift in the radial water canals results in three of them (the trivium) being in contact with the substratum. This phenomenon is not restricted to holothuroids on hard substrata because it occurs in the deep-sea elasipods living on particulate substrata. The increase in area served by a radial canal results from side lateral branches which spread into the interradii.

Suckered tube feet are found generally on the lower surface of dendrochirotids and aspidochirotids. In *Stichopus*, the tube feet are restricted to the radius whereas the lateral canals in *Holothuria* have so many tube feet that the two interradii are filled. The attachment surface of psolids is fringed with the tube feet of the two outer radii, the middle radius being devoid of tube feet in *Psolus* but with varying numbers in other genera (Pawson 1967). The smoothness of the surface of the sole and the continuous fringe of the tube feet implies that the sole is adhesive. The difference in these two modes of providing strong attachment to the substratum is because the tube feet of *Holothuria* are also used for locomotion whereas those of *Psolus* are not.

Synaptid holothuroids attach to the substratum by the anchors in their thin body wall (Figures 3.83 and 3.84). This is effective only in small individuals. *Ophlodesoma spectabilis* up to one-third of a metre in length are found in shallow waters with breaking waves where they curl among the coral rubble and attach by their anchors, whereas large individuals up to 2 m in length live only in the most protected locations (Berrill 1966).

3.5. PROTECTION FROM PREDATION AND PHYSICAL DAMAGE

Many ways by which echinoderms reduce the probability of being preyed upon or damaged physically are not related directly to their structure. Thus

Figure 3.83: (A) *Ophedesoma spectabilis* showing warts on the body wall. (B) Enlarged view of the warts showing the density of the anchors. (C) Position of the anchors and anchor-plate surface forming the warts. ao: Anchor ossicle, po: anchor-plate ossicle. (From Berrill 1966)

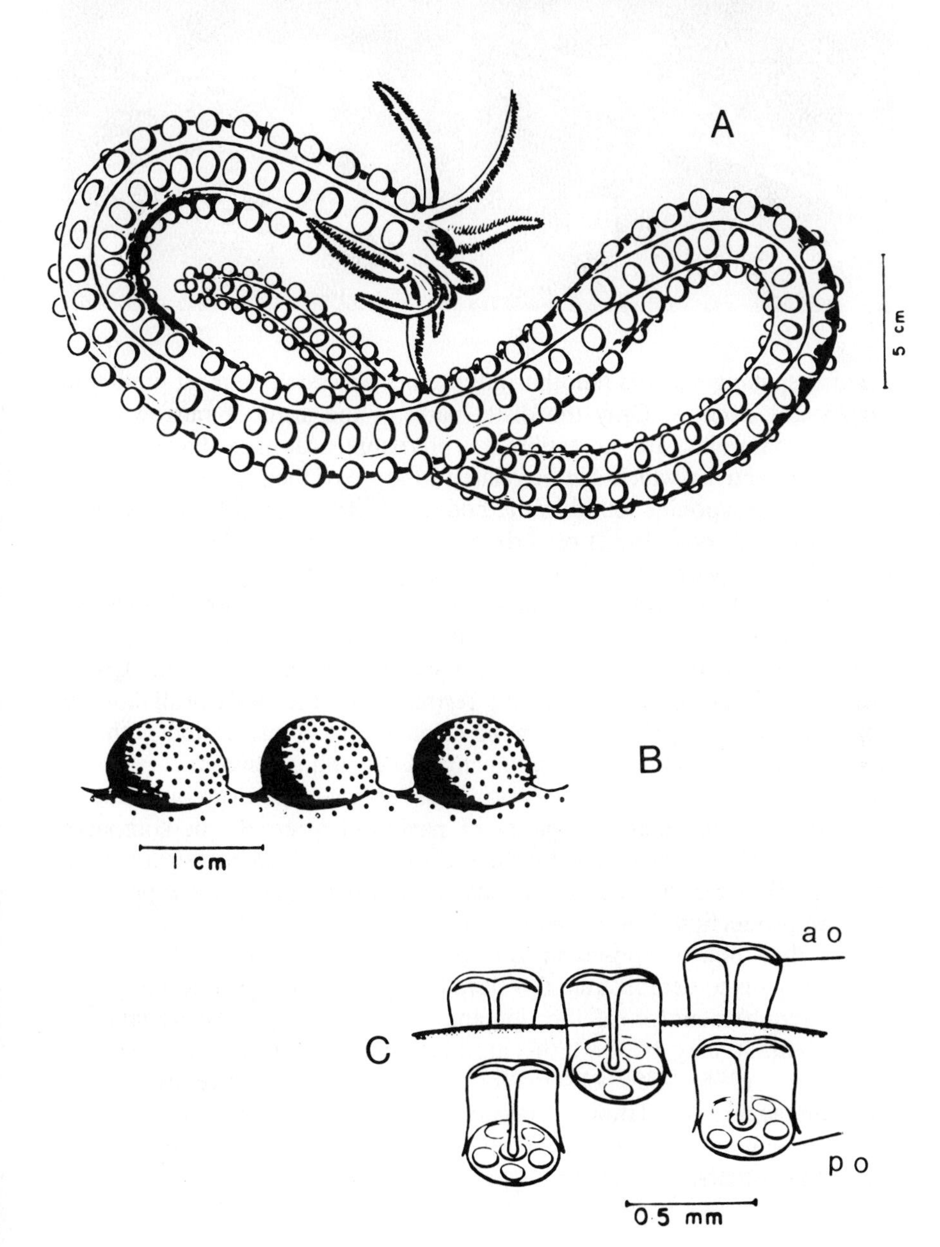

Figure 3.84: (a) Anchor, and (b) anchor plate of *Leptosynapta gallienni* (J.-P. Féral, unpublished)

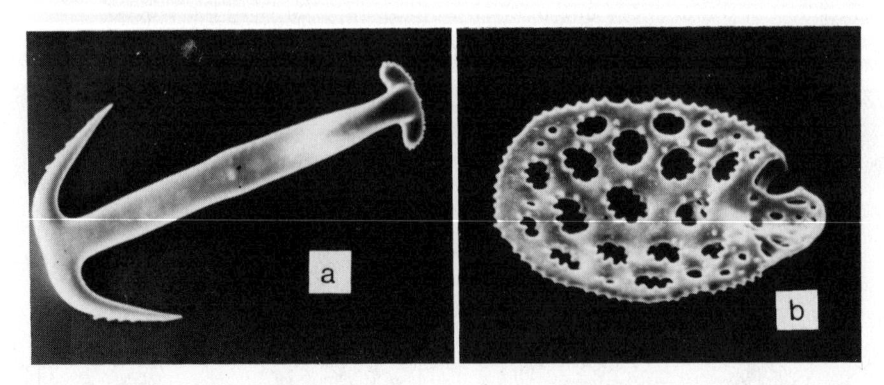

infaunal spatangoids and holothuroids are not accessible to most predation or physical damage. Only the distal parts of the arms of burrowing or cryptic ophiuroids are accessible to predators. All may have diurnal rhythms of activity which can be related to those of their predators (Reese 1966). Toxic saponins found in the body wall and organs of asteroids and holothuroids (Voogt 1982) provide protection against predation in a way not associated with body form. A decrease in the energy return to the predator by a high level of inorganic material or refractory organic material may lessen predation without being associated with body form.

Structural adaptations related to the function of protection against predation or physical damage are restricted to the body wall and its appendages. The skeletal system of echinoderms is associated with the dermis, which is composed of connective tissue and calcareous deposits in varying degrees. The system may be almost completely calcareous in composition as in the marginal plates of paxillosid asteroids, or composed almost completely of connective tissue as in some aspidochirotid holothuroids. Either system can be effective as a physical barrier. It is probable that the production of a primarily calcareous dermis costs less in terms of energy than one composed of connective tissue (Emson 1985), but no measurements to demonstrate this have been made. However, although the energy requirement for calcification and the level of organic material associated with it may be low, the absolute amounts present can be high (Lawrence 1985). Few observations indicate the effectiveness of the skeleton against predation. The presence of regenerating arm tips of crinoids, ophiuroids and asteroids indicates that predation on these body extensions occurs.

3.5.1. Crinoidea

Heavy calcification of the column, cup (calyx), and arms of crinoids seems essential to provide physical support for their mode of suspension feeding (Breimer 1978). The coiling of the arms obviously uses the skeletal elements to protect the ambulacral grooves against predation or physical damage. Only the tegmen shows variation in the degree of calcification which thus could be interpreted as having a protective function.

In the isocrinids, the tegmen is generally highly calcified and is almost completely filled with plates and scales. The plating is not rigid, indicating a role for flexibility. Radial oral plates protect the ambulacra as they pass over the tegmen to the mouth. These oral plates are fully preserved in *Holopus* with the ambulacral grooves running beneath them. Oral plates are present in the pentacrinus of comatulids but are resorbed. The interradii are strongly calcified in isocrinids, a complete pavement of irregularly arranged plates being present. In contrast, the tegmen of adult comatulids may be calcified but is often leathery with spicules only in the interradii (Figure 3.85). Either less protection is needed for the tegmen of these comatulids or other protective devices are present, or the benefits of a flexible tegmen outweigh the potential for predation.

The need for protection of the tegmen of comatulids is indicated by the modifications of the proximal pinnules into oral combs which are either long, slender and very flexible or are rigid and spine-like (Figure 3.86) (Meyer and Macurda 1977). Combs are best developed in the comasterids (H.L. Clark 1921), but can be found in antedonids (Meyer 1972). *Nemaster grandis* lives in exposed habitats and has more comb-bearing oral pinnules than do congeneric species living in more sheltered habitats, a characteristic related to the proposed function of removing foreign particles from the tegmen (Meyer 1973). The proximal pinnules of many tropical Indo-Pacific comatulid genera are stiff and large with sharp spines. These are considered to be defences against predation and to indicate a higher degree of predation pressure in the region (Meyer and Macurda 1977). Spine-like pinnules are typical of species of comatulids that live fully exposed on the reef at Lizard Island (Great Barrier Reef) but not of semicryptic species (Meyer 1985). The importance of these pinnules is indicated by their rapid regeneration.

The sides of the exposed ambulacral grooves of crinoids are raised in repeated scallops which are calcified (Figure 2.1). These lappets border and may cover the ambulacral grooves for protection. The lappets on the pinnules can be closed or opened to hide or expose the tube feet. Those on the arms and tegmen do not necessarily open (Breimer 1978).

Figure 3.85: Variation in the degree of plating of the tegmen of (a) *Antedon multiradiata* and (b) *Actinometra regalis* (from P.H. Carpenter 1888)

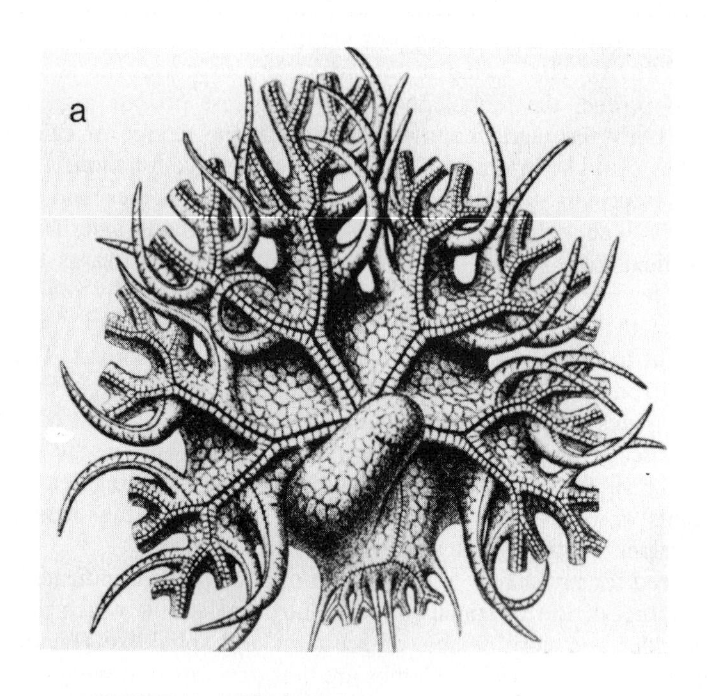

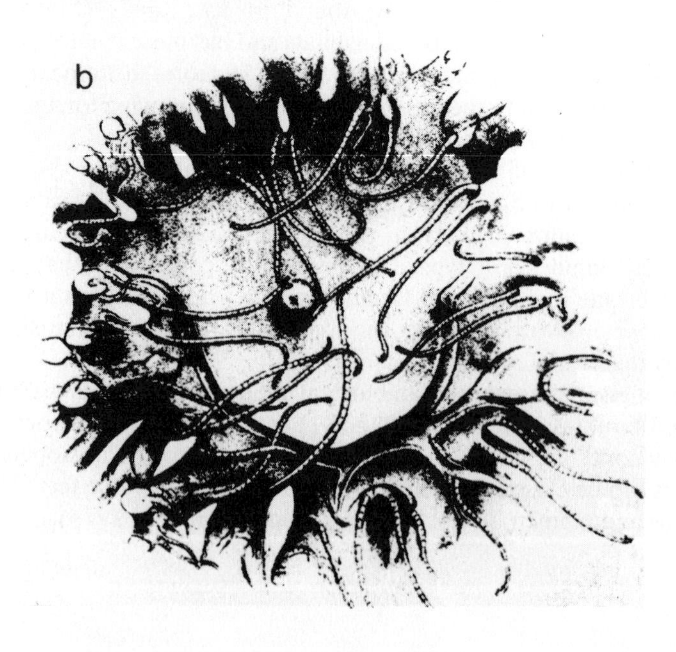

Figure 3.86: Oral combs of the comatulids (a) *Ctenantedon kinziei*, (b) *Florometra serratissima*, (c) *Anthometra adriani*, and (d) *Nemaster discoidea*. Arrowheads indicate distal direction. Bar = 1 mm. (From Meyer 1972)

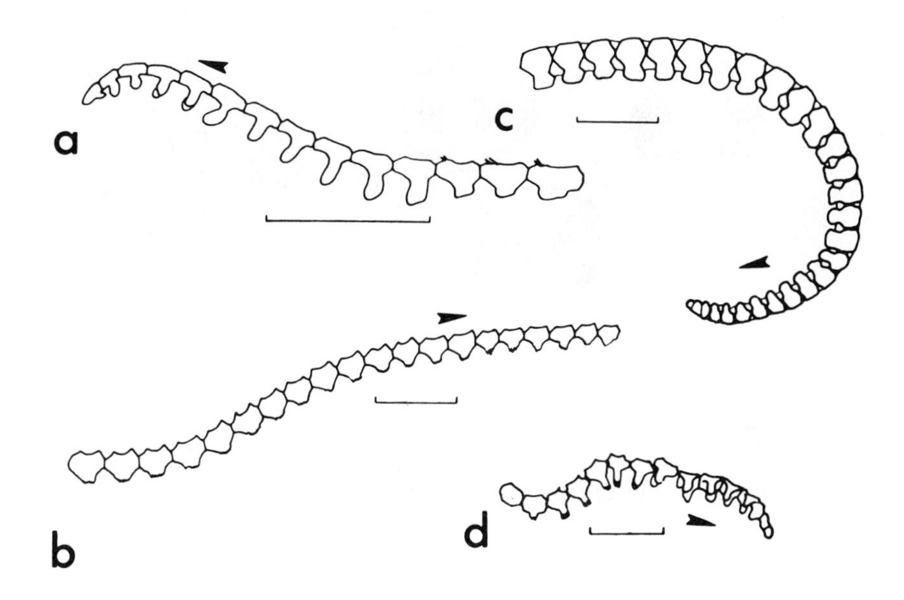

3.5.2. Ophiuroidea

Like the arms of crinoids, the arms of ophiuroids are always heavily calcified because of their supportive role in feeding and locomotion. Evolution of the ophiuroids involved the envelopment of the radial water canal by the ossicles, the gradual enlargement and migration of the adambulacral ossicles to form the lateral arm shields, and variable degrees of development of the dorsal arm shield (Figure 3.87) (Nichols 1969). They can provide protection by coiling.

The vertebrae and the spines found on the lateral shields undoubtedly provide some protection, but the high incidence of regenerating arm tips in ophiuroids suggests that loss by predation occurs (Bowmer and Keegan 1983; Sides 1982). Spines on the arms are quite variable in their number and degree of development between species and along the length of the arm. Because spines function in feeding and establishment of burrows, their development cannot always be considered only in terms of protection. Spine development can vary in congeneric species such as *Ophiocoma* (Figure 3.88) and can be related to the potential for predation (Sides and Woodley 1985). The uppermost spine is largest in *Ophiocoma wendti* which forages in the open; intermediate in size in *O. echinata* which is

Figure 3.87: Evolutionary change in position of radial water canals and arm shields in ophiuroids. (a) *Lapworthura* (Silurian), (b) *Ophiocanops*, (c) *Astroschema*, (d) *Ophiothrix* (from Nichols 1969)

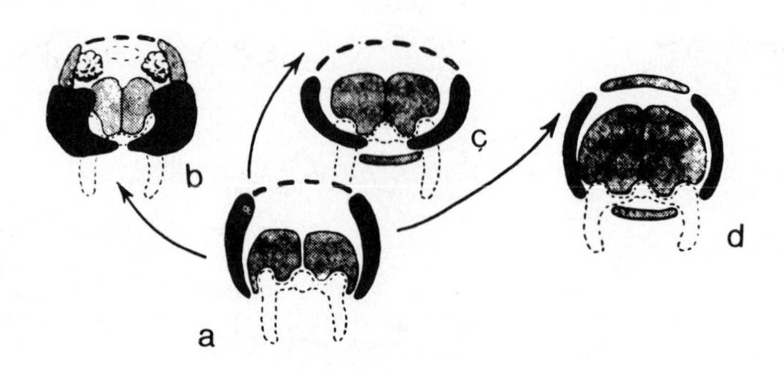

Figure 3.88: Cross-section of the arms of (a) *Ophiocoma wendti*, (b) *O. echinata*, and (c) *O. pumila* (from Sides and Woodley 1985)

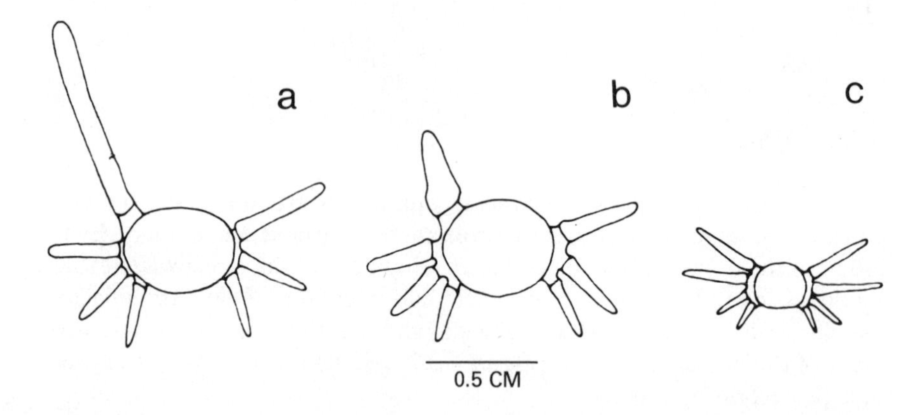

rarely found away from potential shelter; and smallest in *O. pumila* which is never found in the open. Spines protect the tube feet like the lappets of crinoids. The failure of the ophiuroids to modify spines into pedicellariae is considered to be a consequence of their primarily burrowing or cryptic way of life.

The upper surface of early ophiuroids was covered by small scales over a presumably flexible disc (Spencer and Wright 1966). Although some extant species have a skin covering for the upper surface of the disc without any distinct ossicles in or below it, most later ophiuroids have a relatively stout and rigid covering. Calcification of the disc does not confer considerable protection because the disc of infaunal ophiuroids appears in the stomach contents of a number of predators (Turner, Heatwole and Stancyk 1982). The tendency of ophiuroids to autotomise the disc has been interpreted as a response to predation (Emson and Wilkie 1980).

Although ophiuroid tissues do not produce toxins, the external epithelial cells of at least one species of ophiuroid *Ophiocomina nigra* produce copious amounts of mucus when mechanically stimulated. This seems to function to decrease predation (Fontaine 1965). No modifications of the body form are associated with this mucous secretion.

3.5.3. Asteroidea

The upper surface of asteroids, like that of many early ophiuroids, was very weak (Spencer and Wright 1966). The skeleton of the upper body wall of extant asteroids can be open and reticulated as in *Solaster*, flexible and imbricated as in *Asterina* and *Patiria*, or absent except for some axial elements as in *Dermasterias*. Thus the degree to which the skeleton provides protection from predation or physical damage is variable. A thick body wall composed of refractory organic material (Figure 3.89) can be rigid and could be very effective in protection from predation. The level of inorganic material can range from 60 to 70% of the dry weight of the body wall (Table 3.1). The level of insoluble protein, tough connective tissue which could potentially deter predation or make it less rewarding, can vary from 16 to 27%. It would seem that the body wall of *Diplasterias meridionalis* is more costly than that of *Asterina gibbosa*.

Some skeletal elements in the body wall are necessary for support and protection of the papulae when they are present. In paxillosids, the spinelets on the top of the paxillae are raised to expose the papulae and lowered to protect them (Figure 3.90). In forcipulatids, the papulae are simply retracted to the reticulated skeletal elements of the body wall.

The most vulnerable part of the asteroid is the radial water vessel which is exposed in the ambulacral groove of asteroids instead of being internal as in ophiuroids. The ambulacral groove is lined by ambulacral ossicles which form a furrow (Figure 3.91). The row of ossicles next to the ambulacrals becomes continuous in a radial direction and forms a wall overhanging the ambulacral groove (Spencer and Wright 1966). These adambulacral ossicles increase the depth of the groove and shelter the retracted tube feet. The adambulacral ossicles usually bear prominent spines as well (Figures 3.89 to 3.91). The spines rest on the underlying ossicles in a ball-and-socket arrangement and are movable. The ambulacral groove must be open for the tube feet to be protruded and to function. Muscles attached to the ambulacral ossicles allow the asteroid to open the ambulacral groove for functioning, and to close it for protection (Figure 3.92).

Figure 3.89: Partial cross-section near the base of an arm of (a) *Poraniopsis echinaster,* (b) *Porania (Porania) pulvillus,* and (c) *Chondraster grandis* showing differences in the degree of skeleton and connective-tissue development in the body wall of the valvatid family (Poraniidae). ao: Ambulacral ossicle, aao: adambulacral ossicle. Bar = 5 mm. (From A.M. Clark 1984)

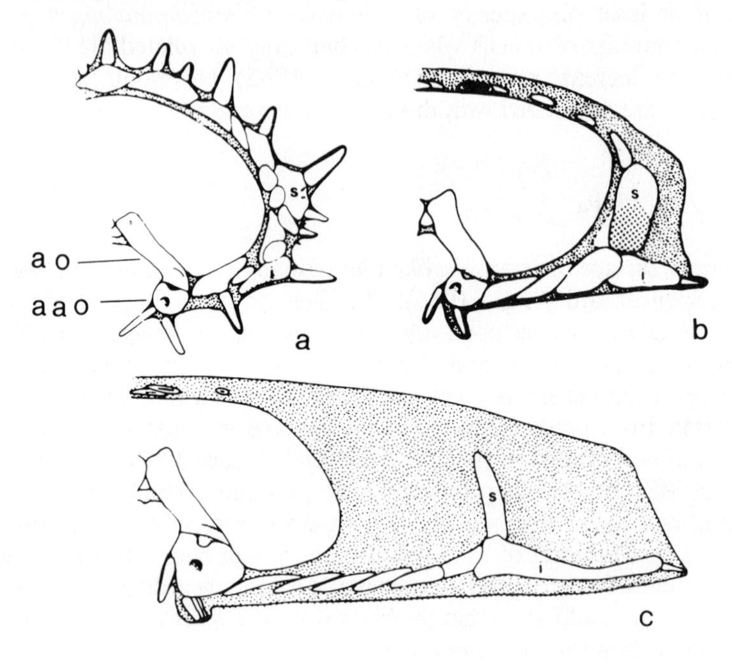

Table 3.1: Level of ash and insoluble protein (% dry weight) in the body walls of asteroids and holothuroids. (From Lawrence and Guille 1982.)

Species	% Ash	Insoluble protein
Asteroidea		
Asterina frigida	70	15
Asterina gibbosa	77	13
Diplasterias meridionalis	75	27
Leptychaster kerguelensis	65	20
Odontaster meridionalis	60	22
Othelia luzonica	63	16
Pteraster affinis	63	20
Holothuroidea		
Actinopyga mauritana	14	59
Afrocucumis africana	42	44
Chiridota rigida	41	35
Cucumaria planci	61	18
Eumolpadia violacea	49	35
Holothuria atra	24	48
Holothuria difficilis	50	38
Labidoplax digitata	24	39

Figure 3.90: Proximal view of a half-arm section of *Luidia clathrata* showing paxillae on the upper surface. (From A.M. Clark 1982)

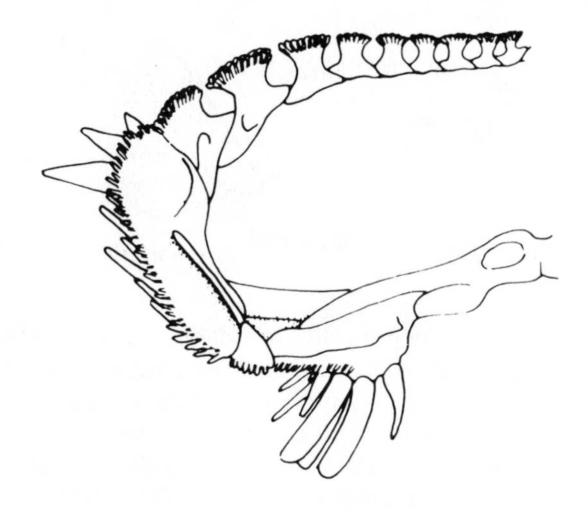

Figure 3.91: Undersurface of the porcellanasterid *Styracaster elongatus* (radius = 18 cm) showing pronounced adambulacral ossicles and spines (from Madsen 1981)

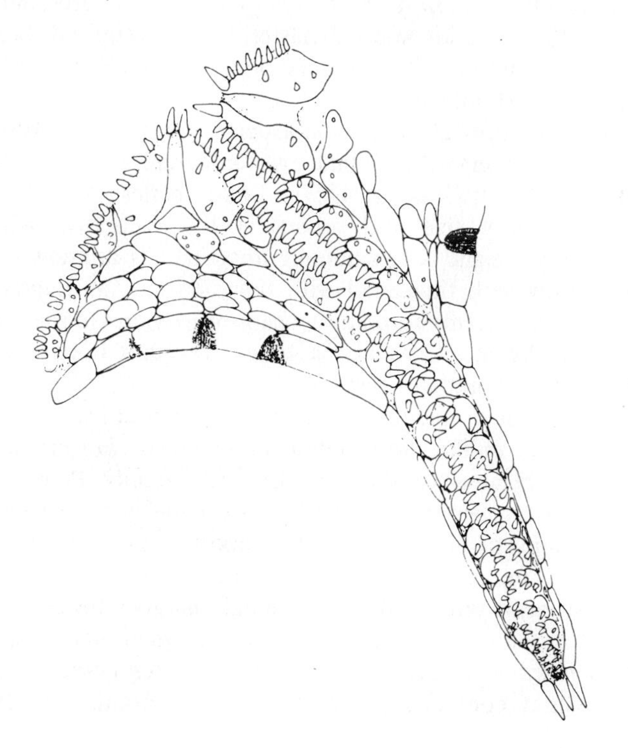

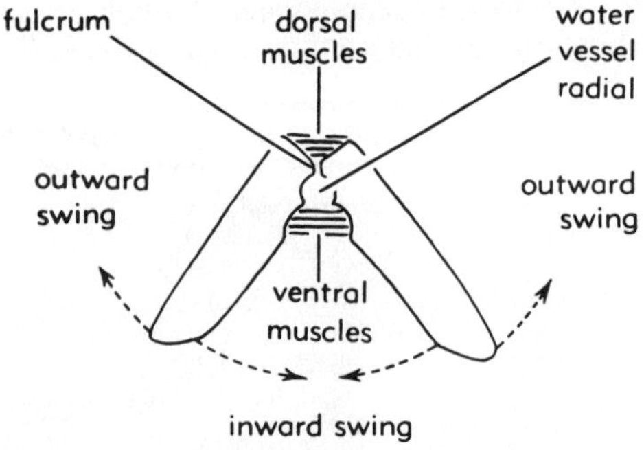

Figure 3.92: Lever action of ambulacral ossicles of asteroids (from Spencer and Wright 1966)

Blake (1983) evaluated the protective role of the structure of furrow and oral areas of valvatids and echinasterids. These include well developed adambulacral and oral ossicles, spines, and thickened dermal tissues. The strong primary skeleton seems most important in the valvatids, and the spines and dermal tissue seem most important in the echinasterids. The extensive development of protective devices by these primarily tropical forms seems related to the high predation pressure in the region. The development of the protective structures is counter to the asteroid's own predatory requirements. It is possible that some groups of asteroids have limited potential to develop structural antipredator structures, and the ones developed by the valvatids and echinasterids have required them to shift from being predators on active solitary prey that would require the open ambulacral groove (Blake 1983).

Most poraniids show at least some aspects of progressive resorption of the skeleton during growth, so that even apparently well calcified large individuals may have hollow marginal and other plates (A.M. Clark 1984). This maintains body size and retains skeletal support. Whether the decrease in body weight is functionally important is unknown. Poraniids also have a body wall thickened with tissue to varying degrees (Figure 3.89). The increase seems to be a mechanism to increase the width while maintaining the height. Certainly, a tough, resistant body wall of connective tissue can be predator resistant.

Mucus is used for protection against predation in at least one species of asteroid, but here, unlike the situation in *Ophiocomina nigra*, structural adaptations are related to the function. The family Pterasteridae are spinulosids with cruciform or lobed plates on the upper surface bearing paxilliform groups of spinelets which support a supradorsal membrane (Figure 3.11). The cavity between the membrane and the body wall beneath it has been termed the nidamental chamber because young are brooded in it in a number of species. However, when mechanically stimulated the membrane of *Pteraster tesselatus* secretes copious amounts of mucus which may contain saponins (Nance and Braithwaite 1979). The mucus deters predation by the asteroids *Solaster dawsoni* and *Pycnopodia helianthoides.* Individuals with the supradorsal membranes removed are readily devoured by these predators. In the defensive response, water and secreted mucus are forced out of the numerous spiracular openings that perforate the supradorsal membrane (Nance and Braithwaite 1981). The requirement of this structural modification for the use of mucus as a protection against predation is related to the respiratory role of the papulae on the body surface. The lack of similar modifications on *Ophiocomina nigra* probably indicates either a lesser amount of mucus production or requirement for a high capacity for movement of oxygen across the arm surface of ophiuroids.

The pedicellariae of asteroids are secondary spinelets attached directly to underlying ossicles (A.C. Campbell 1983). They all have pincer-like action and show a gradation in degree of complexity and presumably in effectiveness. *Sessile* pedicellariae are formed from adjacent modified spines and occur in some paxillosids. These are *spiniform* when simply formed from elongate spines, *pectinate* when formed from short curved spines, or *fasciculate* when formed from a cluster of spines. *Alveolar* pedicellariae are formed from adjacent spines sunk in an alveolus or pit. These are *bivalved* when the spines are valve-like in shape, *spatulate* when broadened at the tip, or *excavate* when the valves are tong-like and fit into recesses of the alveolus. *Pedunculate* pedicellariae are stalked and have two jaws attached to a basal ossicle at the top of the stalk. The pedunculate pedicellariae can be *straight* (*forficiform*) with the two jaw ossicles being parallel, or *crossed* (*forcipiform*) with the two jaw ossicles having a scissor-like arrangement.

The occurrence of pedicellariae is variable in asteroids, showing evolutionary as well as environmental and morphological correlations. Pedicellariae are rare in paxillosids and are only of the sessile type (Figure 3.93). As with the ophiuroids, the lack of development of pedicellariae in the paxillosids has been related to their burrowing habits. Closely related species differ in the number of pedicellariae present. *Astropecten pedi-*

Figure 3.93: Sessile pedicellariae and paxillae of *Luidia atlantidea* (from Madsen 1950)

cellaris has numerous pedicellariae on the upper surface and the adambulacral plates whereas *Astropecten tenellus* has few only on the upper surface and *Astropecten griegi* has none (Fisher 1919).

Sessile and alveolar pedicellariae are never abundant in valvatids, and may be absent. Sessile pedicellariae are present in some species of *Odontaster*, and alveolar in some species of *Linckia*. The presence of the pedicellariae varies with individual. They may be numerous or absent on the dorsal surface of *Odontaster nudus*. Sometimes only one or a few alveolar pedicellariae are found in each interradius and it is difficult to understand their effective mode of action.

Pedicellariae are also rare in spinulosids. They do not occur in *Asterina*, *Patiria*, *Crossaster*, *Solaster* or *Pteraster*. They are rare in linckiids (*Linckia*, *Nardoa*). Their presence is variable in echinasterids, occurring in *Acanthaster* but not in *Echinaster*.

Pedunculate pedicellariae are restricted to the forcipulatids where they are abundant and complex. They never possess more than two valves. The distal ends of the valves are typically in the form of teeth which meet tightly when the valves are closed (Figure 3.94). Lacking stalks themselves, the pedunculate pedicellariae frequently are clustered on spines that extend the pedicellariae further from the body surface in a localised area and provide support for them. Straight pedicellariae can occur as festoons (Figure 3.95) (Fisher 1930) or as clusters of crossed pedicellariae (rosettes) that can encircle spines on the margin or upper surface (Figure 3.96) (Lambert, De Vos and Jangoux 1984). These rosettes rest at the bases of the spines until stimulated, when they are elevated by contraction of longitudinal muscles (Figure 3.97). A prey-catching function has been suggested for these pedicellariae in some species (Chia and Amerongen 1975; Jennings 1907; Robilliard 1971). Although they do respond to tissue fluid and grasp small pieces of tissue, their individual, random and uncoordinated variation in response suggests that they function primarily to protect the body surface from foreign material and small organisms (Lambert *et al.* 1984).

The strength of the crossed valves should be greater than that of the straight jaws because of their shorter valves and the lever action of the lateral processes of the valves (Figure 3.98). (Lambert *et al.* 1984). Contraction of the set of proximal adductor muscles closes the valves. As is usually the case, the adductor muscles responsible for closing the valves are thicker than the abductor muscles. The arrangement of these muscles is such that movement of the prey between the jaws causes stretching of the muscle which results in more firmly locking of the valves (Chia and Amerongen 1975).

Figure 3.94: (A) Skeleton of crossed pedunculate pedicellaria of *Stylasterias forreri* with closed jaws. (B) Distal part of jaw skeleton. B: Basal ossicle, C: canine-like teeth, I: inner distal teeth, M: median teeth, O: outer distal teeth, P: teeth on medial projection, S: muscle scar. (From Chia and Amerongen 1975)

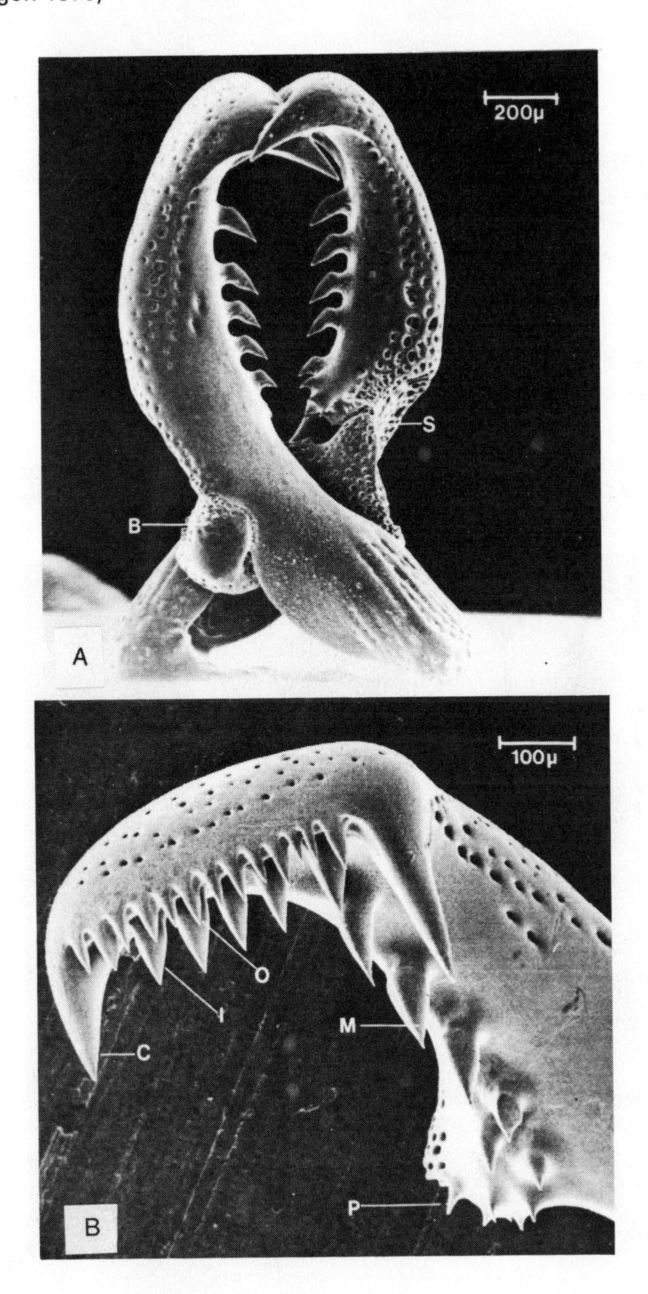

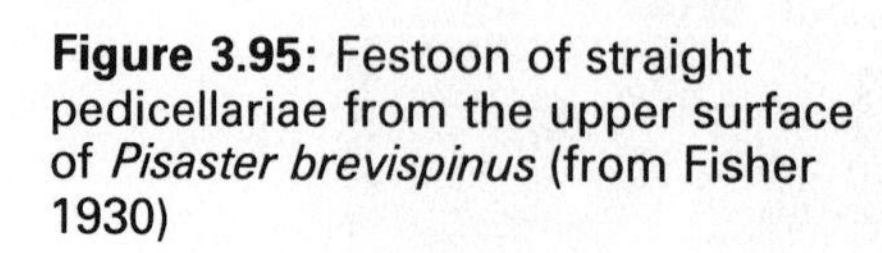

Figure 3.95: Festoon of straight pedicellariae from the upper surface of *Pisaster brevispinus* (from Fisher 1930)

Figure 3.96: Rosette of crossed pedunculate pedicellariae surrounding a spine of *Marthasterias glacialis*. Inset: Arrows indicating right- and left-handed pedicellariae in a single cluster. s: Spine, p: pedicellaria. (From Lambert *et al.* 1984)

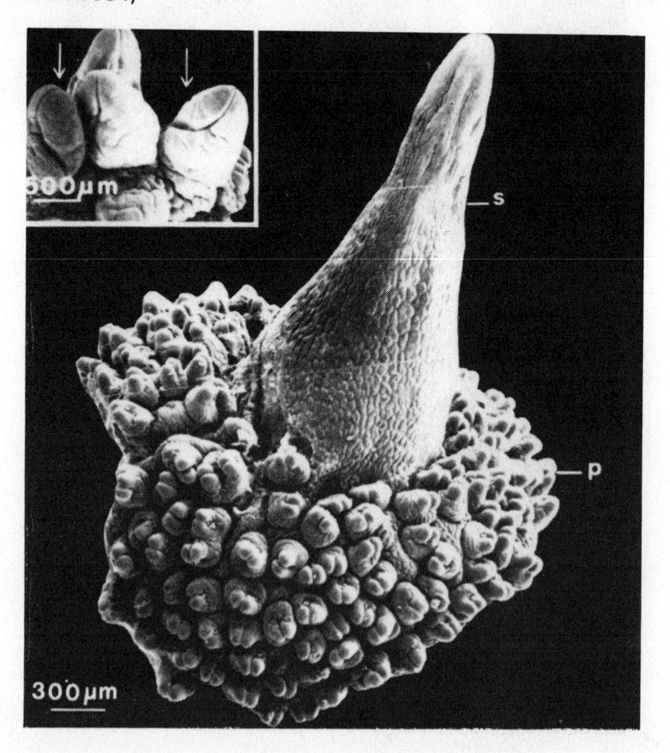

Figure 3.97: Diagrammatic longitudinal section of a rosette of crossed pedunculate pedicellariae of *Marthasterias glacialis* in the quiescent (A) and active (B) state. cs: Central spine, cm: circular muscle, e: epidermis, lm: longitudinal muscles, tm: transverse muscles, p: pedicellaria. (From Lambert *et al.* 1984)

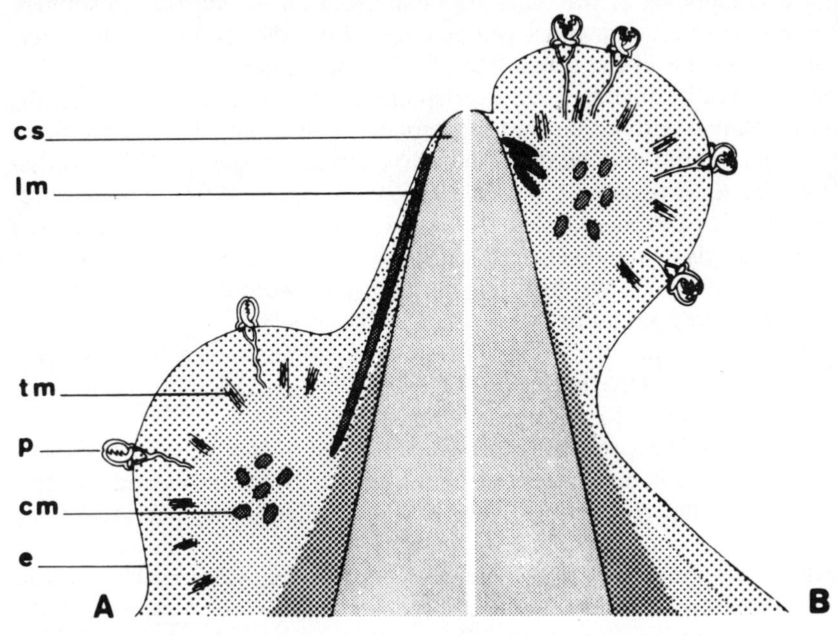

Figure 3.98: Representation of the upper surface of the basal piece of straight (A) and crossed (B) pedicellaria of *Marthasterias glacialis*. ab: Attachment area of abductor muscles, ad: attachment area of adductor muscles, c: cavity. (From Lambert *et al.* 1984)

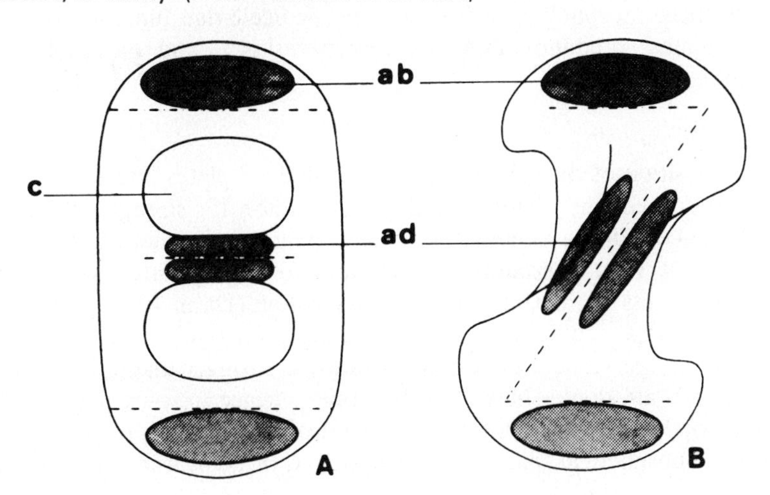

3.5.4. Echinoidea

The echinoid test provides support for the internal organs, spines and tube feet. Support is also necessary in regular echinoids for the stresses involved in the functioning of the Aristotle's lantern. That the supportive strength necessary for these functions can be minimal is indicated by the characteristics of the test of the echinothuriids. The plates can be so little calcified that they are flexible, and can overlap and possess so little rigidity that the body collapses without the buoyant support of sea water. The body form of the echinothuriid may be the result of hydrostatic pressure, similar to that described for asteroids (A. Agassiz 1869; J.S. Pearse 1967). An unawareness of this phenomenon led to bizarre representations of the deep-sea echinothuriids by the nineteenth century naturalists. It is probable, though, that echinothuriids have the catch mechanism between the plates of the test and can assume a rigid form.

It thus seems that the strength of the echinoid test should be interpreted primarily as protective in function, against either predation or impact loading (i.e. an external force acting on the body).

Ebert (1982) hypothesised that the allocation of energy to the body wall of echinoids is inversely related to the longevity of the individual. This implies that the production of the test and spines is an energetic drain which evolutionarily has not been accommodated for. Certainly the regeneration of spines can reduce the rate of body growth (Ebert 1968), just as body growth can reduce gonadal production when food is extrinsically limiting (Lawrence and Lane 1982), but it is not evident that somatic growth is necessarily a stress that reduces longevity when food is not extrinsically limiting.

The strength of the test can be modified by varying the body form, the strength of the individual plates, and the strength of the sutures. In contrast to these three morphological features, the pedicellariae function for protection against predation and probably parasitism, and to cleanse the epithelium.

3.5.4.1. Body form

The composition of the echinoid test by connected plates affects the body form and consequently its potential and limitation for strength. Seilacher (1979) conceived the echinoid test as a mineralised pneu, a tensional balloon made rigid. Certainly the echinoid form is greatly modified by mechanical stress or interference with calcification (Dafni 1983, 1985).

A dome shape provides the greatest strength for rigid structures, and a high dome (large vertical to horizontal ratio) is stronger than a low dome. All echinoids are dome-shaped on the upper surface to some degree. The echinoid body form was primitively spherical, becoming subspherical in cidaroids; subspherical, tall and subcylindrical, or conical in other regular echinoids; depressed and hemispherical in some irregular echinoids; and thin and flattened or domed in others (H.B. Fell 1966; Melville and Durham 1966). The shape of the test can be variable even within a family as in the temnopleurids in which it can be hemispherical (*Salmaciella*), flattened (*Trigonocidaris*), globular, egg-shaped (*Amblypneustes*), or sausage-shaped (*Holopneustes*) (Mortensen 1943a). These deviations in body form must have a biological origin in function, and require the development of other mechanisms to give the body support and protection.

The vertical diameter to horizontal diameter ratio is frequently greater for large individuals than for small individuals (Telford 1985a). The relation between the vertical and horizontal diameters of many species of both regular and irregular echinoids is a straight line (Figure 3.99). The horizontal and vertical diameters of regular echinoids do not completely indicate the body profile. The test shape can vary among populations of a species. Although the vertical diameter to horizontal diameter ratio increases with size for some specimens of *Lytechinus variegatus* and *Tripneustes ventricosus* (Telford 1985a), the ratio decreases for others (Mortensen 1943b). The stress of force exerted by the attachment tube feet of regular echinoids has long been thought to affect the shape of the test (Thompson 1961). Populations of species of regular echinoids tend to have a larger vertical diameter to horizontal diameter ratio in protected habitats where less strength of attachment is required (Lewis and Storey 1984; McPherson 1965; Moore 1935; Nichols 1982). This structural response is opposite to that which would be predicted from a consideration of the body form–strength relationship, a higher dome giving greater strength. That the body form is indeed a response to the strength of tube feet attachment has been demonstrated for *Tripneustes gratilla* (Dafni 1985). Individuals which lacked a solid substratum for attachment grew to an increased vertical diameter to horizontal diameter ratio, an experimental confirmation of the effect of vectorial tensions in the ontogeny of body form. Similar confirmation is shown by the development of abnormal body forms by *T. ventricosus* grown in solutions that affect calcium deposition (Dafni 1983).

The body form of spatangoids is domed, with the vertical diameter to average width and length ratio being similar to that of regular echinoids (Telford 1985a). In contrast, the body form of clypeasteroids and laganoids is usually flattened, an indication that adaptations other than meeting test-strength requirements through the body form are important. A clear distinction exists between the spatangoids and the clypeasteroids and laganoids. The mode of feeding by spatangoids does not require a large horizontal surface for collection. Their infaunal existence does not require flattening to reduce hydrodynamic effects. Consequently, spatangoids maintain a domed body form. The clypeasteroids and laganoids show body flattening, related to their mode of feeding and to their shallow-burrowing or epibenthic mode of existence. The laganoids have the flatter body form,

Figure 3.99: The relation between the vertical and horizontal diameters of (A) regular and (B) irregular echinoids (from Telford 1985a)

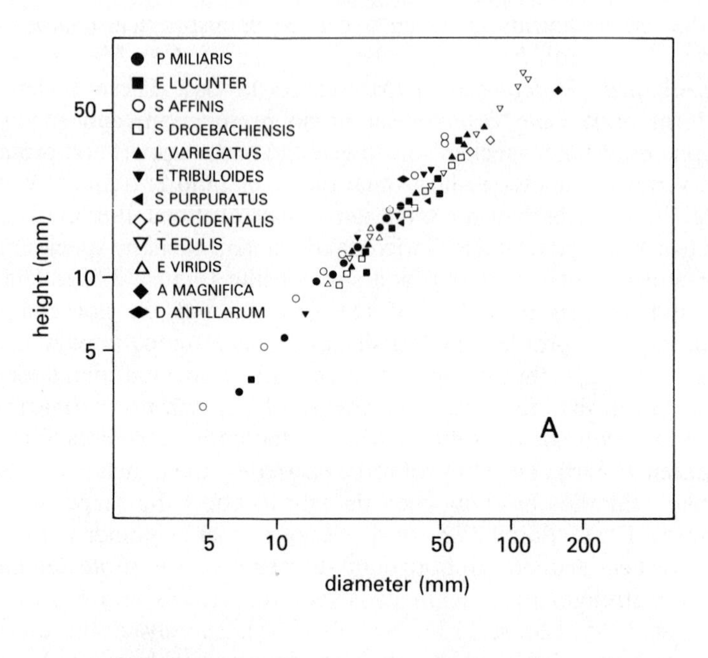

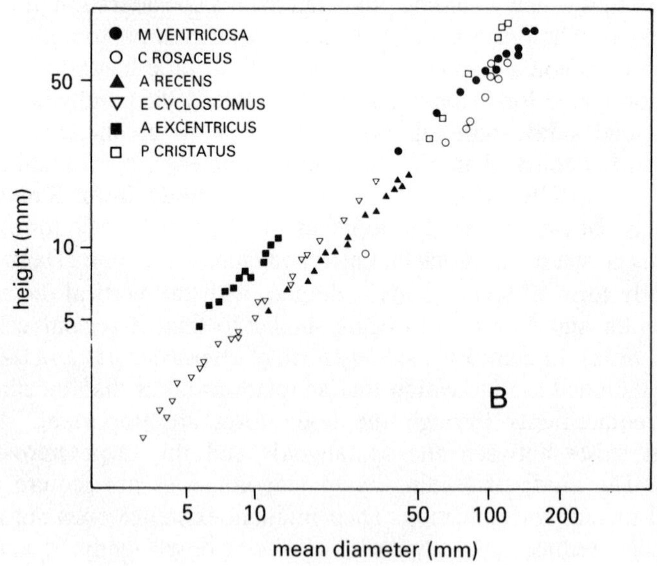

with a thin ambitus because they locomote through the substratum (Seilacher 1979). In contrast, the clypeasteroids have a thicker, more rounded ambitus because they do not. The petaloid regions of the upper surface of the clypeasteroids is dome shaped to varying degrees (Figure 3.100), both raising the petals above the substratum and increasing the surface area for respiratory function. The posterior upper surface of some spatangoids may have a longitudinal ridge rather than being rounded. Similarly, although the plastron of many spatangoids is flat, it has a keel like a boat in many species (Figure 3.101). This should strengthen the lower surface, but may also be necessary to lower the labrum to facilitate feeding. The sharp distinction between the upper and lower body surface with a more or less acute ambitus may be responsible for the increase in body size of irregular echinoids resulting from an increase in plate size rather than from continued increase in plate number as in regular echinoids (Durham 1966).

The adaptiveness of the dome shape to provide strength is primarily as support for an evenly distributed static load, and may not provide protection from unidirectional impact loads (Strathmann 1984). Thus,

Figure 3.100: View of the lateral profile of (a) *Clypeaster altus alticostatus* (Miocene), (b) *C. rosaceus,* (c) *C. euclastus,* (d) *C. reticulatus,* and (e) *C. latissimus.* View of the lower surface of (f) *C. rosaceus,* (g) *C. ravenelii,* (h) *C. europacificus,* and (i) *C. reticulatus.* (From Durham 1966).

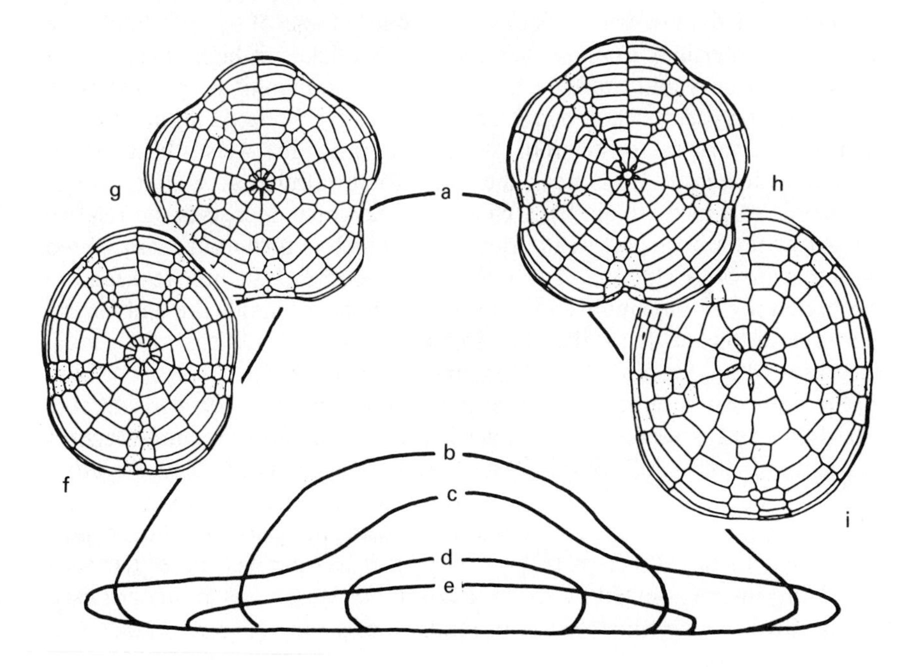

Figure 3.101: (a) Posterior, (b) side, and (c) anterior views of *Spatangocystis challengeri* (from A. Agassiz 1881)

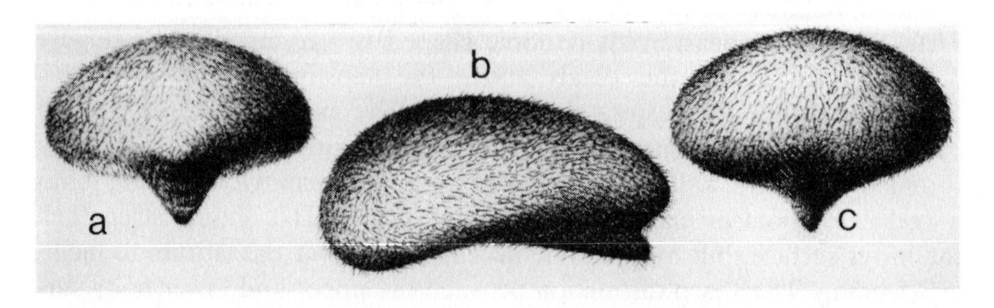

although any strength derived from the dome shape would provide some protection to the test, adaptive features for resisting unidirectional forces must be sought elsewhere.

3.5.4.2. Structural characteristics of the test

Structural characteristics affecting the strength of the test include the thickness of the plates and morphological features such as radial ribs, hoops, compression rings, and tension rings (Telford 1985a). These structural features become more important as the body form deviates from a dome shape.

The strength of a plate is related to its thickness and stereom structure. The tests of deep-sea regular echinoid species are typically much thinner than those of shallow-water species, particularly those of the intertidal. The test of the intertidal *Echinometra lucunter* is thicker in high-energy environments (Lewis and Storey 1981). That of spatangoids is much thinner than that of clypeasteroids, again a difference clearly related to the requirement for structural protection. Within an individual, test thickness in echinoids increases from near the apex to the ambitus or even to the peristome (Telford 1985a, b), a characteristic which decreases the relative weight of the upper part of the dome and which is fortuitously associated with the site of production of new plates. The pores of the tube feet require strengthening of the ambulacral plates. In regular echinoids the ambulacral plates are thicker than the interambulacral ones, and the differential increases with body size (M.L. Moss and Meehan 1967).

Stereom structure of the coronal plates of echinoids shows considerable variability and is related to phylogeny, growth rate and function (Figures 3.102 and 3.103) (A.B. Smith 1980b). Most of the interambulacral plate is galleried or labyrinthic stereom. *Galleried* stereom has long parallel galleries running in one direction only with no pore alignment perpendicular to this direction (Figure 3.104). It has parallel trabecular rods which are interconnected by struts. *Labyrinthic* stereom is an unorganised mesh of trabeculae which seems to function as a filler (Figure 3.104). It is

Figure 3.102: Schematic representation of the stereom structure of echinoid tests (from A.B. Smith 1980b)

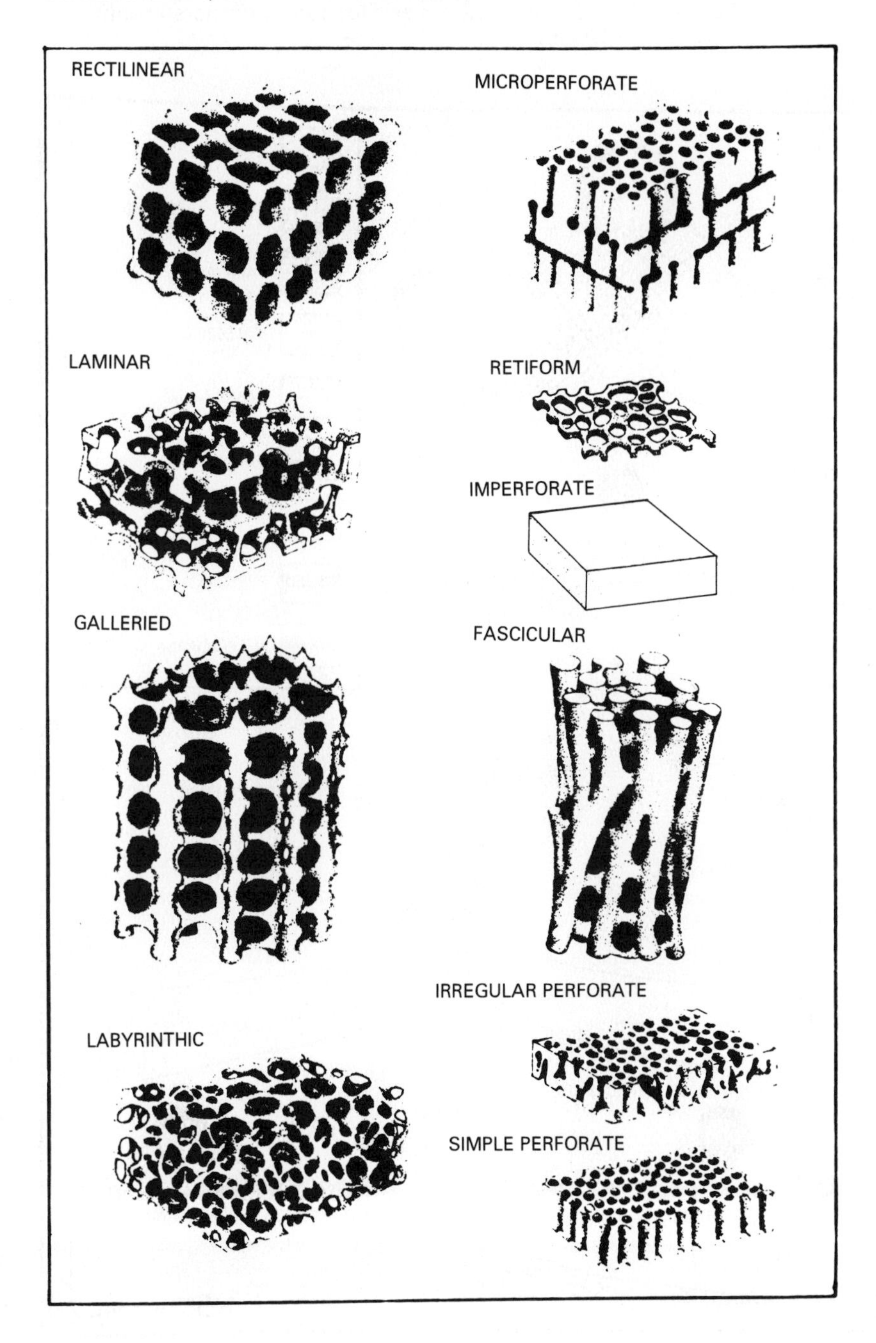

Figure 3.103: Schematic cross-sections of the interambulacral plates of echinoids showing the different types of stereoms associated with the plates and tubercles. See Figure 3.102 for an illustration of the stereom types. (From A.B. Smith 1980b)

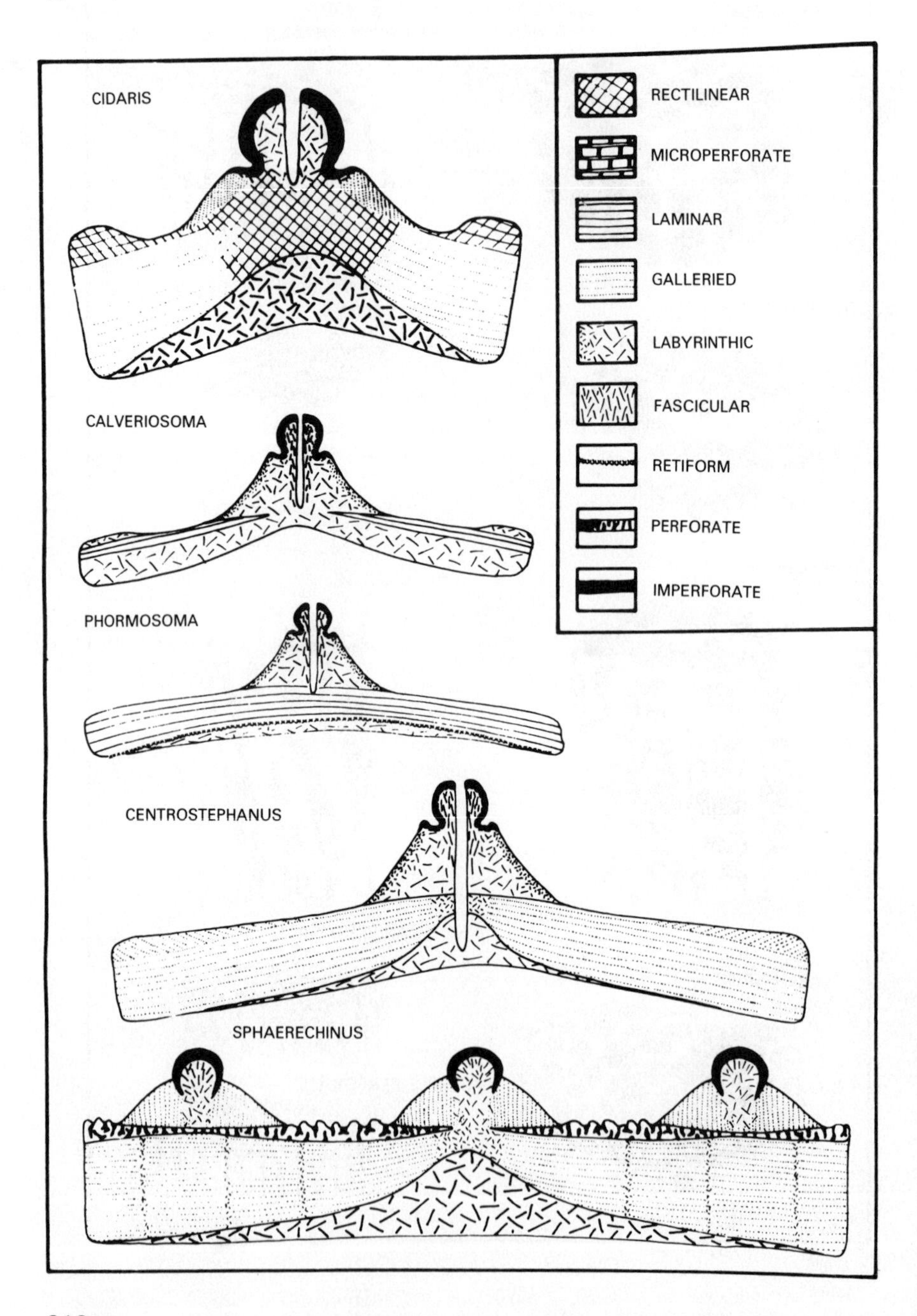

Figure 3.104: (a) Galleried stereom of an ambital interambulacral plate of *Cidaris cidaris.* (b) Labyrinthic stereom of an aboral interambulacral plate of *Calveriosoma hystrix.* (c) Laminar stereom of an aboral interambulacral plate of *Paramaretia peloria.* (From A.B. Smith 1980b)

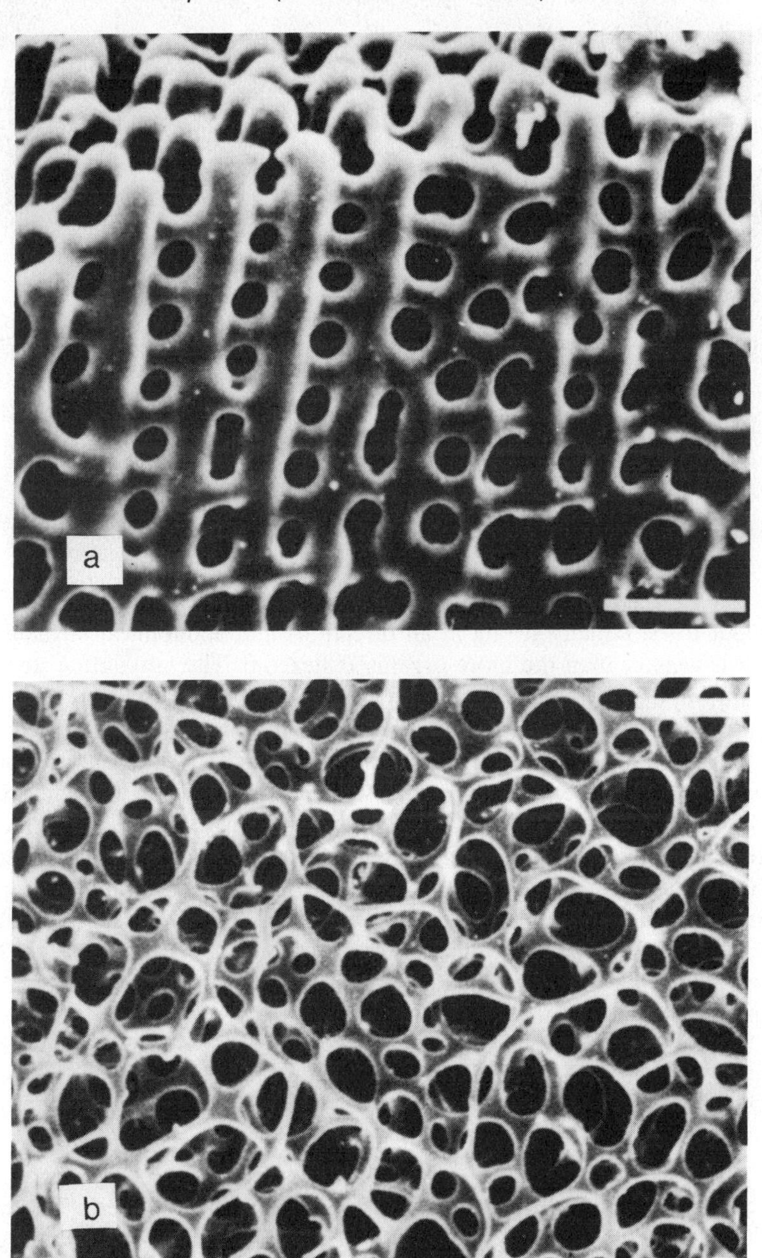

Figure 3.104: *continued*

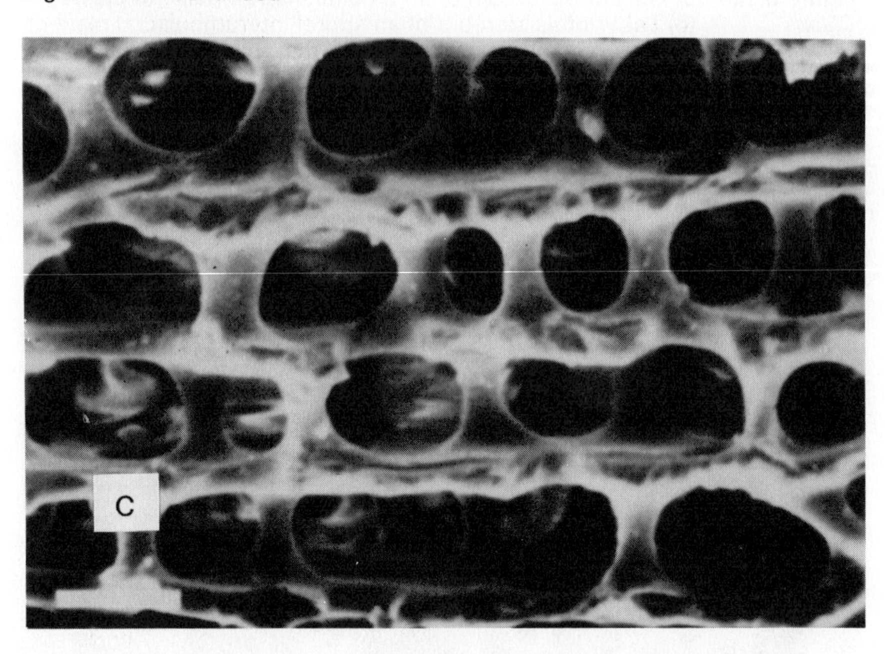

thought to be equally strong in all directions and economical in material, but it is weaker than the more organised stereom. The labyrinthic stereom can vary in density and consequently supposedly in strength. It is relatively open in the plates of the echinothuriid *Calveriosoma hystrix* but is dense in the thick plates of *Colobocentrotus atratus.* Labyrinthic stereom is deposited first in the plates of euechinoids with rigid tests, with later, outward deposition of galleried stereom. The change from labyrinthic to galleried stereom may reflect the onset of rigid plate suturing. No galleried stereom is deposited in euechinoids with flexible tests. *Laminar* stereom is formed from thin sheets of stereom separated by trabecular struts, and makes up the greater part of the plates of spatangoids (Figure 3.104). This arrangement is economical in calcite and allows plate thickening over the large plates characteristic of spatangoids.

The test may have structural features to increase its strength (Telford 1985a). Plates of several genera of echinaceans (*Echinus, Strongylocentrotus, Echinometra, Tripneustes*) are thicker at the sutures so that slender supporting ribs are formed. The plates around the peristome and the periproct are thick, and the ring formed around the very large periprocts of cidaroids and diadematoids is particularly prominent. The concavity of the test elevating the mouth in some clypeasteroids would give increased support to the Aristotle's lantern. In regular echinoids, the necessity to protrude the lantern for feeding seems to override this mechanism for strengthening the test.

Internal supports strengthen the test of flattened clypeasteroid sea biscuits and laganoid sand dollars. They occur only in those regions of the test in which new plates are not being added, i.e. the periphery, until production of new plates ceases (Seilacher 1979).

Characteristics of the test conspicuously separate *Echinocyamus* and *Fibularia*, both members of the family Fibulariidae (Telford 1985b). The former has a low, flattened test and the latter has a high globular or fusiform test. The height to mean diameter ranges from 0.39 to 0.46 for four *Echinocyamus* species and from 0.49 to 0.80 for four *Fibularia* species. *Echinocyamus* species are characterised by internal interradial buttresses that are lacking in *Fibularia* species. In the dome-shaped *Echinocyamus pusillus*, curved buttresses extend inwards from the interambulacral plates to the inflexion point of the test (Figure 3.105). *Echinocyamus australis*, which has only posterior buttresses, has a height to mean diameter ratio (0.69) in the range of the fibulariid species. Species of *Fibularia* and of *Echinocyamus* which lack internal supports have greater test thickness in the transverse than the longitudinal axis.

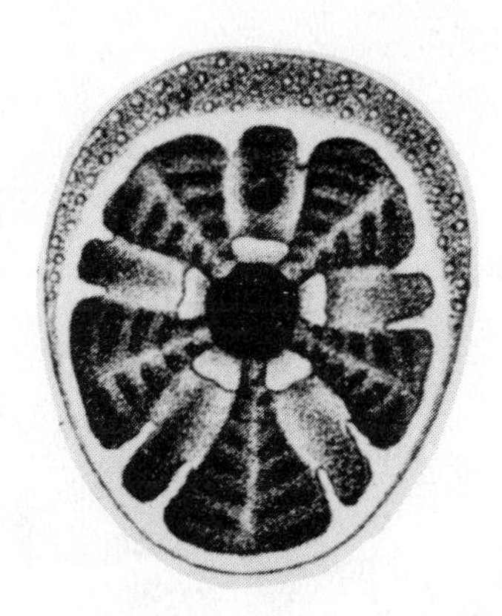

Figure 3.105: View from above showing the internal buttresses in the test of *Echinocyamus pusillus* (from L. Agassiz 1841)

The internal supports of clypeasteroids and laganoids can be so extensive that space available for the filled gut and the gonads is considerably reduced. These can result in almost monolithic tests in the clypeasteroids (Figure 3.106). In the laganoids the thin ambitus is vulnerable to predation despite being essentially solid. Individuals of *Mellita* frequently show irregular edges resulting from predation, possibly by crustaceans.

The echinoid test differs fundamentally from the shell of molluscs in being composed of pieces. Consequently, the suture of adjacent plates is a potential site for strengthening the test. Although the sutures were once thought to be a site of weakness in the test, they are more resistant to

Figure 3.106: Internal buttresses and supports in the clypeasteroid *Clypeaster latissimus* (a) shown from above by X-ray and (b) of *Clypeaster virescens* in profile (from Mortensen 1948b)

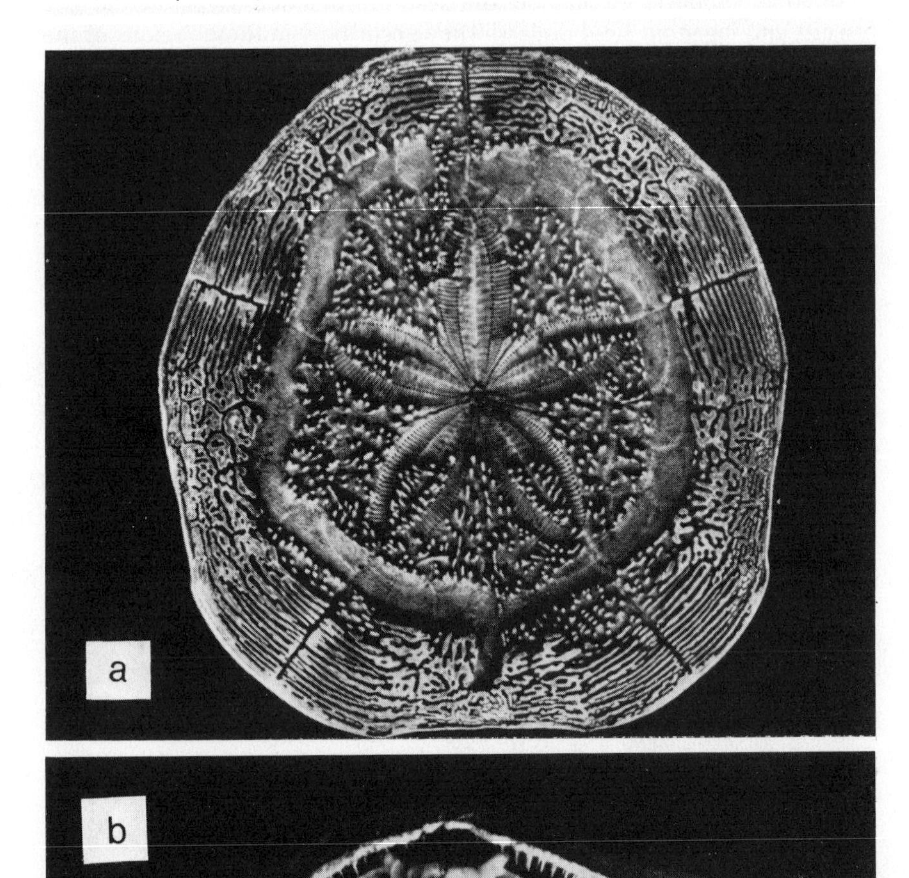

breaking than the plates (Duncan and Sladen 1885). The sutural areas of the test and the sutural connective tissues are coextensive in inflexible echinoid tests; and the strength of the sutures is a result of the arrangement of the plates, the structural features of the plate edges themselves, and the connective tissue binding the plates together (M.L. Moss and Meehan 1967). These authors pointed out that the sutural septa form a framework of dense organised connective tissue which is in many respects analogous to a series of structural beams. They suggested that the sutures act as stress breakers for external loads, transmitting such forces throughout the test.

The evolutionary trend in echinoids has been from ambulacra composed of simple plates at the beginning of the Mesozoic to compound plates (Figure 3.107) (Kier 1974). Compound plates are composed of two or more elemental plates bound together by a single primary tubercle. This development increased the number of sutures as well as the number of vectors which could diffuse stress in different directions. The plates tend to be wider than tall, probably as a result of their ontogenetic development (Seilacher 1979) which involves production at the apical pole followed by differential growth and resorption so that the plates adjust their size and shape to their respective position in the corona (Märkel 1981). Compounding the ambulacral plates resulted in larger plates which could support larger spines (Kier 1974).

Structural elements interlock the plates. The rods that form the galleried stereom of the plate are enlarged at the suture face to produce prominent pegs which interlock into depressions in the adjacent plates (Figure 3.108). (A.B. Smith 1980b). A very rigid test is produced by fusion of trabeculae from adjacent plates as occurs in the cassiduloid *Apatopygus* and the holectypoid *Echinoneus* (Figure 3.108) (Telford 1985a). In the latter, fusion occurs to such an extent that the sutures are obscured.

On a grosser level, small knobs and sockets of adjacent plates fit into one another in some species of the Temnopleuridae (Figure 3.109) (Duncan 1883) and Arbaciidae (Duncan and Sladen 1885). These undoubtedly contribute to the strengthening of the test. The plates break before the sutures open, and transverse sutures that do not have knobs and pits open before vertical sutures that have them. They are not the sole factors, however. The structures are well developed in the delicate and fragile tests of the large *Salmacis* and *Amblypneustes*, but only slightly developed or entirely lacking in the small but strong tests of *Temnotrema* (Mortensen 1943a). Sutural interlocking is variable in the clypeasteroids. The projections from the edge of the plate are short in *Clypeaster* and *Arachnoides* and tend to be long and dense in scutellids (Seilacher 1979). These projections from the plates would increase the rigidity of the test.

The tests of temnopleurids are also sculptured, having a variety of pits, grooves or depressions on the outer surface (Figure 3.110). The pits would seem to weaken the suture, but the grooves and depressions form ridges leading to the tubercles, cross the sutures, and seem to follow tension lines.

In contrast, collagen fibres would connect the plates but provide flexibility. They could provide variable rigidity if they possess the reversible characteristics of the system found in the body wall of holothuroids (Motokawa 1984; Wilkie 1984). These fibres are grouped in bundles which penetrate deeply into each plate, giving off branches that merge with other bundles and looping around the trabeculae (A.B. Smith 1980b). The

Figure 3.107: Evolutionary development of compound plates in euechinoids. A: Grouped primary plates, B: triad of primary plates, C: acrosaleniid, D: diad of primary plates, E: diadematoid, F: holectypid, G: diplopodid, H: arbacioid, I: plesiechinid, J: phymosomatid, K: pyrinid,

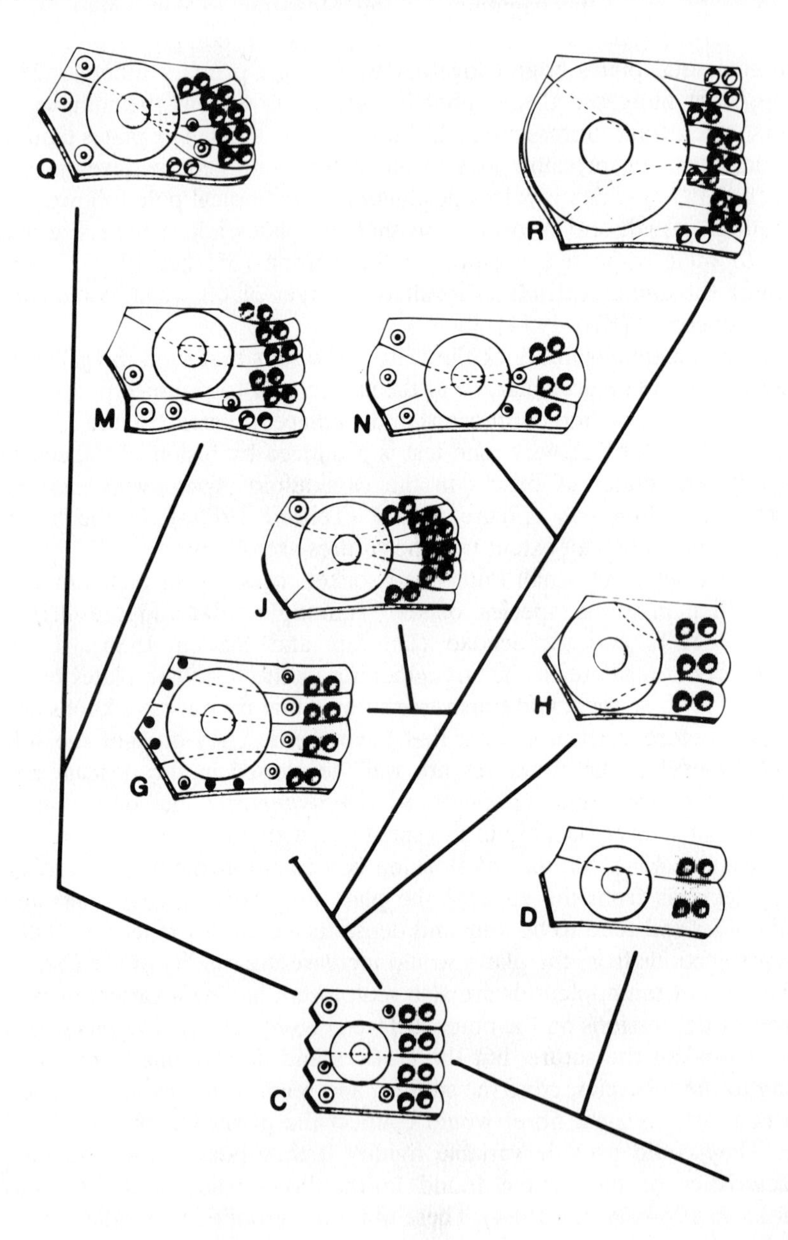

L: discoidid, M: glyptocidarid, N: stomechinid, O: micropygid, P: oligopygoid, Q: echinoid, R: stomopneustid, S: echinothuriid, T: clypeasterid. (From Jensen 1981)

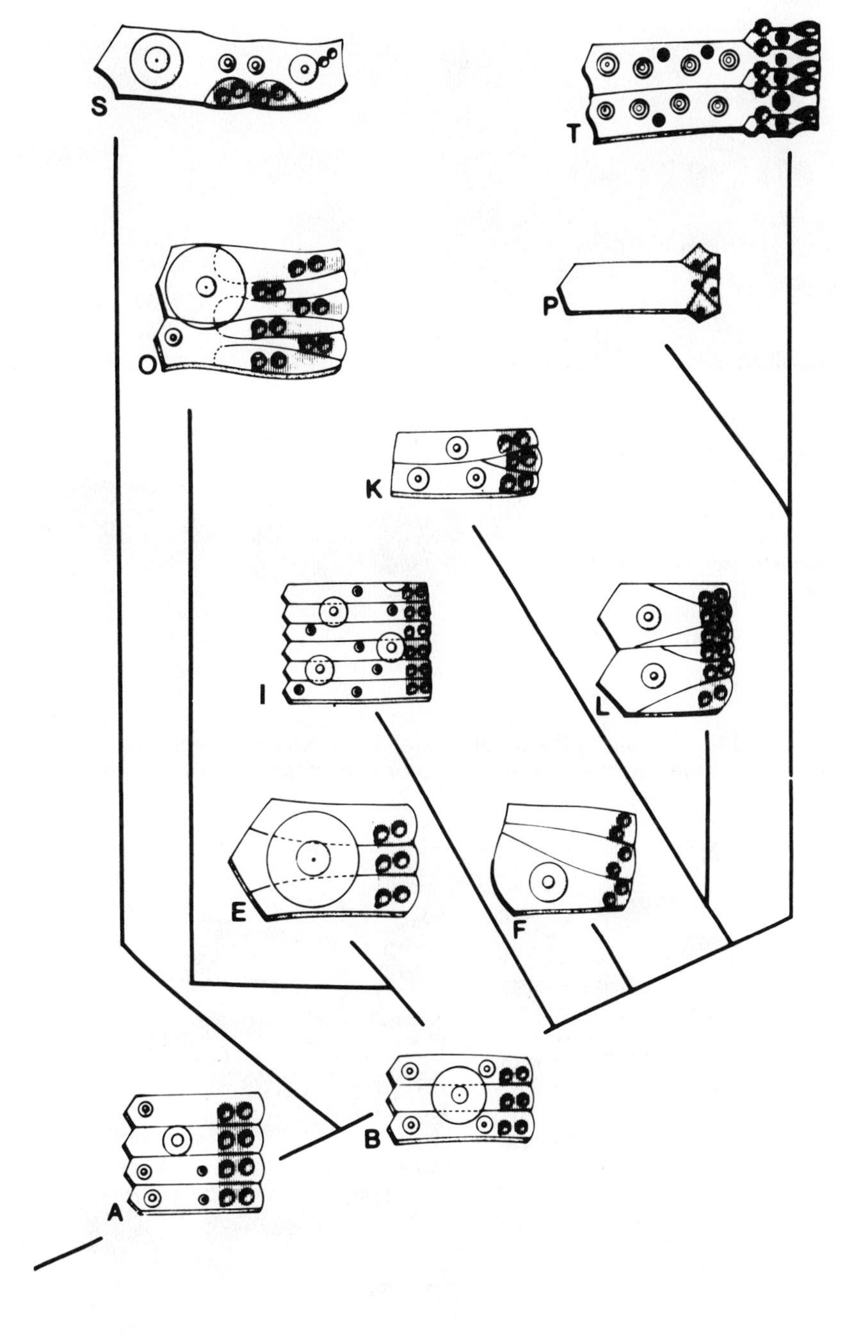

Figure 3.108: (a) Junction between tension and compression sutures in interambulacral region of *Diadema antillarum*. (b) Junction between tension sutures in interambulacral region of *Echinoneus cyclostomus*. (From Telford 1985a)

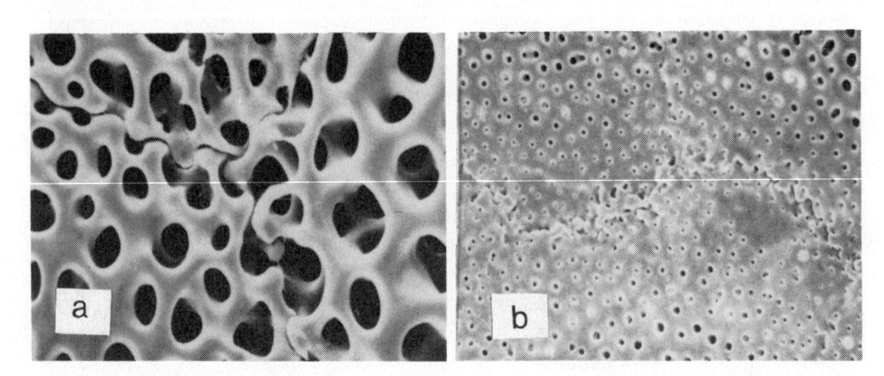

Figure 3.109: Interlocking knobs and sockets of (a) outer edge, and (b) inner edge of interambulacral plates of *Salmacis virgulata* (from Duncan 1883)

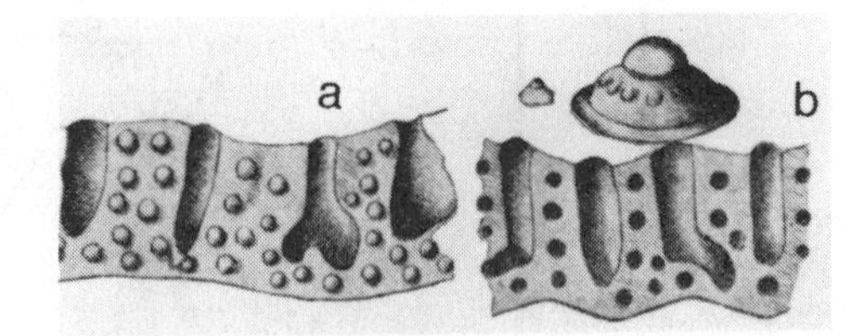

Figure 3.110: Pits along the suture of (a) *Temnotrema pulchellum*, and (b) grooves and depressions of the plates of *Orechinus monolini* (from Mortensen 1943a)

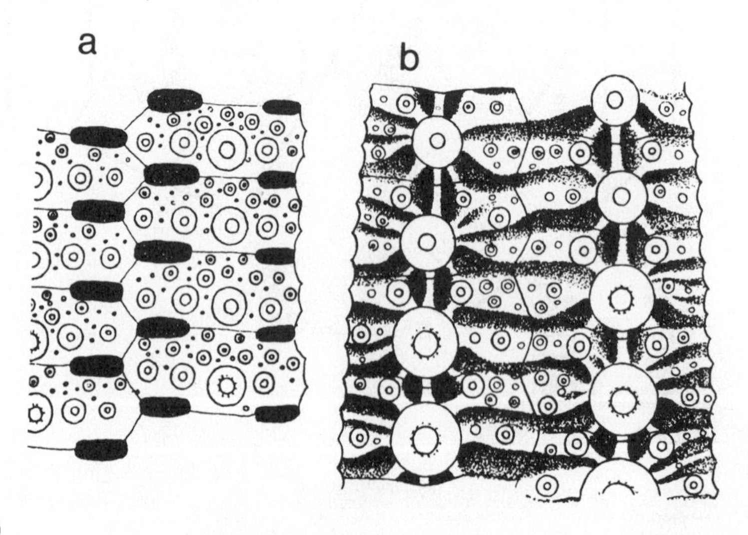

galleries of the stereom of the suture faces run perpendicular to the surface to allow penetration by the fibres. Relatively wide sutures with much collagen allow flexibility in the test of *Diadema*, while narrow sutures with reduced collagen produce the more rigid test of *Arbacia* (Telford 1985a). The radial sutures which are most frequently subjected to tensile forces have the most collagen.

3.5.4.3. Spines

The number and variety of spines of echinoids greatly exceed those of the other classes. The spines are not extensions from the plates, but are skeletal elements attached to them (Figure 3.111). The primary significance of this is that it provides for the orientation of the spine. It also gives a mechanism to autotomise a damaged spine (Swan 1966), although in the cidaroid *Eucidaris tribuloides* autotomy also occurs at a defined level just above the collar (Prouho's membrane) by the action of phagocytes (Märkel and Röser 1983).

The spines differ in gross characteristics and substructure in different echinoid groups (Table 3.2, Figure 3.112). The hollow spine is a retention of the condition of the developing young echinoid (A. Agassiz 1881). The

Figure 3.111: Longitudinal section of spine and tubercle of *Eucidaris tribuloides* (from Märkel and Roser 1983)

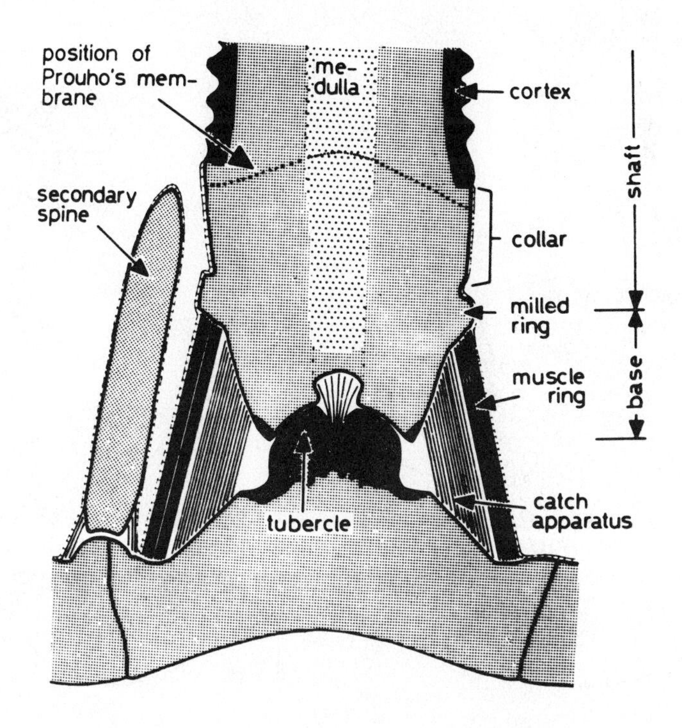

Table 3.2: Characteristics of the spines and tubercles of extant echinoids (compiled from Durham *et al.* 1966)

	Tubercles	Spines
Subclass Cidaroidea		
	Perforate Crenulate or non-crenulate	Solid
Subclass Euechinoida		
Superorder Echinacea	Imperforate Non-crenulate	Solid
Superorder Echinothuriacea	Perforate Usually non-crenulate	Hollow
Superorder Diadematacea	Perforate Usually non-crenulate	Hollow
Superorder Atelostomacea	Usually perforate Usually crenulate	Hollow
Superorder Gnathostomacea	Usually perforate Usually crenulate	Hollow

Figure 3.112: Horizontal cross-sections of spines of (a) *Temnopleurus reevesii*, (b) *T. michaelseni*, (c) *T. decipiens*, (d) *Meoma grandis*, (e) *Eupatagus rubellus*. (From Mortensen 1943a, 1951)

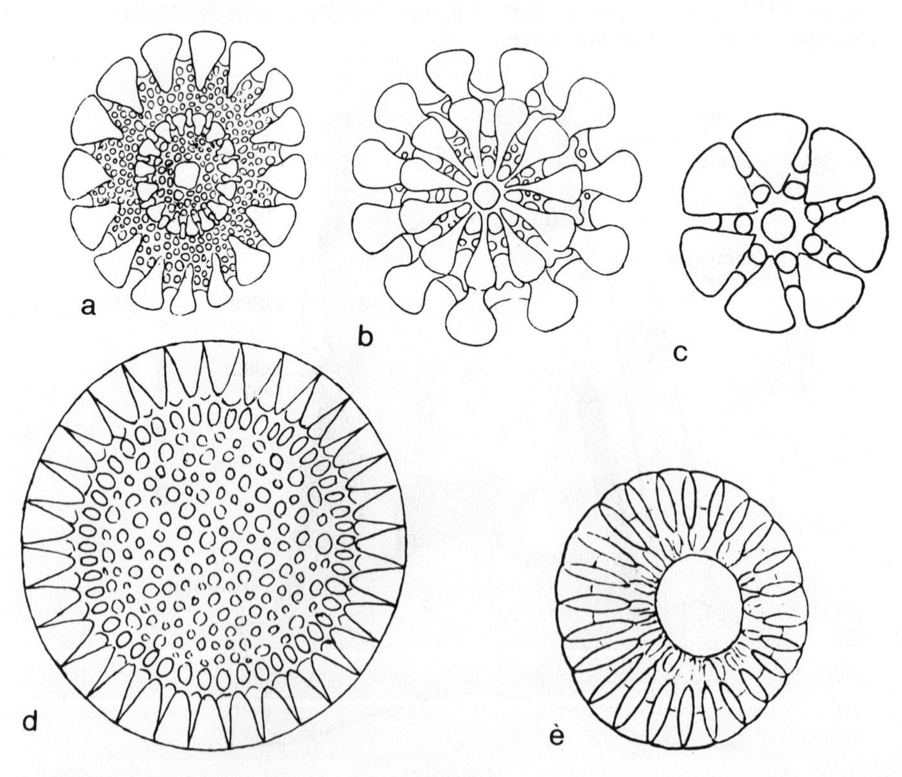

solid spine should be heavier, more rigid and brittle, and less strong per unit weight. The solid spines of *Heterocentrotus* and *Colobocentrotus* have a lower energetic density (Table 3.3) and make up a high proportion of the energy content of the individual (Table 2.3). The absolute cost of the spines of these exposed regular echinoids is much greater than that of spatangoids, although the relative cost may be no greater (e.g. compare *Abatus* and *Echinometra*).

Table 3.3: Level of energy ($kJ.gdwt^{-1}$) in the test and spines of echinoids

	Test	Spines
Subclass Cidaroidea		
Eucidaris nutrix	3.105	5.407
Subclass Euechinoidea		
Superorder Echinacea		
Echinometra mathaei	2.879	2.674
Heterocentrotus sp.	2.817	1.251
Colobocentrotus atratus	2.624	1.561
Lytechinus variegatus	2.821	2.586
Sterechinus diadema	4.005	5.411
Superorder Atelostomacea		
Abatus cordatus	3.695	3.507

The crushing strength of the solid spines of several species has been measured (Currey 1975; Weber, Greer, Voigt, White and Roy 1969), but the use of dried spines may have affected the results. The thin, hollow spines of diadematids are flexible because of the living tissue within the spine and the lack of cleavage planes (Figure 3.113) (Burkhardt, Hansmann, Märkel and Niemann 1983). In addition, the base of the wedges comprising the shaft are at the periphery which produces a high load-bearing capacity with minimum weight. Material is concentrated at the base of the spine so that the strain is constant all along the axis. The transverse bars connecting the wedges are few in the hollow spines in contrast to the numerous ones found in solid, stiff spines.

The spines are set on a raised portion of the test, the *tubercle*, whose size varies with that of the spine it bears. In several groups, the spine is attached to a perforated tubercle by a central elastic ligament which should strengthen the attachment. Perforated tubercles are best developed in long-spined regular echinoids, and function to increase the capacity to hold the spine rigid and prevent its dislocation (A.B. Smith 1980c). The absence of the ligament should be correlated with an increased ability to autotomise the spine. The imperforate tubercle of the Echinacea is correlated with their solid spines, but the cidaroids have both perforate tubercles and solid spines.

Figure 3.113: Longitudinal section of the skeleton of (a) the base and (b) the middle part of the shaft of a primary spine of *Diadema setosum*. c: Cylinder, mr: milled ring, s: septum, t: thorn, tb: transverse bar, X: damage to spine. (From Burkhardt *et al.* 1983)

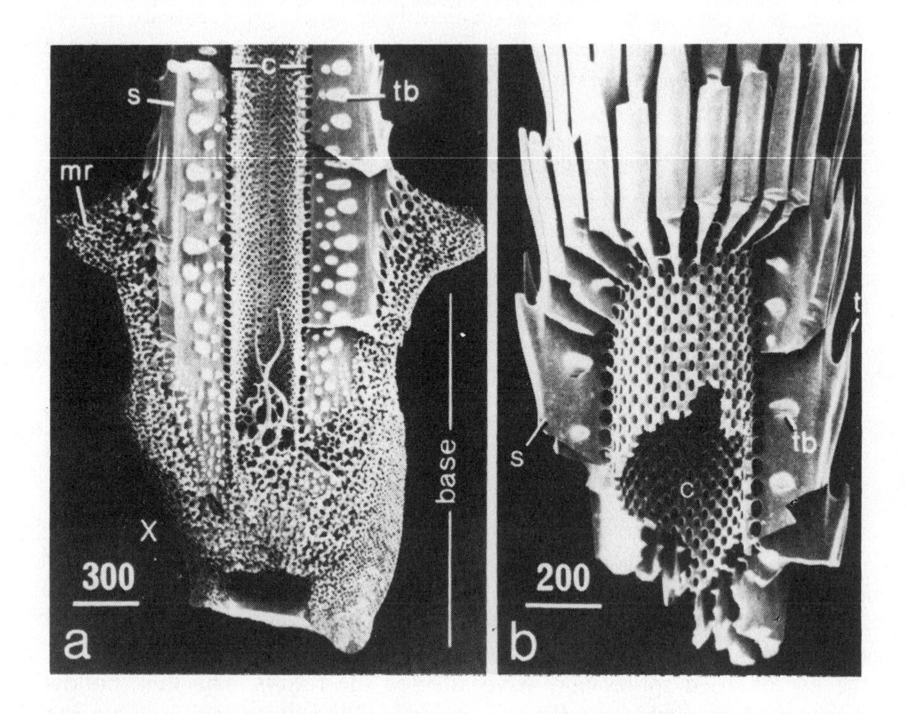

In the absence of the central ligament, the spines are attached only by an outer ring of muscles and an inner ring of fibrous tissue (the *catch apparatus*) attached to the circumference of the spine (Figure 3.111). The catch apparatus is extremely important in the functioning of the spine because it is capable of becoming rigid and locking the spine in a given position (Motokawa 1984, Wilkie 1984). The ligament is composed of collagen bundles with minute muscle fibres lying parallel and close to the bundles (D.S. Smith, Wainwright, Baker and Cayer 1981). The ligaments on the side of a contracting spine muscle shorten but do not buckle. In an undisturbed individual, spine ligaments are in all phases, from mobile to catch. Mechanical stimulation causes the ligaments to catch.

Crenulated edges of the tubercle match the milled edge of the base of the spine. They interlock when the spine tilts and help hold the spine in position (A.B. Smith 1980b). Crenulation of the tubercles (Figure 3.114) is common in the temnopleurids, although the degree of crenulation is variable (Mortensen 1943a). It is distinct in *Salmaciella*, indistinct in

Figure 3.114: Crenulation of the tubercle of (a) *Salmaciella dussumieri*, and (b) *Asterechinus elegans* (from Mortensen 1943a)

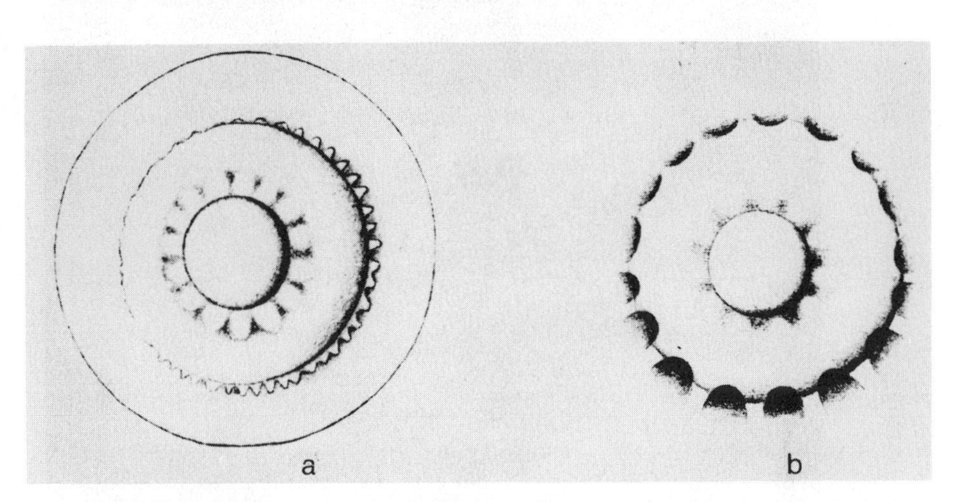

Mespilia, and absent in some species of *Pseudechinus*. The absence of perforated tubercles and crenulation in Arbacioida and Echinoida may be because most species in these groups have relatively small spines or have no requirement for strong movement of the spines. Perforation and crenulation of the tubercules are two separate evolutionary trends with the transitions occurring more or less independently at different rates in different families (H.B. Fell 1966). The major primary spines on the upper surface of shallow-burrowing spatangoids which may be subjected to extreme stress or predation may have extreme depressed areoles which lack crenulation in contrast to the locomotory primary spines on the lower surface (Figure 3.115) (Mortensen 1951).

The type of stereom in the tubercle seems correlated with the amount of stress encountered by the spine (Figure 3.103) (A.B. Smith 1980b). The stereom may be galleried with the galleries oriented perpendicular to the test as in *Cidaris* and *Sphaerechinus*, or be labyrinthic stereom as in the irregular echinoids, the echinothuriid *Calveriosoma*, and the diadematid *Centrostephanus*. Much of the tubercle of *Cidaris* is strong rectilinear stereom. The portion of the tubercle directly below the mamelon is usually labyrinthic stereom, indicating that vertical stress is not as important as angular stress. In most instances, there is an inner dome underlying the tubercle filled with labyrinthic stereom. The galleries of the overlying galleried stereom usually angle upwards along the sides of this dome suggesting that the stress placed on the spine is spread over a large area.

The spines of Ordovician echinoids were small and undifferentiated (Kier 1966). Most Permian and most Triassic cidaroids had evolved a large

Figure 3.115: Views of the upper surface of *Lovenia elongata* (a) showing the long protective spines and (b) the extremely depressed aurioles of these protective spines (from Mortensen 1951)

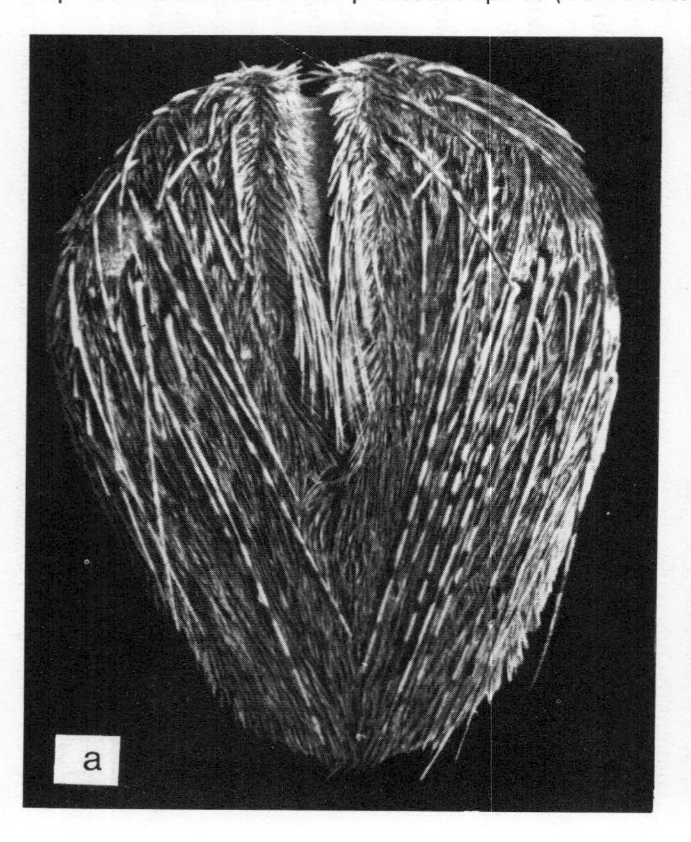

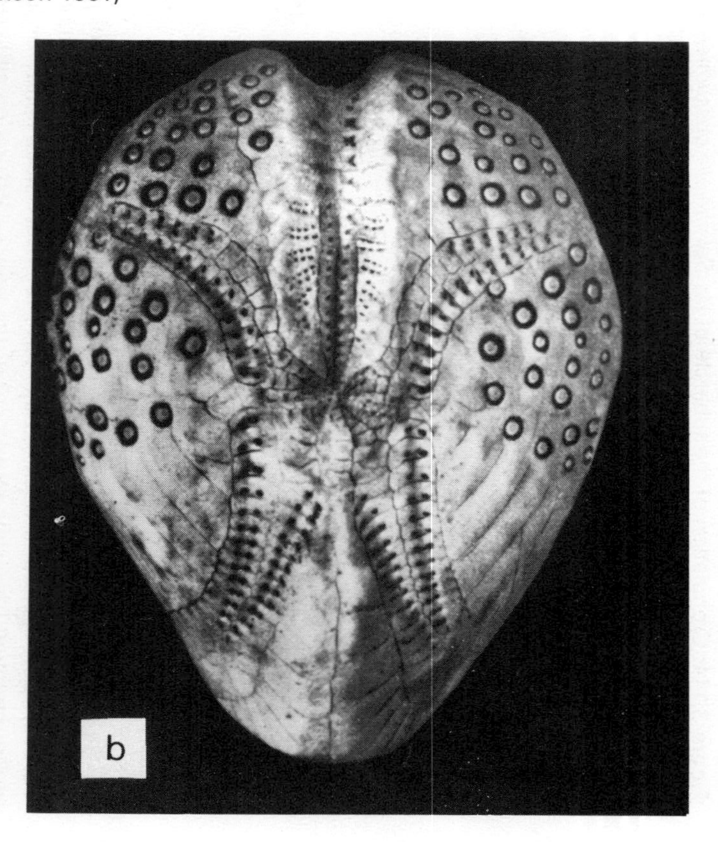

primary spine on each interambulacral plate. Euechinoids have shown a subsequent general evolutionary trend towards an increase in number associated with a decrease in size of primary spines on interambulacral plates and the development of primary spines on ambulacral plates (Kier 1974).

A protective function for the spines is inferred on the basis of their structure and location, but few studies have documented this role. It is now clear that spines provide protection not only from predation, but also from impact loading and from abrasion of the test epithelium. They also function to wedge the individual into crevices for maintenance of position. The protective role of the spines of burrowing irregular echinoids seems primarily to create a barrier between the individual and the particulate substratum, except for defensive spines of some shallow-burrowing spatangoids. The role of the spines on the upper body surface is indicated by the increase in their density with a decrease in the sediment grain size (A.B. Smith, 1980c).

Both regular and irregular echinoids can be ingested whole by fish and paxillosid asteroids. For example, an intact *Cidaris cidaris* was found in the stomach of the fish *Lophius piscatorius* (Mortensen 1932) and an intact *Arbacia punctulata* was found in the stomach of the fish *Opsanus pardus* (Shirley 1982). As usual in prey species that are exposed, individuals have a refuge in size. Juvenile echinoids are more susceptible to predation than the larger adults. When the prey are ingested whole, this is usually assumed to be a result of body size becoming too large for ingestion. However, the asteroid *Luidia clathrata* ingests small *Mellita quinquiesperforata* by preference, not because of an inability to ingest large individuals (McClintock 1984).

Spines could decrease predation simply by increasing the effective size of the individual. This would be an energetically inexpensive mode of providing a refuge in size at a much lower energetic cost than an increase in the size of body itself and would avoid the biological consequences to the individual of an increase in size. An increase in effective body size requires long spines.

Spines that are long relative to the body size are found primarily in the cidaroids and diadematids. Long spines also occur in representatives of the salenids, arbaciids, and echinometrids. Almost all irregular echinoids have extremely short spines which would not increase the effective body size. Either the sizes of potential predators for euechinoids are so small that a large body size of the echinoids is not necessary, or the benefits of short spines outweigh the disadvantages. Long spines would be inappropriate for the way of life of most irregular echinoids. Those protective spines found on the upper body surface of some shallow-burrowing spatangoids such as *Lovenia* (Figure 3.115) are retracted posteriorly, flush with the body, and raised when the individual is disturbed. Long spines would probably be disadvantageous for regular echinoids by decreasing their mobility or being inappropriate for their habitat. The long spines of the diadematids do not hinder locomotion, but the individuals are susceptible to currents.

Most predation on echinoids is not by ingestion of whole individuals, but by breaking the test. The internal organs are most preferred by predators, but fish often eat the test and spines also. Spines can provide protection from this kind of predation by their ability to wound predators or to prevent the predator from reaching the test. The ability to wound a predator requires that the spine have a sharp edge or be pointed, and that force be applied. Ornamentation of the shaft seems limited to echinoids with long spines and may be grotesque and irregular in cidaroids (Figure 3.116), or may occur as outwardly pointing barbs in cidaroids, salenoids and diadematoids. The outwardly pointing spines seem designed to deter ingestion by predators, not to facilitate penetration (D.P.B. Smith 1975). This conclusion is supported by the observation that the primary spines of diadematids tend to bend rather than penetrate unless the tip is normal to a substantial force. It is the fine secondary spines that penetrate easily. The primary spines of cidaroids and *Heterocentrotus* species are too thick to possess a sharp point, and a pointed tip would not be functional in the use of the spines to wedge individuals into crevices.

The spines of most other euechinoids are of varying, but typically short to moderate, diameter, and taper only quite near the distal end. This provides a point for penetration as well as mechanical support. The slender end of these spines frequently breaks off in the flesh of the predator. The

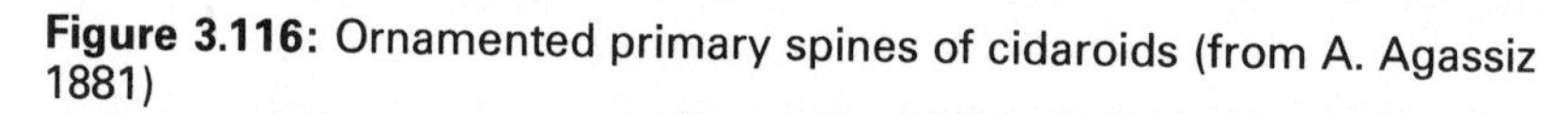

Figure 3.116: Ornamented primary spines of cidaroids (from A. Agassiz 1881)

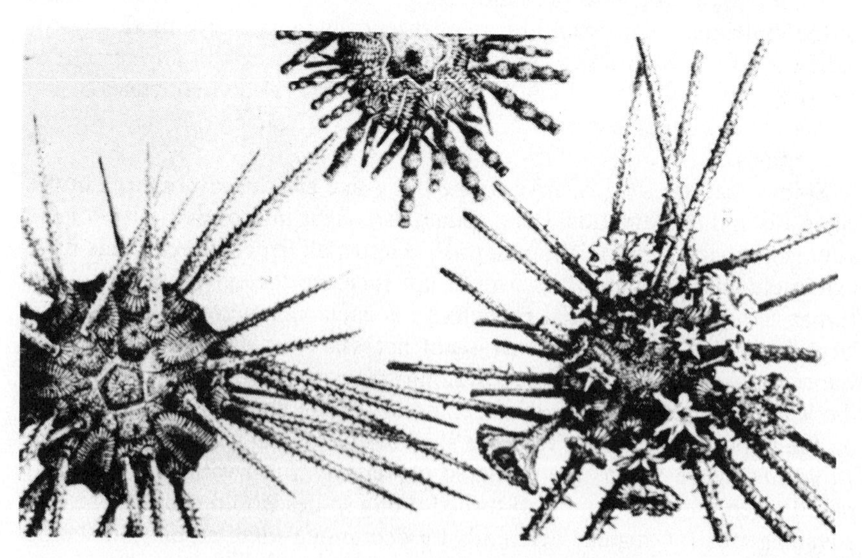

effectiveness of the spines of asthenosomids is increased by the presence of poison glands. The effectiveness of the spines in these roles depends on whether the predator is a fish, a mammal, a crustacean, a gastropod, an anemone or an asteroid. The pointed spines would be most effective against predators that would use soft tissues to produce force in the feeding process. Even short spines can be effective in such a situation, and the real question seems to be why some species have long spines and not why most species have short spines.

Thirty-four species of reef fishes eat echinoids, and six species use them as the primary food (Randall 1967). Several species eat the cidaroid *Eucidaris tribuloides*, indicating that its spines are not effective in deterring predation (McPherson 1968). The long spines of diadematids are effective in decreasing predation by fish. Several species of fish are predators on *D. setosum* (Fricke 1971). Lethrinid and labrid fish penetrate between the spines of the echinoid and bite the test. The fish is not deterred by the penetration by the spines of its thick lips and skin of its bony head. That the spines are protective, however, is indicated by several observations. The spines of *D. setosum* show a 'shadow reflex', pointing quickly to the area of the test which becomes shaded. Attacking fish circle the echinoid and attack between spines that have not become focused. Small *D. setosum* are attacked more frequently than large ones, indicating the effectiveness of the spines.

The labrid *Cheilinus fasciatus* grasps large *Diadema setosum* by the mouth region and breaks the echinoid against hard surfaces (Figure 3.117). The balistid *Balistapus undulatus* grasps the spines of *D. setosum* with the mouth, inverts the echinoid, and attacks the less protected mouth region.

Figure 3.117: *Chelinus trilobatus* feeding on *Diadema setosum* (from Fricke 1971)

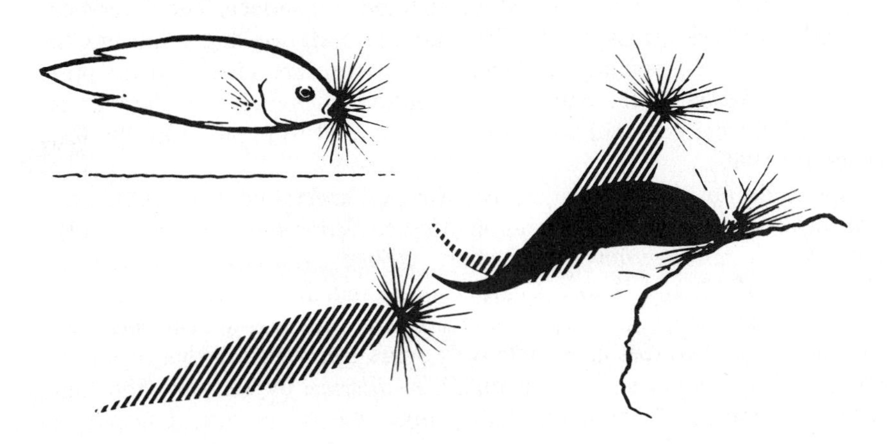

Balistes fuscus expels jets of water from its mouth to expose the undersurface of *D. setosum.* The short spines of small *Echinometra mathaei* are no deterrent to attacks by the fish, but the diadematid *Echinothrix calamaris* and the echinometrid *Heterocentrotus mammillatus* are not attacked. The massive spines and test of *H. mammillatus* seem to be too strong to be broken by biting.

The effectiveness of echinoid spines in preventing predation by fish can be seen with the labrid wrasse *Pimelometopon pulchrum* which preys upon strongylocentrotids (Tegner 1980). *Strongylocentrotus purpuratus* has a maximum horizontal test diameter of *c.* 70 mm and has short spines relative to the diameter, whereas *S. franciscanus* can have a diameter of 140 mm and has long spines relative to the diameter. The fish attacks *S. purpuratus* from any angle and the entire test can be ingested. In contrast, large *S. franciscanus* are inverted by the fish and the vulnerable peristome is attacked by the fish's teeth.

The sea otter *Enhydra lutris* preying on echinoids has similar problems in penetrating the test without being punctured by the spines. Small individuals of *Strongylocentrotus* species are crushed between the teeth, but larger ones are held on the chest and cracked by being struck with a rock (Estes 1974).

The exoskeleton of the chelae of crustaceans is not penetrated by echinoid spines. The latter thus function only as a physical barrier between the test and the crustacean predator. The ability of the lobster *Panulirus interruptus* to feed on *Strongylocentrotus franciscanus* and *S. purpuratus* is a direct function of both lobster and echinoid sizes (Tegner and Levin 1983). Small individuals were simply fractured into pieces or attacked through the oral surface and peristome. Even moderately large *S. purpuratus* could be eaten by large lobsters by placing the ambitus of the echinoid directly to the mandibles. In contrast, the same-sized lobster did not attempt to breach the spine barrier of large *S. franciscanus,* but instead inverted the echinoid and attacked through the oral surface. The difference in mode of attack seems due to differences in body size and spine length. The spines of the spatangoid *Meoma ventricosa* are ineffective in preventing predation by the crab *Calappa flammea.* One claw of the crab is used to hold the echinoid while the other is used to break away the test (Chesher 1969).

Spines do not deter predation by gastropod species on some echinoids. The helmet shell *Cassis* preys upon *Meoma ventricosa* (Chesher 1969). *Cassidulus caribaearum* (Gladfelter 1978). *Lytechinus variegatus* (Engstrom 1982) and *Diadema antillarum* (Randall, Schroeder and Starck 1964). Spines affect the way in which *Cassis* preys on *C. caribaearum.* Boring is most frequent in the relatively spine-free ventromedial region of the test. The helmet shells feed on *D. antillarum* by elevating the foot anteriorly, moving forward and falling upon the echinoid, and rasping a hole for feeding. The spines of *D. antillarum* do not penetrate the gastropod foot, but whether this is because of the toughness of the foot or because the spine tips are avoided is not clear. The triton *Charonia tritonis* feeds on *Heterocentrotus mammillatus*, ingesting the secondary spines presumably in the process of breaking into the test (Ebert 1971).

Asteroids usually apply only sufficient pressure to hold an echinoid prey immobile, and the spines would not be expected to play a role in preventing predation. Certainly the small spines of the clypeasteroids ingested by paxillosid asteroids are too small and dense to penetrate the gut wall. *Asterias vulgaris* applies its mouth to *Strongylocentrotus droebachiensis* and digests the test epithelium until the spines fall off (Himmelman and Steele 1971). *Oreaster reticulatus* mounts both the regular echinoid *Tripneustes ventricosus* and the irregular echinoid *Meoma ventricosa* and feeds by everting the stomach over the echinoid's test (Scheibling 1982). *Dermasterias imbricata* feeds on *Strongylocentrotus purpuratus*, which is partially ingested (Rosenthal and Chess 1972), and *Pycnopodia helianthoides* completely ingests it (Fisher 1928). However, *Meyenaster gelatinosus* can crush the test of *Loxechinus albus* without being deterred by the spines (Dayton *et al.* 1977). *Strongylocentrotus franciscanus* uses its long spines to fend off attacking *P. helianthoides* (Moitoza and Phillips 1979). When an arm of the asteroid comes down between the echinoid's spines, the spines close on the arm and pinch it. This generally results in successful deterral of the asteroid's attack.

A similar lack of applied pressure in feeding occurs with the anemone *Urticinopsis antarcticus*. Its principal prey is the echinoid *Sterechinus neumayeri* whose spines are ineffective against the gentle grasp of the anemone's tentacles (Dayton, Robilliard and Paine 1970). Debris held on the upper surface of the echinoid can be released to the anemone, allowing the individual to escape.

The spines could also function to decrease the effect of unidirectional impact loading (Strathmann 1981). This phenomenon does not result from the modes of predation on echinoids, but may occur in high-energy environments where objects may be moved about. *Strongylocentrotus droebachiensis* and *Loxechinus albus* can detach from the substratum in response to the presence of predators (Dayton 1975; Dayton *et al.* 1977), and certainly experience unidirectional impact loading. It is probable that the use of spines to protect the individual from dislocation by wedging the individual into crevices should be considered impact loading also, although of a sustained rather than abrupt nature. The spines of regular echinoids absorb energy or spread loads because they break or because the connective tissue attaching the spine to the test is stretched or torn (Strathmann 1981). This role of the spines is adaptive for echinoderms with rigid tests which rest directly on a hard substratum, and the appearance of spines on the flexible primitive echinoids may have been a necessary precondition for the development of a rigid test (Strathmann 1981). *Echinocyamus pusillus* usually uses its tube feet to hold pebbles and shell fragments against the spine tips (Nichols 1959). This would disperse applied forces over a greater number of spines (Telford 1985b).

3.5.4.4. Pedicellariae

The pedicellariae of echinoids are different from those of asteroids and have evolved independently. In contrast to the asteroids, the pedicellariae of echinoids have a stalk attached to a small tubercle on the test by a ball-and-socket joint like a spine. The head of a pedicellaria typically has three valves (or jaws), although two, four or five may occur. The length of the stalk is variable, and a tubular neck may separate the stalk and head and provide flexibility to the distal end. There are four major types: globiferous, ophicephalous, triphyllous, and tridentate (A. Agassiz and Clark 1907 to 1909; A.C. Campbell 1983; H.L. Clark 1912 to 1917; Cuénot 1948; Döderlein 1906; Jensen 1981, 1982; Mortensen 1928 to 1951). The occurrence of types of pedicellariae varies considerably with superorder and order, within genera, and even with individual. As with the spines, the structure of the pedicellariae implies a protective function, but little documentation of their effective action exists. Their small size and behaviour indicate that they function primarily to keep the test clean and to repel small organisms, although the poisonous pedicellariae may function against macropredators.

Globiferous pedicellariae (Figure 3.118 and 3.119) can be either large (up to 1 cm in height) or small (to 0.25 cm) (A. Agassiz and Clark 1907 to 1909; H.L. Clark 1912 to 1917). They have valves with poison glands or glandular tissue which vary in position on the pedicellaria (Jensen 1982). Glands or glandular tissue are found within the valve of the globiferous pedicellariae of cidaroids and on the back of the valves of echinids. The glands are found on the back of the valves of the diadematid *Centrostephanus*, but the heads of the globiferous pedicellariae in other diadematids degenerate, producing the so-called 'claviform' pedicellariae with glands around the stalk. Both the valves and the distal part of the stalks are embedded in glands in the spatangoids. Only the valves of the echinid globiferous pedicellariae are grooved. The stalk of the globiferous pedicellariae supports the head and usually lacks a neck.

Globiferous pedicellariae are absent in the clypeasteroids and laganoids, except for the genus *Fibularia* (Mortensen 1948b). This is generally attributed to the burrowing habit of individuals in this group, but globiferous pedicellariae do occur in some spatangoids. The correlation with burrowing does not explain the absence of globiferous pedicellariae in arbaciids. When present in spatangoids, they are usually on or beside the bare ventral ambulacrum. The large globiferous pedicellariae of cidaroids occur mainly around the primary spines of the upper surface and rarely in the

Figure 3.118: (1) Inner and (2) lateral views of a valve of a globiferous pedicellaria of *Psammechinus miliaris*. ad: Adductor muscle insertion, b: blade, f: foramen, k: keel, s: subterminal teeth, vt: venom tooth. (From Oldfield 1976)

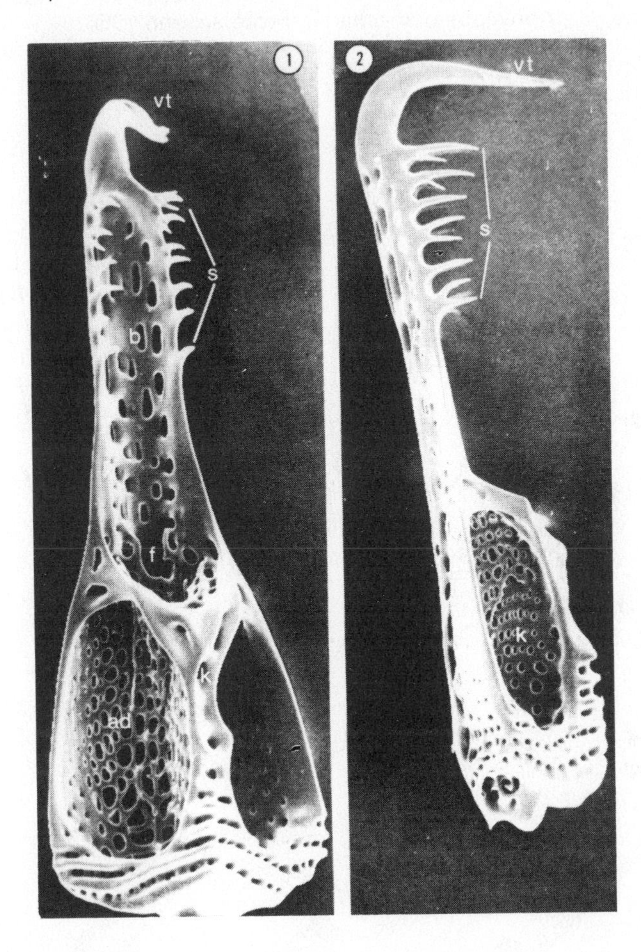

ambulacra, and may be absent in individuals or entire groups of species (A. Agassiz and Clark 1907 to 1909; H.L. Clark 1912 to 1917) whereas the small ones occur over the entire test except the ambulacra. They are particularly abundant around the mouth. Globiferous pedicellariae reach their greatest development in the Toxopneustidae and may form such a dense cover to the test and be of such colours that Mortensen (1943a) likened individuals to a flower garden.

The globiferous pedicellariae are stimulated to release venom by specific chemosensitivity (von Uexkull 1899; A.C. Campbell and Laverack 1968).

Figure 3.119: Pedicellariae of *Psammechinus miliaris* in (a) the quiescent state, (b) stimulated by an asteroid tube foot, and (c) stimulated mechanically. A: Primary spine, B: secondary spine, C: tube foot, D: globiferous pedicellaria, E: ophicephalous pedicellaria, F: tridentate pedicellaria, G: triphyllous pedicellaria. (From Jensen 1966)

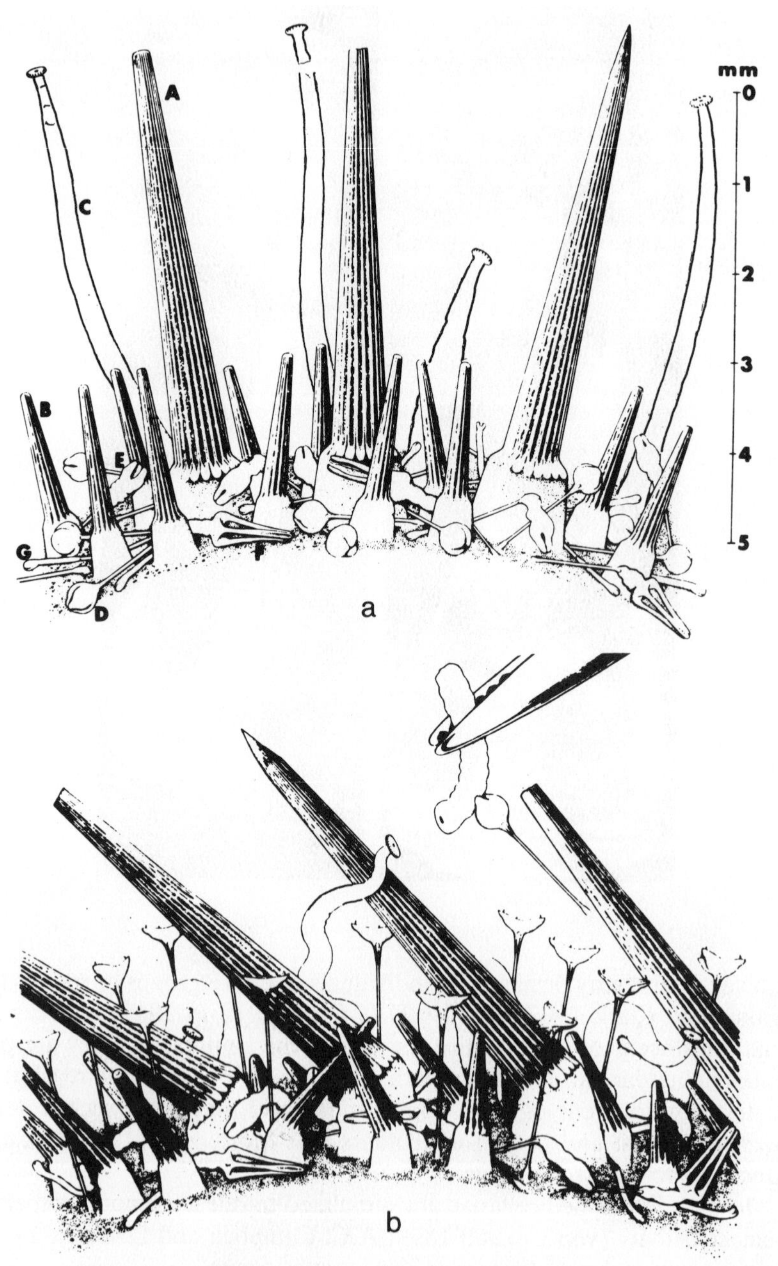

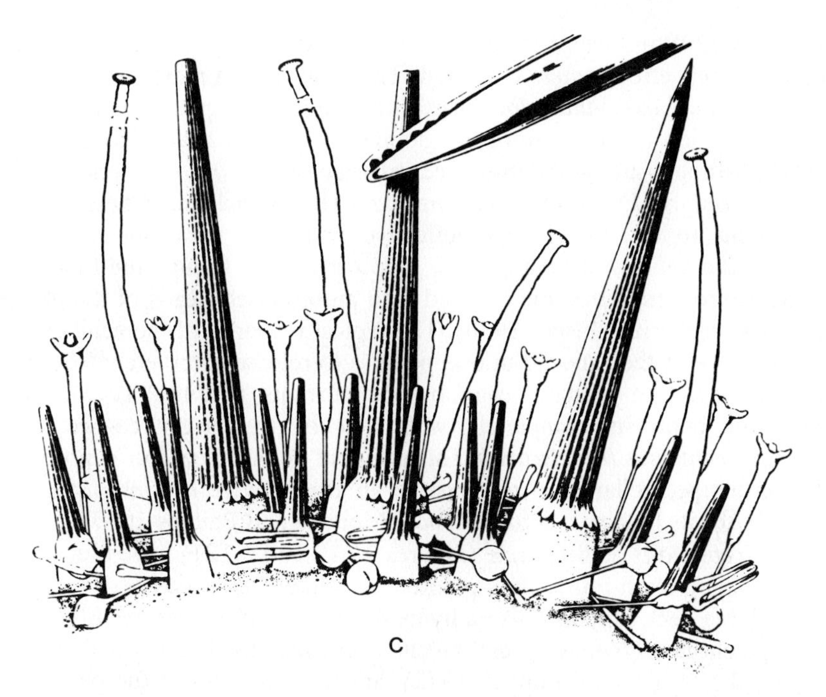

Commensals (planarians, polychaetes, crustaceans) on echinoid tests are not attacked by the pedicellariae but foreign representatives of these groups are attacked (A.C. Campbell 1983). The heads of globiferous pedicellariae autotomise after use (A.C. Campbell 1976). Crushing *Tripneustes ventricosus* in the field causes the heads of the globiferous pedicellariae of other individuals in the area to autotomise and to be released into the surrounding sea water (pers. obs.), presumably as a defence reaction.

Globiferous pedicellariae are used against predatory asteroids. *Paracentrotus lividus* attacked by *Marthasterias glacialis* does not use its spines in defence, instead bending them tangential to the test and raising the globiferous pedicellariae in a defence which can be successful (Prouho 1890). The globiferous pedicellariae of *Strongylocentrotus purpuratus* attack the tube feet of attacking *Astrometis sertulifera* (Jennings 1907). The effectiveness of the globiferous pedicellariae seems related to the response of different echinoid species to predatory asteroids. The approach of *M. glacialis* results in a fleeing response by *Strongylocentrotus droebachiensis*, while *Psammechinus miliaris* does not flee but flattens its spines, and defends itself with its globiferous pedicellariae (Figure 3.119). *Strongylocentrotus purpuratus* exposed to *Dermasterias imbricata* retracts the tube feet, flattens the spines, erects the globiferous pedicellariae and usually flees (Rosenthal and Chess 1972).

Although the globiferous pedicellariae attach to the asteroid's epidermis

(Figure 3.120) and cause localised retraction of papulae and retraction of the arm, their effectiveness is limited. The asteroid is a major predator on the echinoid. *Loxechinus albus* flees from *Meyenaster gelatinosus*, but can deter predation through use of the globiferous pedicellariae (Dayton *et al.* 1977). Although the globiferous pedicellariae of *S. purpuratus* respond to the presence of *Pycnopodia helianthoides* (Rosenthal and Chess 1972), they seem to be completely ineffective with this voracious predator. *Pycnopodia helianthoides* ingests *S. purpuratus* even though the tube feet of the asteroid may become covered with autotomised heads of the globiferous pedicellariae (Fisher 1930). The typical response of the echinoid to the presence of the asteroid in tide pools is to release completely from the substratum to be washed away by the currents (Dayton 1975). This behaviour contrasts completely with that of the larger co-occurring *Strongylocentrotus franciscanus.* This echinoid neither flees nor exposes its globiferous pedicellariae in response to an attack by *P. helianthoides*, but instead uses its long spines for defence (Moitoza and Phillips 1979).

Ophicephalous pedicellariae (Figures 3.119 and 3.121) have blunt jaws with grasping teeth rather than pointed ones like globiferous pedicellariae. The flexible neck functions like a hydroskeleton to orient the head (Hilgers and Splechtna 1976). A special mechanism exists for holding active prey (Figure 3.122) (A.C. Campbell 1972). Small teeth occur at the periphery of the jaws and larger ones in the centre of the gripping area. The forces set up by organisms held in the ophicephalous pedicellariae are transferred by

Figure 3.120: Globiferous pedicellariae of *Strongylocentrotus purpuratus* attached to the arms of the asteroid *Dermasterias imbricata* (from Rosenthal and Chess 1972)

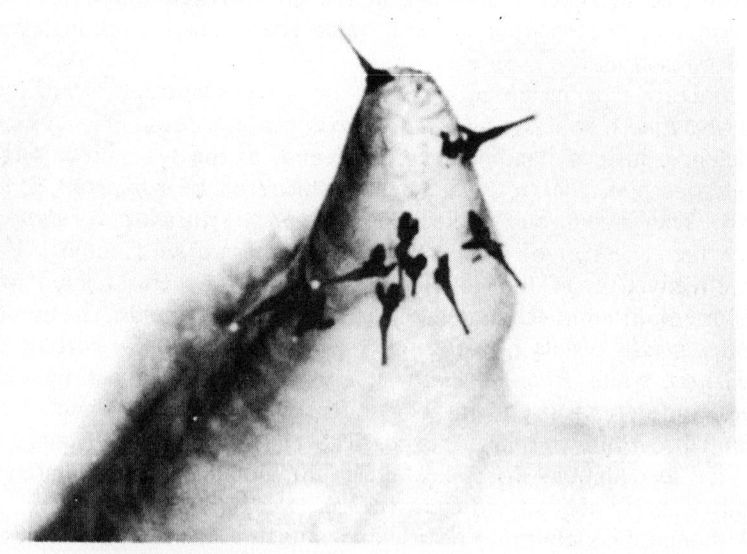

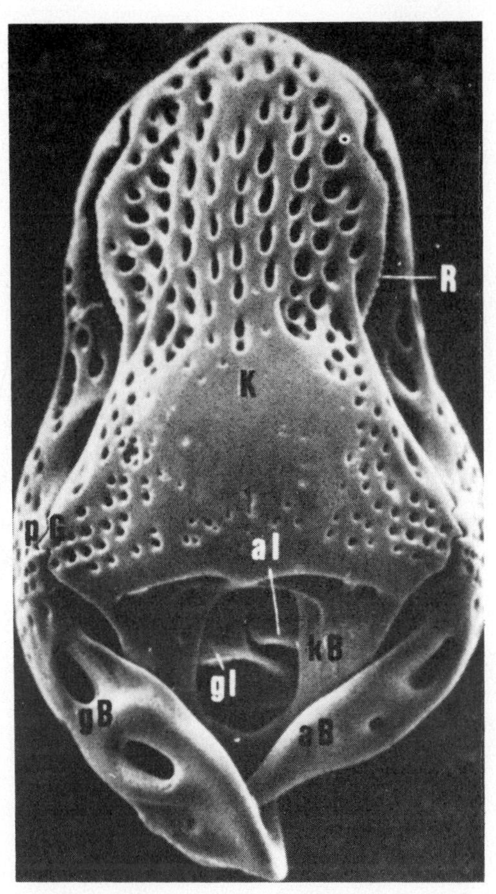

Figure 3.121: Ophicephalous pedicellaria of *Sphaerechinus granularis*. al: Inner handles of the valves with asymmetrical handle, K: valve, R: edge of valve, aB: asymmetrical handle, gB: large handle, gI: inner handle of valve with large handle, kB: small handle, pG: peripheral articulation. (From Hilgers and Splechtna 1976)

the collagen fibres which link the stalk ossicle with the valves back to the valves themselves. Thus the gripping power of the ophicephalous pedicellariae is not limited to the power of the adductor muscles alone.

Ophicephalous pedicellariae occur in all euechinoids, but not in cidaroids, although their presence is sporadic within groups (A. Agassiz and Clark 1907 to 1909; H.L. Clark 1912 to 1917; Mortensen 1928 to 1951). They are few in clypeasteroids, where they occur mainly on the upper surface. In young spatangoids they are often abundant, but they may be absent in adults. The ophicephalous pedicellariae are most abundant on the lower surface of spatangoids, and most are on the posterior part, particularly in the ambulacral region. The regular euechinoids have ophicephalous pedicellariae primarily around the mouth.

The ophicephalous pedicellariae of an active echinoid are fully extended and sweep across the surface of the test (A.C. Campbell and Laverack 1968). The jaws are held open, but shut rapidly if mechanically stimulated on their inner surface and remain shut for some time if any object is caught.

Figure 3.122: Lines of force acting in an ophicephalous pedicellaria when a captive organism held by the gripping teeth (q) tries to free itself from the test in direction l. Forces generated are transferred to the handles of the valves (r) by the collagen of the stem 2. The action of the handles (r) and the cog teeth (t) force the valves closer together in direction 3. (From A.C. Campbell 1972)

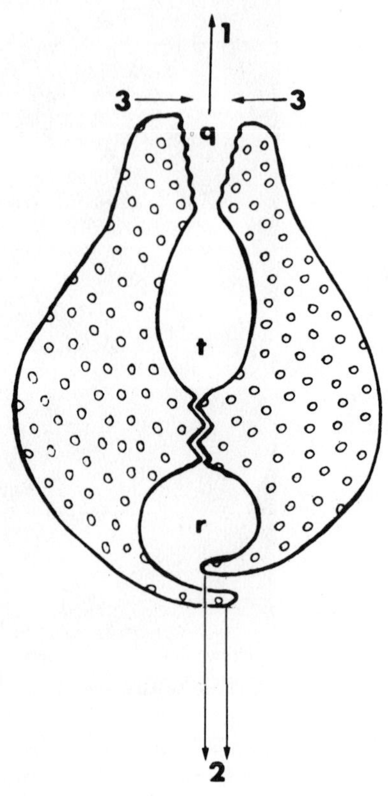

The ophicephalous pedicellariae are not stimulated by chemicals, in contrast to the globiferous pedicellariae. They remain quiescent in the presence of predatory asteroids but are extended to their full length in response to mechanical stimulation (Figure 3.119).

The *triphyllous* pedicellariae (Figure 3.119) are small, ranging between 0.1 and 0.5 mm in length (A. Agassiz and Clark 1907 to 1909; H.L. Clark 1912 to 1917). In contrast to the other types of pedicellariae, their short, broad jaws do not meet distally and lack teeth. The triphyllous pedicellariae are lacking in cidaroids, but are present in all other groups. They are scarce in scutellids and are reduced to small forms which are bivalve rather than trivalve (Mortensen 1948b). The triphyllous pedicellariae differ from the other types in being continuously active, with seemingly spontaneous jaw movements which are increased by mechanical but not chemical stimulation (A.C. Campbell 1973). The structure and behaviour of the triphyllous pedicellariae indicate that they function to keep the test clean. No obvious correlation exists between the occurrence of triphyllous pedicellariae and the biology of an echinoid, and there is no indication that those echinoids that lack the triphyllous pedicellariae suffer from the deprivation.

Tridentate pedicellariae typically have three valves which differ in shape from that of ophicephalous pedicellariae but seem to have the same basic features for functioning (Figure 3.119) (A.C. Campbell 1972; Hilgers and Splechtna 1976; Jensen 1982). They typically have a broad proximal part and a long jaw. Only the tips of the teeth grip in contrast to the ophicephalous and triphyllous pedicellariae. Although the tridentate pedicellariae may be short (0.2 mm), they may be much longer than any of the other types, reaching more than 5 mm in length (A. Agassiz and Clark 1907 to 1909, H.L. Clark 1912 to 1917). Often the stalk is short, and most of the length is the long flexible neck. The neck functions as a hydroskeleton (Chia 1969) like that of the ophicephalous pedicellariae.

Tridentate pedicellariae occur in all extant groups of echinoids except some groups of cidaroids and arbaciids (A. Agassiz and Clark 1907 to 1909; H.L. Clark 1912 to 1917; Jensen 1982). Their abundance in the echinoids seems to be complementary to that of globiferous pedicellariae. They have a uniform distribution over the body of spatangoids, but are mainly on the lower surface of clypeasteroids.

Tridentate pedicellariae like ophicephalous pedicellariae respond to mechanical but not chemical stimuli (A.C. Campbell and Laverack 1968; A.C. Campbell 1972). Usually quiescent on the surface of the test, they are erected and the jaws are opened when stimulated. Commensals on the test do not activate the tridentate pedicellariae (A.C. Campbell 1973). It seems that the tridentate pedicellariae of regular echinoids are stimulated by contacts with swimming organisms (A.C. Campbell 1973; Jensen 1966). The tridentate pedicellariae found in the anterior ambulacrum of the spatangoid *Moira atropos* actively help the spines to move particles down the groove, but are not very efficient in action (Chesher 1963).

3.5.4.5. Spicules

Spicules can occur in the internal tissues of echinoids, as in the gut and gonads. Small, coarse, semi-lunar spicules stud the gut wall of *Salmaciella dussumieri.* A complete mail of spicules occurs in the gut and gonads of *Amblypneustes grandis* (Figure 3.123). The occurrence of the spicules is such that they probably do not provide support as they do in the tube feet. Whether they are adaptive by decreasing the energetic density of the organs or by damaging the gut of a predator, or whether they simply result from a relaxation of the control of gene expression is unknown.

Figure 3.123: A: Spicules from the (a, b, c) gonad and (d) gut wall of *Amblypneustes grandis*. B: Spicules from the (a) gut wall, (b) tube feet, (c) buccal membrane, and (d) gonad of *Salmaciella dussumieri*. (From Mortensen 1943a)

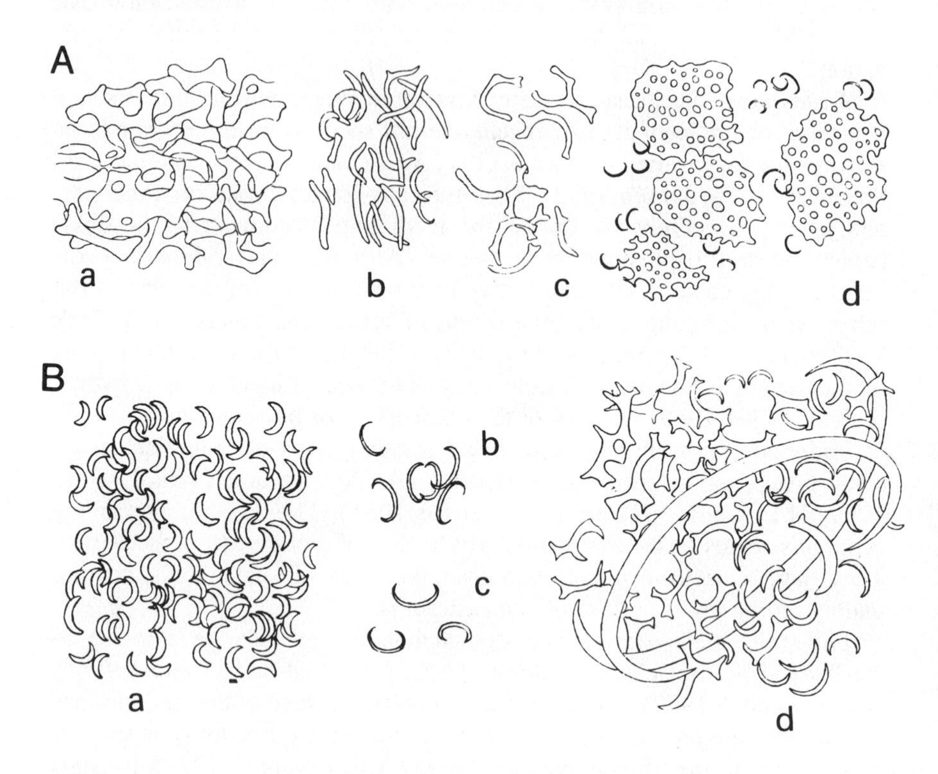

3.5.5. Holothuroidea

Few morphological features of holothuroids can be related directly to protection from predation. Except for the feeding tentacles and pedicels, there are no extensions of the body that are vulnerable to predation. Consequently, the individual holothuroid must be ingested whole or a portion of the body must be bitten off. The small species *Cucumaria miniata* is ingested intact by asteroids (Mauzey *et al.* 1968).

The lack of reports of predators biting off a portion of the holothuroid body might simply be the result of the fact that the holothuroids are not a preferred food because of the presence of toxins in the body or because of the presence of spicules and high amounts of refractory structural organic material in the body wall which makes the energetic return to the predator minimal, rather than being a protective structural characteristic of the holothuroids.

The spicules in the body wall of most species are usually microscopic and would seem ineffective protective elements although they may have sharp points. The anchors characteristic of the body wall of synaptids (Figure 3.84) suture wounds in the body wall of *Opheodesoma spectabilis* (Fankboner 1979). A wounded individual folds its body towards the wound, causing the anterior and posterior edges to press together. Peristaltic muscular waves from both anterior and posterior directions produce additional force on the edges of the wound. The pointed hooks of the anchors set in the opposite walls and suture the wound.

Some ossicles possess spines but these are processes from the ossicle and not separate elements in contrast to those of asteroids and echinoids. These spines are never large. Some elasipods such as *Deima* have crowded, overlapping ossicles in the body wall, papillae and pedicels to form a test which is rigid, rough and very brittle (Figure 3.124) (Théel 1882). Because of the

Figure 3.124: (A) Overlapping ossicles of the body wall of *Deima validum*, and (B) lateral view of the individual showing outline of ossicles in the body wall (from Théel 1882)

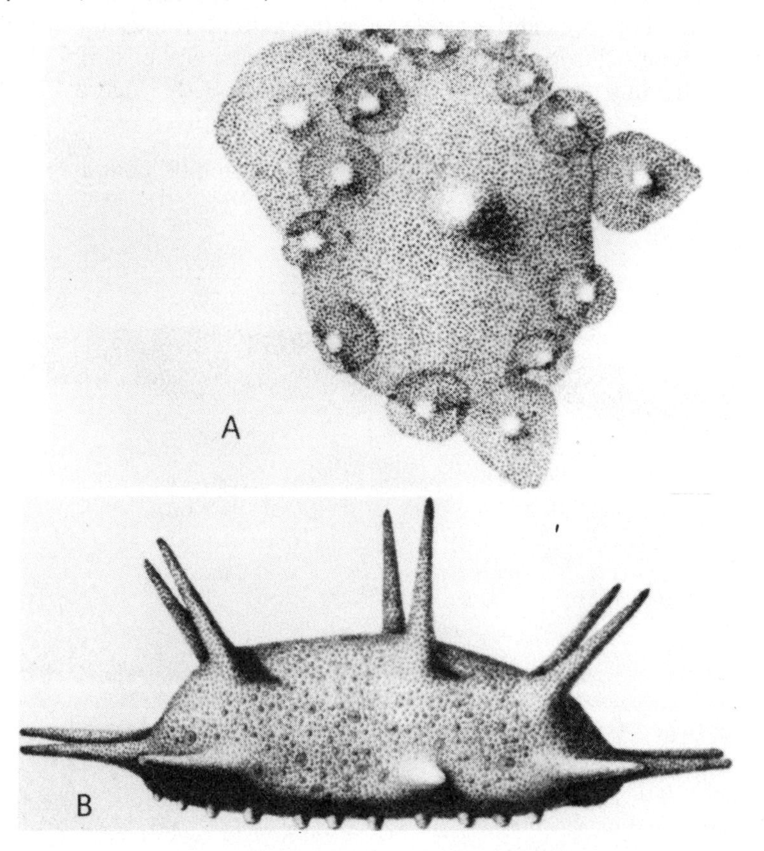

fragility, this skeletal development may be more related to the requirements of body form and of the water vascular system than to protection from predation. The larger, imbricating plates which give a body form to psolids (Figure 2.51) are also very fragile, and a role in protection from predation has not been documented. Enlarged ossicles, the anal teeth, seemingly protect the anal opening of actinopygids, but there are no reports on predatory entry by the anus for those holothuroids which lack them.

For most holothuroid species, the connective tissue comprising the greater part of the mass of the body wall is the primary deterrent to predation. The resistance it possesses even in the relaxed state is exaggerated when the body contracts. The contracted turgid body not only makes a more compact body wall which increases the difficulty for a predator to obtain a bite, but increases stiffness of the body wall due to the change in state of non-contractile proteins (Figure 3.125) (R.S.H. Stott, Hepburn, Joffe and Heffron 1974).

The tentacles would seem to be the part of the holothuroid body most vulnerable to predation, particularly in dendrochirotids in which the tentacles are both long and extended into the water. It is thus not surprising that in dendrochirotids the body wall immediately posterior to the tentacles, the introvert, can be drawn down into the body (Figure 3.126). In

Figure 3.125: *Holothuria scabra* in the (A) relaxed, and (B) contracted states (from Stott *et al.* 1974)

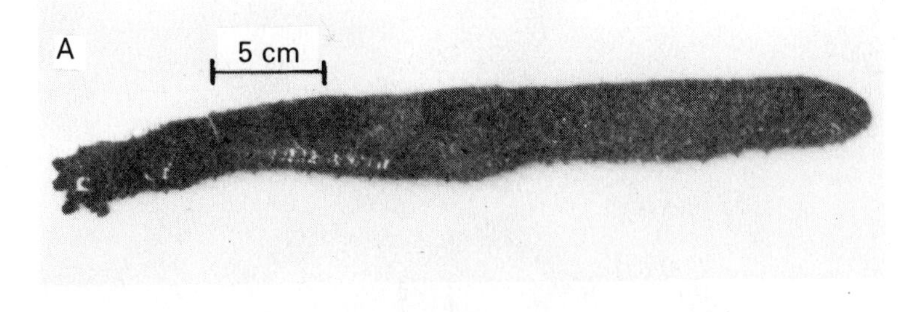

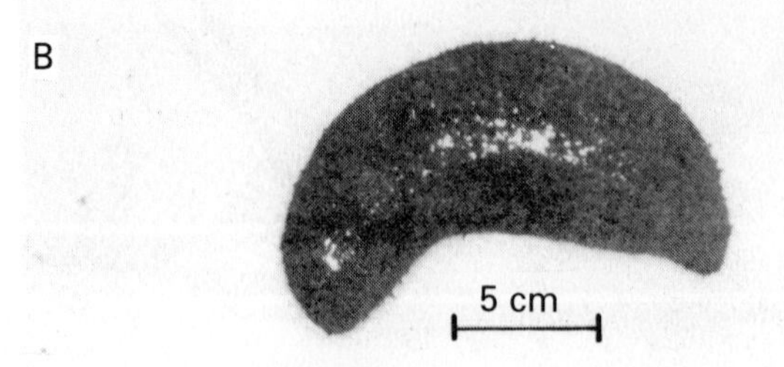

elasipods and aspidochirotids, a more or less developed extension of the body wall usually occurs at the oral end into which the tentacles can be retracted. This is variable even in the same family, because the tentacles of *Oneirophanta* are not retractable into its oral cavity whereas those of *Deima* are (Théel 1886). There is no structural feature to protect the retracted tentacles of apodids or molpadids. Because of their lack of spines at the ambulacra, holothuroids are unique in the echinoderms in the absence of any structural protection for their tube feet.

The most unusual protective devices of holothuroids are the Cuvierian tubules found in some species of the aspidochirotids *Holothuria* and *Actinopyga* (Figure 3.126). This is correlated with the postulated higher degree of fish predation in the lower latitudes (Bakus 1973, 1974). The Cuvierian tubules are elaborations from the gut, being attached to the stem or bases of the respiratory trees. Individuals artificially stimulated raise the

Figure 3.126: *Holothuria forskali* dissected from the dorsal side. o: Perforated cloacal wall; oag: left respiratory tree; t: Cuvierian tubules attached to the stem of the respiratory tree; t′: Cuvierian tubules protruded through the anus; vb: basal vesicles of the Cuvierian tubules. (From Cuénot 1948)

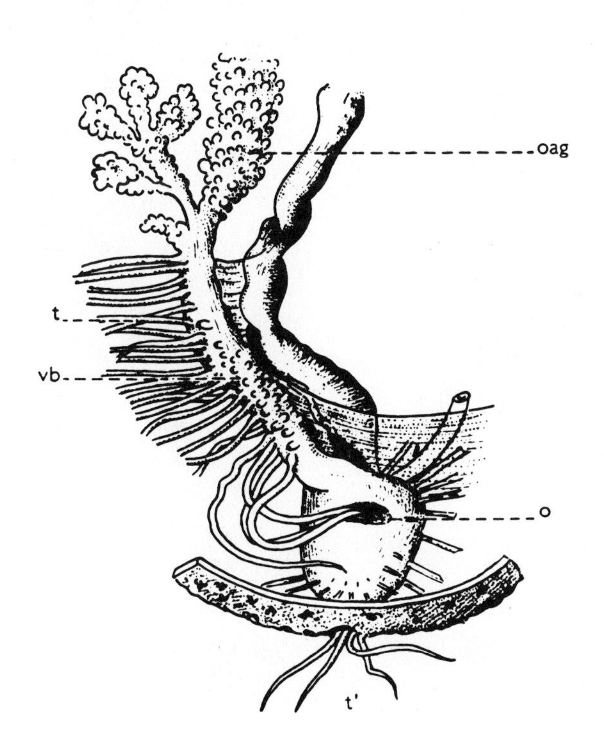

anus, point it in the direction of the stimulation, and contract the body wall to expel the tubules through the anal opening (Bakus 1968; Hyman 1955). The Cuvierian tubules of *Holothuria* species are sticky whereas those of *Actinopyga* are not (Hyman 1955). Toxic properties have been demonstrated for the Cuvierian tubules of *Actinopyga agassizi* (Nigrelli and Jakowska 1960) and *Holothuria difficilis* (Bakus 1968).

4

Reproduction

4.1. ENERGETICS OF REPRODUCTION

As with the acquisition of resources, the use of resources for reproduction can be analysed in terms of bioenergetics. This approach allows the comparison of acquisition of resources with their use by employing the same energy units. It also allows a comparison of the ways in which energy resources are used for reproductive activities. The various ways in which this is done give insight into those factors that determine fitness.

Along with survival, reproduction is a component of fitness. As such, and according to the maximisation principle, reproduction should be maximised (Calow 1984a, b). Exactly how reproduction is maximised is under constraints imposed both by the environment and by the characteristics of the organism. Spatial and temporal variations in the physical and biological features of the environment can affect the acquisition of nutrients and survivorship. Characteristics of the organism determine how these nutrients can be used.

At least some acquisition of nutrients for growth is essential. Otherwise, each successive reproduction would continue to decrease the mass of the offspring. Subsequent somatic growth is considered adaptive and the rate and limit are important variables among species. Subsequent somatic growth also has implications for reproduction beyond simply affecting the reproductive output.

Both somatic growth and reproduction require nutrients. The rate at which nutrients can be acquired is limited by the characteristics of the organism (intrinsic limitation), and may or may not be limited below this level by the environment as well (extrinsic limitation). A limitation of the acquisition of nutrients is implicit in most theories of life history. Because of the limitation in resource acquisition, trade-offs between somatic growth and reproduction are assumed. They should be those that maximise fitness.

Nutrients are also required for maintenance activities for survival of the individual. This includes physiological processes at both the organismal and sub-organismal levels as discussed in Chapter 3. Maintenance activities constitute a third use of resources, and trade-offs between the use of nutrient resources for survival, growth and reproduction are assumed to occur in such a way as to maximise fitness.

Echinoderm species vary greatly in size and, although few data are available, in their rates of growth and survival. All seem to have a sigmoidal growth curve although the growth period and ultimate size vary. These characteristics can also vary greatly within a species, growth and size being determined by food availability.

Most echinoderms seem to become reproductively capable at least by the second year of life, and the reproductive output continues to increase with an increase in size. Species may be either iteroparous or semelparous. Survival may be for only one to a few years, but some species are estimated to live twenty to thirty years. Constraints on reproductive output seem associated with the small perivisceral coelomic cavity in crinoids, ophiuroids and some echinoids.

Trade-offs between allocation of resources for somatic growth and reproduction are apparent with extrinsic limitation of resources, but not with intrinsic limitation. Growth and reproduction can occur simultaneously. Trade-offs between allocation of resources for maintenance activities and for somatic growth and reproduction are little known. High temperatures can increase use of resources to such an extent that somatic growth is affected. An effect of reproduction on survival indicating a trade-off in the allocation of resources between maintenance and reproduction has been postulated (Ebert 1975, 1982), but not demonstrated.

4.2. SEXUAL REPRODUCTION

The vast majority of the echinoderms have sexual reproduction only and are gonochoric with male and females of equal size. These gonochoric species typically have a sex ratio of one (Table 4.1) and are rarely hermaphroditic (Table 4.2). They produce small eggs which are broadcast into the sea and fertilised externally to develop as planktotrophic larvae (Emlet, McEdward and Strathmann 1987) or are associated with brooding. Among those characteristics of body form that affect the quantity of gametes that can be produced are the location and space available for gonadal development and, in the case of brooding species, for the developing embryos and juveniles. Body size itself is an extremely important character affecting both the reproductive output and the relative amount of material and energy used for reproduction (*reproductive index* and *reproductive effort*).

Table 4.1: Sex ratios in gonochoric echinoderms

Species	Ratio (male:female)	Reference
Echinoidea		
Arbacia punctulata	1.03:1	Shapiro 1935
Cassidulus caribaearum	1:6, less than 30 mm length	Gladfelter 1978
Dendraster excentricus	1:1	Timko 1975
Echinocardium cordatum	1:0.98	Moore 1935
Echinolampas crassa	1:0.89	Thum and Allen 1976
Echinometra mathaei	1:0.83[a]	Pearse 1969
	1:1.09	Pearse and Phillips 1968
Eucidaris tribuloides	1:1.16	McPherson 1968
Evechinus chloroticus	1:0.84[a]	Dix 1970
	1:0.96	
Heliocidaris erythrogramma	1:1.34[a]	Dix 1977
Strongylocentrotus purpuratus	1:0.68[a]	Gonor 1973
	1:0.80[a]	
	1:1.06	
Strongylocentrotus purpuratus	1:1.05	S.L. Baker 1973
Temnopleurus toreumaticus	1:1.018	Ikeda 1931
Tripneustes ventricosus	1:1.5[a], 80–100 mm HD	McPherson 1965
	1:1, 45–79 mm HD	
Asteroidea		
Acanthaster planci	1:1.11	Conand 1985
Asterias amurensis	1:0.93	Kim 1968
Asterias rubens	1:2.41	Vevers 1949
	1:0.98	
Asterias rubens	1:0.85, 3.11 mm R	Vevers 1949
	1:2.33, 19.61 mm R	
Asterias rubens	1:0.88	Jangoux and Vloebergh 1973
Asterias rubens	1:1.17	Pelseneer 1926 in Lowe 1978
Asterias vulgaris	1:1.69[a]	Lowe 1978
Astrostole scabra	1:0.82	Town 1979
Coscinasterias calamaria	1:1	Crump and Barker 1985
	1:4	
	55:1	
Echinaster sepositus	1:0.87	Delavault 1960
Henricia perforata	1:0.37[a]	B.N. Rasmussen 1965
Henricia pertusa	1:0.89	B.N. Rasmussen 1965
Henricia sanguinolenta	1:1.04	B.N. Rasmussen 1965
Leptasterias hexactis	1:1	Menge 1970
Leptasterias tenera	1:1, 10 to 14 mm R	Hendler and Franz 1982
	1:1.8, 25 to 29 mm R	
Luidia clathrata	1:1.13	J.M. Lawrence unpubl.
Oreaster reticulatus	1:0.87	Scheibling 1981a
Patiria miniata	1:1.4	J.M. Lawrence and A.L. Lawrence unpubl.
Pisaster ochraceus	1:1.03	J.M. Lawrence and A.L. Lawrence unpubl.
Ophiuroidea		
Amphioplus thromboides	1:0.79	Stancyk 1970
Ophioderma brevispinum	1:0.82	Stancyk 1970

Table 4.1: *continued*

Species	Ratio (male:female)	Reference
Ophiolepis elegans	1:0.93	Stancyk 1970
Ophiophragmus filograneus	1:1.09	Stancyk 1970
Ophiothrix angulata	1:0.96	Stancyk 1970
Holothuroidea		
Parastichopus californicus	1:1	Cameron and Fankboner 1986
Holothuria atra	1:8.5, 100 g wet wt. 1:0.07, 1000 g wet wt.	Harriott 1982
Microthele fuscogilva	1:0.88	Conand 1981
Microthele nobilis	1:0.98	Conand 1981
Thelenota ananas	1:0.93	Conand 1981

[a]Ratios significantly different from 1:1. Other values not significantly different or uncalculable. If more than one ratio is given, different populations or different sizes were studied.

4.2.1. Body form

No direct evidence indicates the presence of the gonads in the arms or pinnules or in the theca of fossil crinoids (Ubaghs 1978). In recent crinoids, gonads are usually located in the specialised genital pinnules on the arms, more rarely in the arm itself, but never in the theca (Ubaghs 1978). Gislén (1924) concluded that the pinnules developed for feeding and that the gonads appeared there subsequently. The gonads of some recent forms (*Metacrinus, Notocrinus, Comatula*) occur in the arms and not in the pinnules. Similarly, ontogeny indicates a late migration of the gonads into the arms and pinnules. Thus originally the gonads may have been in the voluminous anal sac, separated from the other viscera in the theca. N.G. Lane (1984) considered the ultimate shift of the gonads to the arms to be a response to the vulnerability of the energy-rich anal sac to predators. The trade-off would have been the decrease in attractiveness of the gonads as food for predators because of their dispersion among the calcareous arms and pinnules and the decrease in feeding efficiency.

Several modifications exist for the use of the arms and pinnules for reproductive and feeding functions. Although gonads appear on pinnules far out on the arms in most comatulids, those on distal pinnules are less developed and gradually do not occur (Gislén 1924). In certain types, the gonads are restricted to proximal genital pinnules (Figures 2.3, 4.1). This specialisation should decrease the interference of gonadal development with the function of the feeding pinnules and the accessibility of the gonads to predators. Development of massive gonads in the arms would not be compatible with suspension feeding so the gonads are numerous and small. The comasterids are bilaterally symmetrical because the mouth is near the

Table 4.2: Incidence of hermaphroditism in gonochoric echinoderms

Species	Incidence	Reference
Echinoidea		
Arbacia punctulata	1 of 2350 individuals	Shapiro 1935
Echinarachnius parma	1 of 2000 individuals	Herold 1969
Echinocardium cordatum	1 of 358 individuals	Moore 1935
Echinus esculentus	1 of 3000 individuals	Moore 1932
Evechinus chloroticus	None in 4670 individuals	Dix 1970
Lytechinus variegatus	As high as 15.1%	Moore, Jutare, Bauer and Jones 1963
Lytechinus variegatus	None in 952 individuals	Brookbank 1968
Mellita quinquiesperforata	None in 446 individuals	Brookbank 1968
Strongylocentrotus intermedius and *S. nudus*	None in 1500 individuals	Fuji 1960
Strongylocentrotus purpuratus	20 of 10000 individuals	Tyler in Boolootian and Moore 1956
Strongylocentrotus purpuratus	None in 1354 individuals in one locality; 2 of 1354 individuals in another locality	Gonor 1973
Tripneustes gratilla	1 of 550 individuals	Kidron, Fishelson and Moau 1972
Tripneustes ventricosus	As high as 27%	Moore, Jutare, Jones, McPherson and Roper 1963
Ophiuroidea		
Ophiocten sericeum	2 of 749 individuals	Thorson 1934
Microphiopholis gracillima	2 of 1187 individuals	Singletary 1970
Asteroidea		
Echinaster sp., type II	1 of 177 individuals	R.E. Scheibling and J.M. Lawrence (unpubl.)
Echinaster sepositus	As high as 23%; type and incidence varies with season	Delavault 1960
Echinaster sepositus	5 of 144 individuals; no difference in incidence between spring and autumn	Cognetti and Delavault 1960
Luidia clathrata	1 of 2000 individuals	P.F. Dehn (pers. comm.)

margin of the tegmen. The anterior arms near the mouth are specialised for feeding whereas the two posterior arms opposite are shorter and thicker and may lack ambulacra (P.H. Carpenter 1884; Gislén 1924). Although shorter and with fewer total pinnules, more of the pinnules have gonads which are also more developed than those on the feeding arms.

The location of the gonads in the arms means that an increase in the number of arms provides the potential for an increase in the number of gonads. The number of genital pinnules increases with the size of the individual. The number also varies with the number of arms, which can reach

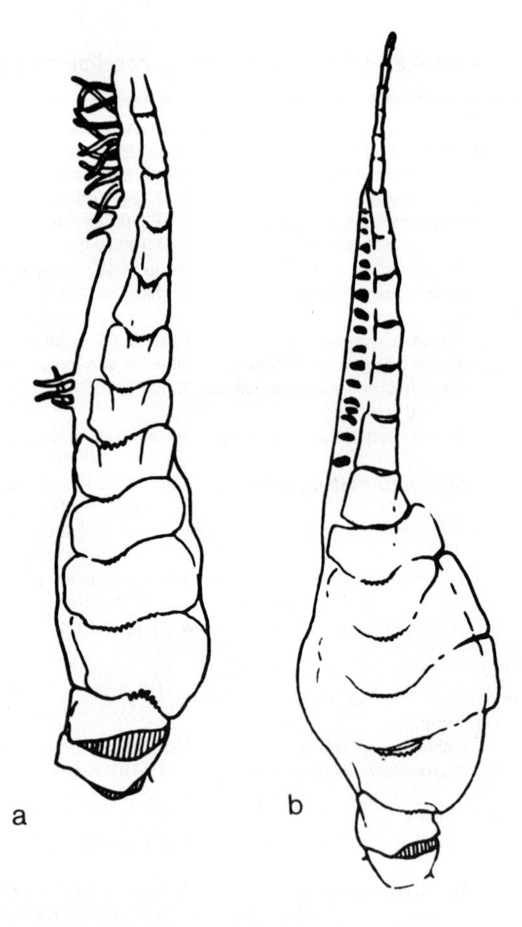

Figure 4.1: Genital pinnule of (a) a male, and (b) a female *Isometra horrida* (from John 1938)

several hundred in some species. The number of arms is usually considered in terms of feeding, but obviously affects reproduction as well.

The genital pinnules containing ovaries are typically larger than those containing testes (John 1938). If the number of genital pinnules is the same in both males and females, the reproductive effort of the latter would be greater.

In ophiuroids, the gonads are attached to the coelomic wall of the bursae. The gonads are restricted to the coelomic cavity of the disc of most ophiuroids, but extend into the arms of the primitive *Ophiocanops* and in asteroschematids and euryalids (Spencer and Wright 1966). Space for gonadal development is thus limited to the size of the disc, but the disc may swell in response to gonadal development (Lyman 1882). Mortensen (1933) attributed the occurrence of gonads in the widened dorsal spaces in the arms of *Ophiocanops fugiens* to the lack of space in the diminutive disc. As in crinoids, the functional role of the ophiuroid arm in locomotion and feeding requires it to be slender so that an increase in volume to accommodate the gonads is not possible as in asteroids.

In asteroids, the gonads may be restricted to the interradial portion of the broad disc or the proximal part of the arm, but generally extends into the arms. It is probable that the gonads were originally interradial and five in number. The gonads on either side of the ambulacra could still be considered interradial. The presence of the gonads in the arms is possible because of the broadening of the arm associated with the skeletal support in the body wall.

The arm form of most asteroids (broad with considerable coelomic space) contrasts with that of crinoids and ophiuroids (slender with minimal coelomic space). This can be correlated with the potential for development of organs involved in reproduction and nutrient storage. In asterinids such as *Asterina* with short arms and a large interradial angle, development of the gonads is restricted to the interradial portion of the arm immediately adjacent to the disc. The gonad here is a relatively compact tuft. In asteriids such as *Asterias* or *Pisaster*, the gonads are found in the long, slender arms which are set off from the disc. The gonads are serial in the Luidiids and some genera of several other families, and discrete gonads with separate gonoducts are found all along the arms. In several *Leptychaster* species the gonads are serial in males but not in females. In *Freyella*, the distal parts of the arms are narrow and specialised for feeding (Figure. 2.28). Here the gonads are restricted to the enlarged proximal portion of the arms. This regional specialisation shows again the relation between arm form (broad vs. slender) and function (reproduction and nutrient storage vs. feeding). The arms of multiarmed asteroids are usually slender (e.g. *Solaster, Pycnopodia, Acanthaster*).

The increase in number of arms results in an increase in gonad number and the potential for an increase in reproductive output. The gonads and the pyloric caeca both occupy space in the arm coelom. However, without a concommitant increase in arm length, addition of arms results in a decrease in the volume of the arms rather than an increase (Lawrence 1987b). *Luidia clathrata*, with five arms, has a calculated arm volume greater than that of similar-sized *Luidia senegalensis*, with nine arms. This is because the volume of the arm is associated with the square of the radius of the arm base and only linearly with the length. Because the increase in arm number increases the amount of material and energy necessary for the body wall without an increase in reproductive capacity, either the multiarmed condition is maladaptive or it has a trade-off with another function, possibly the capacity to feed.

The reciprocal relation in the sizes of the gonads and the pyloric caeca typically results in a lack of restriction by the pyloric caeca on gonadal development. However, when individuals are well fed, the pyloric caeca may not decrease in size as the gonads develop (Lawrence and Lane 1982). This illustrates the difference between intrinsic and extrinisic limitation on nutrient acquisition (Lawrence 1985). A minimal amount of food is necessary for maintenance. Below this threshold, gonadal growth does not occur. Above it, gondal growth can occur and somatic growth is limited (extrinsic limitation). Above a second threshold, both gonadal and somatic growth can occur to the extent limited by the ability to acquire food (intrinsic limitation).

Depot storage is considered to be a means of transferring productivity acquired in one place or time for use at another when conditions for reproduction are good but not by direct acquisition of nutrients (Pianka 1976). This does not really answer the question as to why the nutrients are not deposited directly into gametes. It is possible that nutrients reserves are more readily used for other purposes if stored in a somatic reserve, but gonads of echinoderms are resorbed when individuals are starved (Lawrence 1975; Lawrence and Lane 1982). It is also possible that gametes cannot be stored for long periods of time, and that they must be produced close to the spawning period. Prolonged storage of gametes can occur in echinoids (J.S. Pearse 1969).

In echinoids and holothuroids, the gonads develop in the large perivisceral coelom along with the gut. The test is not distendible in echinoids as is the disc in ophiuroids or the body wall of holothuroids. Regular echinoids generally have low rates of feeding when the gonads are most developed (Lawrence 1975), and Leighton (1968) suggested that this is due to the physical restriction of space at this time. Five gonads, one in each interradius, occur in regular echinoids. In irregular echinoids, an evolutionary posterior migration of the anus has resulted in the loss of one or two gonads. However, this does not seem to be a restriction on the degree of development (reproductive output) of the gonads in these echinoids. Compression of the body form and the development of extensive internal supports in clypeasteroids can reduce the space available for gonadal development (Figure 3.106).

4.2.2. Body size

Simplistically, the larger the individual, the greater the reproductive output possible. For each class the range in body size is quite large. Consequently it seems that factors other than *individual* reproductive output have determined body size, and that the reproductive characteristics of a species are functional for its specific body size.

The *reproductive output* (the amount of gametes produced), the *reproductive index* (the amount of gametes produced per unit body weight), and the *reproductive effort* (the amount of gametes produced per total amount of production) are all of interest. The interspecific variation in body size results in interspecific variation in reproductive output and reproductive index (Tables 4.3A, B). Within the same functional morphological type, the reproductive output shows the expected increase with increase in body size. However, the reproductive index is more variable. It is not clear whether this ratio represents the ability of a particular body form to support gonadal production, or simply an empirical value. For example, echinoids vary greatly in the numbers and sizes of their spines. Spines contribute little to the body weight in *Lytechinus* in contrast to their contribution in *Heterocentrotus.* Because of this, the reproductive index calculated on the basis of a weight is less in *Heterocentrotus* than it would be if the spines were not so well developed. This may be the reason for the lack of a high correlation between body weight and reproductive index in echinoids.

Within a species, the relation between size and reproductive output is more direct. A certain minimal body size is necessary before an individual becomes reproductive (Table 4.4). This may be the body size that is necessary for the individual to be sufficiently functional in terms of feeding to support gonadal growth. The attainment of adult size occurs exponentially and results from allocating most of the acquired resources to growth rather than to reproduction. The strategy seems to be to provide for an ever-increasing reproductive capacity until the adult size is reached.

This would be appropriate for iteroparous species, but semelparous species should grow rapidly to maximise reproductive output for their single spawning. Thus the ophiuroid *Amphiura filiformis* and the spatangoid *Brissopsis lyrifera* are fast-growing, short-lived semelparous species, whereas *Amphiura chiajei*, *Leptopentacta* (= *Cucumaria*) *elongata* and *Echinocardium cordatum* are long-lived, slow-growing iteroparous species (Buchanan 1967). However, survival of *E. cordatum* in populations that grow rapidly and reproduce is not less than that of individuals in populations that grow slowly and do not reproduce because of low nutrient availability. This indicates that allocation of large amounts of nutrients to growth and reproduction does not necessarily result in a decrease in longevity.

Reproductive effort in terms of the relative amount of energy used for gonodal and somatic production has been calculated for a few echinoid species (Table 4.5). As with the reproductive index, the reproductive effort generally increases with body size, and can reach levels above 60% but never approaches 100%. Allocation of energy to somatic growth continues in even the largest individuals.

A major strategy of echinoderms to cope with extrinsic food limitation involves their ability to cease growth. *Asterias rubens* can remain as tiny individuals for months in a 'waiting stage' before assuming a growth phase that leads to a size at which reproduction occurs (Nauen 1978). Some populations of *Echinocardium cordatum* may persist for years as small individuals without ever becoming reproductive, apparently due to a low acquisition of nutrients (Buchanan 1967).

Table 4.3A: Reproductive output and reproductive index of echinoderms

Species	Wet body weight (g)	Reproductive output (g wet wt)			Reproductive index (g wet gonad wt × 100 / g wet body wt)			Reference
		M	F	M/F	M	F	M/F	
Asteroidea								
Order Valvatida								
Oreaster reticulatus	1000	80	160		8	16		Scheibling 1979
Order Spinulosida								
Acanthaster planci	1500	180	225		12	17		Conand 1985
Echinaster type I	1.9	0.04	0.08		5.1	10.6		Scheibling and Lawrence 1982
	3.7	0.29	0.34		14.4	16.7		
	5.8	0.40	0.66		15.9	25.1		
Echinaster type II	0.32	0.03	0.04		16.2	20.9		Scheibling and Lawrence 1982
	0.64	0.09	0.11		25.8	29.7		
	1.81	0.35	0.38		33.3	32.5		
Order Forcipulatida								
Asterias rubens	30	5.7	10.5		19	35		von Bismarck 1959
Astrostole scabra	500	95	95		19	19		Town 1980
Leptasterias hexactis	8	1.5	0.40		19	5		Menge 1975
Pisaster giganteus	250	12.5	7.5		5	3		Farmanfarmaian, Giese, Boolootian and Bennett 1958
Pisaster brevispinus	500	40	30		8	6		Farmanfarmaian *et al.* 1958
Pisaster ochraceus	500	60	85		12	17		Farmanfarmaian *et al.* 1958
Echinoidea								
Superorder Echinacea								
Order Arbacioida								
Arbacia lixula	50			3.5			7	Fenaux, Malara, Cellario, Charra and Palazzoli 1977

Order Echinoida								
Allocentrotus fragilis	75			4.5			6	Giese 1961
Strongylocentrotus droebachiensis	50			7			14	Percy 1971
Strongylocentrotus purpuratus	50			11.5			23	Lawrence 1966
Order Phymostomatoida								
Stomopneustes variolaris	450			77			17	Giese, Krishnaswamy, Vasu and Lawrence 1964
Superorder Gnathostomacea								
Order Laganoida								
Mellita quinquiesperforata	15			0.3			2	J.E. Moss and Lawrence 1972
	60			1.2			2	
Holothuroidea								
Order Aspidochirotida								
Actinopyga echinites	300	19.6	25.1		6.1	7.8		Conand 1982
Holothuria mexicana	125	12.5	18.8		10	15		Engstrom 1980
Holothuria floridana	75	6.8	9.0		9	12		Engstrom 1980
Microthele nobilis	1200	42.3	77.2		2.89	5.04		Conand 1981
Microthele fuscogilva	1500	13.8	36.0		0.83	2.4		Conand 1981
Thelenota ananas	2000	25.1	37.9		1.12	1.58		Conand 1981

Table 4.3B: Fecundity of echinoderms

Species	Body size	Fecundity (number of eggs)	Reference
Asteroidea			
Asterina phylactica[a]	6 mm diameter	*c.* 40	Emson and Crump 1979
	9 mm diameter	*c.* 60	
	12 mm diameter	*c.* 80	
Echinaster sp. I	1.935 g wet body wt	954	Scheibling and Lawrence 1982
	3.653 g wet body wt	3908	
	5.759 g wet body wt	7609	
Echinaster sp. II	0.316 g wet body wt	248	Scheibling and Lawrence 1982
	0.644 g wet body wt	766	
	1.813 g wet body wt	2677	
Leptasterias hexactis[1]	2 g wet body wt	*c.* 550	Menge 1974
	6 g wet body wt	*c.* 950	
	10 g wet body wt	*c.* 1050	
Leptasterias polaris[a]	30 g wet body wt	*c.* 150	Himmelman, Lavergne, Cardinal, Martel and Jalbert 1982
	60 g wet body wt	*c.* 300	
	100 g wet body wt	*c.* 1500	
Ophidiaster granifer	20 mm diameter	*c.* 400	Yamaguchi and Lucas 1984
	26 mm diameter	1100–1600	
Ophiuroidea			
Axiognathus squamata[a]	1.5–2.0 mm disc diameter	2–5	Rumrill 1982
	2.0–2.5 mm disc diameter	2–6	
	2.5–3.0 mm disc diameter	2–7	
	3.0–3.5 mm disc diameter	2–7	
Holothuroidea			
Cucumaria curata[a]	20 mg wet body wt	*c.* 5	Rutherford 1977b
	60 mg wet body wt	80	
	100 mg wet body wt	150	
	140 mg wet body wt	200	
	180 mg wet body wt	280	
Echinoidea			
Abatus cordatus[a]	27 mm length	13	Magniez 1980
	42 mm length	109	

[a]Brooding species.

Table 4.4: Size at first sexual maturity in echinoderms

Species	Sex	Size	Reference
Asteroidea			
Acanthaster planci	M/F	400 g wet body wt 22 cm diameter	Conand 1985
Asterina gibbosa[a]	M	7–14 mm diameter	Emson and Crump 1979
Asterina phylactica[b]	M/F	5 mm diameter	Emson and Crump 1979
Leptasterias hexactis		2 g wet body wt	Menge 1974
Leptasterias tenera		5 mm diameter	Hendler and Franz 1982
Luidia senegalensis		75 mm radius	Halpern 1970
Pisaster ochraceus		70–90 g wet body wt	Menge 1974
Echinoidea			
Cassidulus caribaearum	M F	9 mm length 15 mm length	Gladfelter 1978
Goniocidaris umbraculum	F	17 mm diameter	Barker 1985
Lytechinus variegatus	M F	55 mm diameter 52 mm diameter	Moore, Jutare, Bauer and Jones 1963
Strongylocentrotus intermedius	M F	15 mm diameter 20 mm diameter	Fuji 1960
Strongylocentrotus purpuratus		24 mm diameter	Gonor 1972
Ophiuroidea			
Ophiolepis paucispina[b]	M/F	3.2 mm disc diameter	Hendler 1979
Holothuroidea			
Holothuria floridana	M F	88 mm length 87 mm length	Engstrom 1980
Holothuria mexicana	M F	90 mm length 103 mm length	Engstrom 1980
Microthele fuscogilva	M/F	324 mm length 900 g drained wet body wt	Conand 1981
Microthele nobilis	M/F	227 mm length 580 g drained wet body wt	Conand 1981
Thelenota ananas	M/F	300 mm length 1150 g drained wet body wt	Conand 1981

[a]Protandric. [b]Simultaneous hermaphrodite.

When extrinsic food limitation is less extreme, various species of echinoids, asteroids and holothuroids do not reach their potential growth rate or size but do reproduce (see Lawrence and Lane 1982). In this situation somatic and gonadal growth do seem to have an inverse seasonal relationship. The apparent strategy here is to divert nutrients from growth to reproduction, even though reproductive output is less because of the small size.

Because echinoderms have the capacity to resorb their body wall, they can respond to extrinsic food limitation by a decrease in body size to a level supportable by food availability (see Lawrence and Lane 1982). Ebert (1967, 1968) proposed that echinoids thus have an 'optional' size set by local conditions. Because of the effect of food availability on individual

Table 4.5: Effect of size on the annual energy budget and reproductive effort of echinoids. C: consumption, R: respiration, P: production. Values are in kJ. Reproduction effort is gonadal production per total production

Species	Size	C	R	P somatic	P gonadal	P total	Reproductive effort	Reference
Mellita quinquiesperforata	1[a]	160	30	16	0	16	0	J. Lane 1977
	10	431	67	18	5	23	22	
	30	810	120	25	43	68	63	
Strongylocentrotus droebachiensis	3[a]		9	7	0	7	0	Propp 1977
	8		27	18	12	30	40	
	17		64	17	32	49	65	
	31		107	9	58	67	87	
Strongylocentrotus intermedius	6[b]	368		44	28	72	39	Fuji 1967
	19	709		25	48	73	66	
	48	925		15	63	78	81	

[a]g dry body wt; [b] g wet body wt.

size, biomass and density of echinoderms are independent (Buchanan 1967; Ebert 1968). This has important implications in considering the reproductive potential of populations.

After adult size is reached, an increase in body size continues without an increase in reproductive capacity in the echinoids (*Tripneustes ventricosus* (Moore, Jutare, Jones, McPherson and Roper 1963), *Moira atropos* (Moore and Lopez 1966), and *Strongylocentrotus purpuratus* (Gonor 1972), and in the ophiuroids *Ophiothrix spiculata* and *Ophiopteris papillosa* (Rumrill 1982) (Figure 4.2). In this situation reproductive output remains constant and reproductive effort decreases. These species have a size above which the ability to acquire nutrients and/or to use them for reproduction does not increase further. The body size at which relative gonad production decreases is small in *Strongylocentrotus intermedius* and is related to decreases in the relative sizes of digestive structures and functioning (Figure 2.36) (Fuji 1967). Consequently, the significance of continued somatic growth past the body size at which reproductive output and effort continue to increase must be sought elsewhere. The gonads of extremely large *Diadema setosum* and *Echinometra mathaei* have much connective tissue and show little gametogenic activity. As a result, the gonads of very old or large individuals may become reproductively senile.

Figure 4.2: The relation between the ripe gonad index (10 × gonad volume. test volume^{-1}) and test length of *Moira atropos* (from Moore and Lopez 1966)

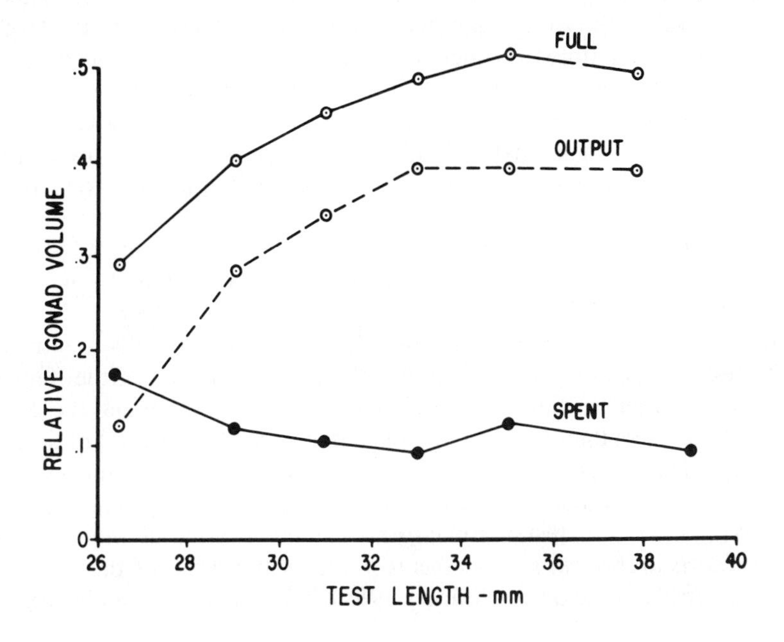

Energy requirements for reproduction seem to affect gonadal development. In some species, the minimal body size for the appearance of the gonads in males is less than for females (Table 4.4). Ovaries are energetically richer than testes even when they are of the same size (Lawrence and Lane 1982), and more sperm can be produced per unit gonad volume than can eggs. Species that show sequential hermaphroditism are protandric, with the male being smaller and preceding the female. Again, this is not related to a difference in functional morphology except perhaps as related to the requirements of growth and the functional state of the individual at different sizes. Small, juvenile stages may be less able to take in sufficient energy to support oogenesis, or the space available for gonads may be insufficient to produce a functional number of eggs but be adequate to produce a functional number of the much smaller sperm.

In gonochoric echinoderms, adult males and females are generally similar in size (Hyman 1955), although there has been little documentation of this. The female spatangoid *Echinocardium cordatum* is larger than the male (Moore 1936), and several species of asteroids and ophiuroids are androphorous (Hyman 1955). No sexual behavioural differences such as sexual competition or attraction which might involve sexual size dimorphism are known to exist in echinoderms.

The size of ripe testes and ovaries may be very similar in a species, but the ripe ovaries are frequently larger than the testes (Table 4.3). This is usually the result of less development of the testes, but in *Leptasterias tenera* approximately half of the testes of mature males actually degenerate or are absent (Hendler and Franz 1982). The higher energetic level of the ovary than that of the testis in addition to the larger size indicates that fitness is increased by the female allocating a larger amount of acquired nutrients to reproduction than does the male. The increased fitness resulting from a smaller allocation to reproduction to the male might result in a smaller food requirement or, in terms of limited resources, a greater amount of energy available for maintenance to increase survival.

Rarely are the testes larger than the ovaries. In the spatangoid *Abatus cordatus* the petals of the female are greatly depressed to form brood chambers (Magniez 1983) (See also *Abatus philippi*: Figure 4.3). The petals in male *A. cordatus* are depressed less. Coelomic space is so reduced in the females that the male gonad is 50% larger than that of the female. Despite the larger gravimetric reproductive output by the male, the energetic output is less than that of the female due to the higher level of lipid in the ovary. The testes of *Leptasterias hexactis* are larger than the ovaries, but the functional consequence of this is not known (Menge 1975). Similarly the mature testes are larger than the ovaries in the echinoid *Goniocidaris umbraculum*, and this is probably the result of the presence of more nutritive phagocytes (Barker 1985). Why the testes should require

Figure 4.3: Inner surface of the upper test of (A) male, and (B) female *Abatus philippi*, showing sexual dimorphism in the degree of depression of the lateral ambulacra that function as marsupia in the female. Only three gonads occur. (From Thomson 1876)

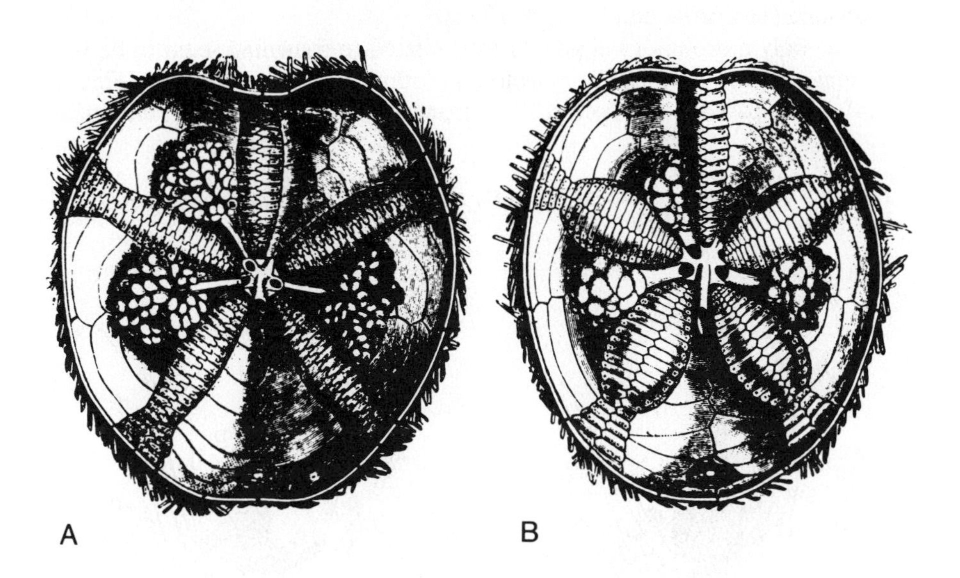

more nutritive phagocytes than the ovaries is not known, but their occurrence still contributes to a greater energetic demand on the male.

One might expect behaviour that increases the probability of successful fertilisation to result in smaller-sized males. Male *Archaster typicus* cover females during spawning and are slightly smaller (Komatsu 1983). Male *Astrochlamys bruneus* are dwarf and are borne upon the upper surface of the larger females (Mortensen 1936). Female *Ophiodaphne materna* carry smaller males over the mouth and 'so far from being an affectionate mother nursing its young in a self-sacrificing manner, being a passionate mistress living in continuous close embrace with its male lover' (Mortsensen 1933).

These examples indicate that the size of males can respond evolutionarily to the probability of successful fertilisation. That most echinoderm species have not reduced the body size of males but have decreased the size of the testes instead indicates an increased fitness resulting from the larger size. Because the large size of the male does not seem to be involved in reproduction, its value must be in survival of the individual.

Even in the absence of pseudo-copulation, aggregations could increase the probability of fertilisation. Relatively large numbers of sperm per egg are necessary for fertilisation, and sperm have limited viability. Fertilisation of eggs of *Strongylocentrotus droebachiensis* is low when males spawn at a distance of about 2 m (Pennington 1985). Although spawning aggregations have been reported for echinoderms, relatively few workers actually document the phenomenon (Pennington 1985). Holothuroids may raise their anterior end to release their gametes into the water column (Cameron and Fankboner 1986), and the ophiuroid *Hemiopholis elongata* raises its disc in spawning (Heatwole and Stancyk 1982).

The only morphological adaptations related to spawning seem to be the genital papillae in some ophiuroids, holothuroids and echinoids (Figure 4.4). The papillae are protrusible from the body wall of *Holothuria*

Figure 4.4: Genital papilla of (A) male and (B) female *Tripneustes gratilla* (from Tahara *et al.* 1958)

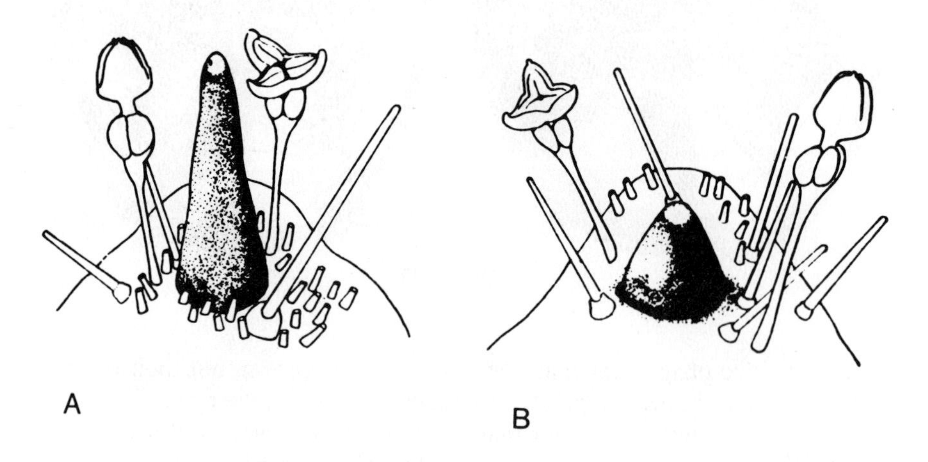

marmorata (Mortensen 1937) and the clypeasteroid *Dendraster excentricus* (Timko 1975) at spawning. In echinoids, sometimes only the males have papillae (Table 4.6). When both sexes do, those of the male are slenderer and longer. The genital papillae increase the distance that the sperm reach during spawning (Pawson and Miller 1979). The gonopores of female echinoids are wider than those of males, but do not necessarily indicate the egg diameter. Mortensen (1938) described the sausage-shaped appearance assumed by the eggs of the asteroid *Fromia ghardaqana* as they pass through the gonopores.

For sexual reproduction to be successful, it is necessary that sufficient gametes be produced for the subsequent events of fertilisation to result in successful recruitment. The number of gametes that is sufficient for this to occur is not just a function of individual body size, but also of other factors that increase the probability of successful fertilisation. Modifications at the individual level of organisation include hermaphroditism, egg–sperm interaction, abbreviated or direct development and brooding. Modifications at

Table 4.6: Sexual dimorphism in genital papillae of echinoids

Species	Reference
1. Male genital papillae short and conical; female without genital papillae	
Amblypneustus ovum	Chia 1977
Anthocidaris crassispinus	Tahara, Okada and Kobayashi 1960
Brissus meridionalis	Chia 1977
Dendraster excentricus	Chia 1977
Echinus esculentus	Swann 1954
Hemicentrotus pulcherrimus	Tahara *et al.* 1960; Okada and Tahara 1970
Paracentrotus lividus	Swann 1954
Mespilia globulus	Tahara, Okada and Kobayashi 1958
Metalia sternalis	Chia 1977
Psammechinus microtuberculatus	Swann 1954
Psammechinus miliaris	Marx 1929; Swann 1954
Pseudocentrotus depressus	Tahara *et al.* 1960
Sphaerechinus granularis	Swann 1954
Temnopleurus toreumaticus	Tahara *et al.* 1960
Toxopneustes pileolus	Tahara *et al.* 1958
2. Male genital papillae are long and tubular, female without genital papillae	
Clypeaster australasiae	Chia 1977
Echinocardium mediterraneum	Hamann 1887
Laganum depressum	Chia 1977
Laganum laganum	Chia 1977
3. Male and female genital papillae are short and conical	
Brissopsis luzonicas	Chia 1977
Moira lethe	Chia 1977
Tetrapygus niger	Pawson and Miller 1979
4. Male genital papillae are long and tubular, female genital papillae are short and conical	
Coelopleurus floridana	Pawson and Miller 1979
Diadema setosum	Tahara *et al.* 1958
Echinocyamus pusillus	Marx 1929
Echinocardium cordatum	Tahara *et al.* 1960
Echinometra mathaei	Tahara *et al.* 1958; Tahara and Okada 1968
Echinostrephus aciculatus	Tahara *et al.* 1958
Maretia planulata	Chia 1977
Maretia ovata	Chia 1977
Spatangus purpureus	Chia 1977
Tripneustes gratilla	Tahara *et al.* 1958
5. Male and female genital papillae are long and tubular	
Arachnoides placenta	Chia 1977
Clypeaster japonicus	Tahara *et al.* 1958
6. No genital papillae in males or females	
Allocentrotus fragilis	Pawson and Miller 1979
Arbacia lixula	Swann 1954
Arbacia punctulata	Harvey 1956; Pawson and Miller 1979
Echinus affinis	Pawson and Miller 1979
Echinus alexandri	Pawson and Miller 1979
Echinus gracilis	Pawson and Miller 1979
Plesiodiadema antillarum	Pawson and Miller 1979
Podocidaris sculpta	Pawson and Miller 1979
Salenocidaris varispina	Pawson and Miller 1979

the supra-organismal level include behaviour (copulation or aggregation for spawning, synchrony of spawning) and population size and density. Species with a small body size must have mechanisms to increase the probability that the limited number of gametes produced results in recruitment. The biologically meaningful analysis is that of the population. Menge (1975) explicitly included the densities of the populations of the asteroids *Leptasterias hexactis* and *Pisaster ochraceus* in evaluating the effect of size on their reproductive success and concluded that the small *L. hexactis* could not function as a broadcast spawner because of its limited fecundity. If this is true, the numerous species of echinoderms that have body sizes much smaller than *L. hexactis* and are broadcast spawners must have mechanisms to ensure successful reproduction.

Superimposed on these functional morphological constraints on reproduction are environmental ones. Changes in temperature and salinity can affect activity processes such as locomotion and feeding and the rate of metabolic processes with effects on reproductive output (Lawrence 1985; Lawrence and Lane 1982).

Another major environmental control on reproductive output and effort is food availability. After the first threshold of food requirements for maintenance and survival, reproductive output is determined by the extrinsic limitation of food (Lawrence 1985; Lawrence and Lane 1982). The extrinsically limited resources may be used for maintenance over reproduction. With extrinsic limitation of food, *Pisaster giganteus* allocates acquired nutrients to somatic maintenance and growth over gonadal development (Harrold and Pearse 1980). This response was considered appropriate for a long-lived iteroparous species. Similarly, the pyloric caeca appear before the gonads in the regenerating arms of *Asterias vulgaris* (King 1898). *Luidia clathrata* allocates food resources to the pyloric caeca of intact arms rather than to the regeneration of arms (Lawrence *et al.* 1986). Although the rapid regeneration of the lost arms would increase the locomotory and feeding ability of the individual, it does not seem to compensate for the decrease in nutrient storage (pyloric caeca) and reproduction (gonads) that would occur.

4.3. BROODING

Brooding is the association of embryos and juveniles with an adult, usually the female parent. Whether brooding involves only the embryo or the juvenile also varies, as does the size at which the offspring assumes independent existence. The advantage of brooding is considered to be the increased probability of survival of the offspring to the stage of independent existence, but the costs that must be considered include decreased fecundity and dispersal, as well as an increase in reproductive effort.

Brooding occurs in representatives of all extant echinoderm classes. Brooding is not common, however, indicating that the costs outweigh the advantage.

The decrease in fecundity associated with brooding results from the size of the offspring at the assumption of independent existence. The responsibility for acquiring nutrients to grow to the appropriate size of the postmetamorphic individual is given to the larva for species with planktotrophic reproduction. Large eggs are produced by species that brood. The production of large eggs has been interpreted to be necessary to provide energy for development during the non-feeding brooded period, but may have its significance in the production of a large offspring (Lawrence, McClintock and Guille 1985).

The term viviparous has been used for intrabursal, intraovarian or intracoelomic brood care in echinoderms (Turner and Dearborn 1979) in the sense of 'live birth' and without the connotation of nutrient transfer. Other workers have restricted the terms viviparous to intraovarian or intracoelomic development and the term brooding to intrabursal or marsupial development. Thus Mortensen (1936) did not consider brooding to include the occurrence of young on the outer surface of the parent.

Morphological adaptations for brooding are minimal or seemingly nonexistent in some brooders, but others have adaptations for holding and protecting the developing embryos. Some of these seem related to features promoting water circulation, presumably for respiratory gas exchange, waste elimination, or provision of dissolved organic nutrients for nutritional purposes. The need for such circulation may limit the size of brood masses (Strathmann and Chaffee 1984).

Although most comatulids are broadcast spawners (John 1938), species with external or internal brooding exist. In species with external brooding, the eggs are characteristically stuck to the pinnules by secretions of cement glands, as in the genus *Antedon.* Female *Antedon bifida* draw the arms together and fold the pinnules along the longitudinal axis in a brooding behaviour which lasts until the larvae hatch (Lehaye and Jangoux 1985b). This would obviously interfere with feeding, and must be considered a reproductive cost not experienced by the male.

Internal brooding occurs in many Antarctic crinoids, primarily in *Isometra, Notocrinus* and *Phrixometra.* Brood chambers are located in the arms at the bases of the pinnules adjacent to the ovary in *Notocrinus* and in the pinnules in the others (Figure 4.5). Modifications of the hard parts occur only in *Isometra* species, as segments of the genital pinnules of females are greatly expanded to anchor over and protect the ovary and brood pouch (John 1938).

Development is direct in these brooding species and, as the eggs are larger, fewer are produced than in non-brooding species. Usually only a few eggs are found in a marsupium. The number of embryos in each

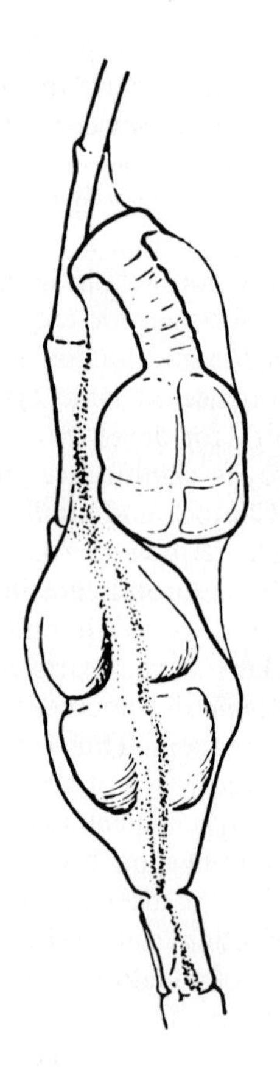

Figure 4.5: Genital pinnule of *Kempometra grisea* from the underside showing the ovary containing large eggs and the brood pouch containing a pentacrinoid larva (from John 1938)

marsupium and the degree to which they develop is variable (John 1938). Thus, *Notocrinus mortenseni* can have up to 92 embryos in all stages of development in a marsupium. Because a large individual can have 23 genital pinnules, nearly 2000 young can be produced. These embryos are ciliated and presumably free-swimming when released. This is the usual situation, but swimming may be for a short distance only. In *Isometra vivipara,* the embryo swim only to the female's cirri, practising adelphophagia by devouring their siblings on the way. In contrast, *Notocrinus virilis* has only one to three embryos per marsupium and its embryos develop no ciliated bands. Presumably they drop to the substratum and develop directly. *Phrixometra nutrix* has only two embryos per marsupium and broods the young through metamorphosis. Brooding in crinoids does not seem to be related to the body size of the species.

Brooding in ophiuroids is usually associated with the bursae, although young may be held on the parent's body. Young *Astrothorax waitei* leave the bursae and cling by the arms to the upper surface of the parent's disc (A.N. Baker 1980). The largest juvenile can be one quarter the size of the parent. Similarly, young *Ophiothrix fragilis* attach to the spines of adults and then enter the bursae of both males and females, indicating a potential cost to the male as well as to the female parent (J.E. Smith 1941). Young *Gorgonocephalus chilensis* are brooded on the body of parents, and embryos are held beneath the disc of *Ophiophycis gracilis* (Mortensen 1936).

The bursae are sac-like invaginations in the oral wall of most ophiuroids, one adjacent to each side of the arm bases. The gonoducts open into the bursae. The flagellated epithelium of the bursae and pumping movements of the disc maintain water currents in the bursae. No special modifications of the bursae have been reported, but a stalk from the bursal sac attaches to the embryos in the hermaphroditic *Axiognathus* (= *Amphipholis*) *squamata* and *A. japonica* (Figure 4.6). H.B. Fell (1946) concluded that the stalk serves as an anchor only, but Oguro, Shōsaku and Komatsu (1982) and Walker and Fineblit (1982) stated that nourishment of the larva must occur in the bursa. The growth of the embryo within the bursa stretches the bursal wall which becomes thin, and the embryo occupies most of the disc coelom. No specific morphological features seem to be associated with

Figure 4.6: Gonads and developmental stages in the bursa of *Axiognathus* (= *Amphipholis*) *japonica*. A: Testis and ovary; B: mature testis and ovary; C: ovulation of ovum and release of sperm; D: fertilised ovum; E: two-cell stage; F: gastrula; G: embryo connected to the bursal wall; H and I: embryonic stages; J and K: juvenile after metamorphosis, free of attachment to bursal wall. b: Bursa; e: embryo; ea: embryonic attachment; fm: fertilisation membrane; gc: genital cleft; j: juvenile; o: ovary; oo: ovulated ovum; ov: ovarian ovum; pvc: perivisceral coelom of parent; t: testis. (From Oguro *et al.* 1982)

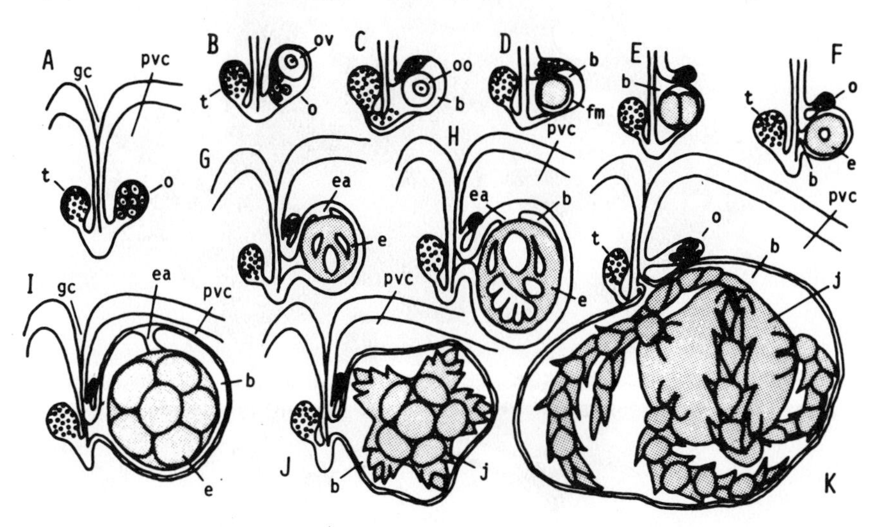

intraovarian development in ophiuroids. No placenta exists, but each juvenile is surrounded by its own ovarian sac (Turner and Dearborn 1979).

Mortensen (1936) noted a tendency to intraovarian development in Antarctic ophiuroids. He could see no advantage in this as adelphophagia occurs in both bursal and intraovarian development. The availability of nutrients from other routes is probably the important factor. Mortensen concluded that gonochoric species are usually primitive, but pointed out that very specialised forms are gonochoric as well. Intrageneric variability in the occurrence of brooding exists. Mortensen (1936) listed seven brooding and five non-brooding species of the genus *Amphiura* from the Antarctic region.

Whether brooding occurs in the bursae or the ovary, the developing young often fill the disc space. In *Ophiochondrus stelliger*, the young fill the space to such an extent that the stomach is reduced to a network among them (Mortensen 1936). Young brooded within the ovary of *Ophionotus hexactis* reach extremely large sizes before becoming free-living (Figure 4.7) (Mortensen 1921). Brood size in *O. hexactis* reaches 54 young per individual, with a mean of *c.* 25, so that space in the disc can be nearly

Figure 4.7: A female *Ophionotus hexactis* with the upper part of the disc removed. Four young in different stages of growth are contained in the ovary. (From Lyman 1882)

filled with developing young (Turner and Dearborn 1979). The decrease in the capacity to acquire nutrients would be a cost to the female parent.

Mortensen (1936) recorded the broods of various ophiuroids and the sizes of the parents of Antarctic and sub-Antarctic ophiuroids. The number of developing young in each bursa ranges from one or two (*Ophioscolex nutrix* with a 10-mm disc, and *Ophiacantha vivipara* with a 18-mm disc) to six to eight (*Ophiura meridionalis*) to ten (*Ophiura rouchi* with a 7-mm disc). The number is always small though variable, and probably depends considerably on the size at which the young leave the parent. An *Ophiolebella biscutifera* with a disc of 7 mm produces only 3 young per individual, but these young can have a disc diameter of up to 2 mm. Most brooding species have disc diameters of 10 mm or less, but several species have disc diameters of 16 mm or more. In contrast, most non-brooding species have disc diameters or more than 10 mm. It seems for these species that a small size is conducive to the development of brooding, but not requisite.

Emson, Mladenov and Wilkie (1985) concluded that small body size in ophiuroids results in a reproductive constraint because of the small number of eggs that can be produced. They suggested that abbreviated development of larvae, direct development often associated with viviparity or brooding, and supplementation of sexual reproduction with asexual reproduction (fission) were the potential alternatives to species of small body size. Emson *et al.* stated that evolution to a small body size has been infrequent in ophiuroids. Is this because of the difficulty in evolving the brooding habit or with its success, or is the small body not successful because of problems associated with feeding or predation?

The relation between interspecific ophiuroid body size and fecundity can be seen in contrasting the maximal fecundity of the viviparous *Axiognathus* (= *Amphipholis*) *squamata* and the ovoviparous *Ophioplocus esmarki*. *A. squamata* (disc diameter *c.* 2 mm) has a maximal fecundity of seven 105-μm diameter eggs. *O. esmarki* (disc diameter *c.* 18 mm) has a maximal fecundity of 2360 320-μm eggs (Rumrill and Pearse 1985). Although Hendler (1975) found a direct relation between body size and the number of embryos brooded in *A. squamata*, Rumrill (1982) found that small adults brooded nearly as many young (two to five) as large adults (two to seven). Some of the variability in fecundity could result from the continuous, non-synchronous release of eggs. Rumrill also noted that small females brooded offspring of a size equivalent to that of the offspring brooded by large females, and concluded that no distinct selective reproductive advantage results from these small ophiuroids attaining a large body size. A close relation exists between body size and number of young brooded by *O. esmarki* for the early stages of development but not for the later stages. Rumrill assumed that the lack of relation for the later stages resulted from the release of juveniles, but did not consider the possibility of mortality. The production of large eggs for brooding decreases fecundity. In contrast to *O. esmarki*, *Amphiodia occidentalis* (disc diameter = 5 mm) produces a maximum of nearly 68000 eggs, which have planktotrophic development.

All hermaphroditic ophiuroids are brooders, and most brooding ophiuroids are hermaphroditic (Mortensen 1936; Hendler 1975). Some, such as *Ophiolepis kieri*, are protandric hermaphrodites, whereas others, such as *Ophiolepis paucispina*, are simultaneous hermaphrodites (Hendler 1979). Hendler proposed the evolution of small *Ophiolepis* species through paedomorphosis, facilitated by adaptations for brooding and selection for hermaphroditism. Protandric hermaphroditism would increase the number of gametes produced per individual at small body sizes.

Brooding in asteroids ranges from behavioural postures to considerable morphological adaptations. The most common behavioural method of brooding by asteroids is that of holding the developing embryos beneath the oral surface. This method is found in the order Spinulosa (families Asterinidae and Echinasteridae) and the order Forcipulatida (family Asteriidae). Various species of *Leptasterias* brood their young (Fisher 1930). *Leptasterias hexactis* forms a humped posture with only the distal ends of the arms attached to the substratum (Chia 1969). In *L. arctica*, the basal parts of the arms are brought so close together that the adjacent marginal spines interdigitate (Fisher 1930). A similar posture is assumed by brooding *L. pusilla*, but *L. aequalis* is more closely attached to the substratum with the disc only slightly elevated (R. H. Smith 1971). Repetitive attachment and detachment of the parent's tube feet from the embryo mass results in cleaning and ventilation of the mass. *L. polaris* remains flattened with the arms curved and the larvae attached to the substratum (Himmelman *et al.* 1982). In *L. tenera*, the embryos are taken into the parent's stomach for a period of time and then released into the brood chamber formed by the disc for the remainder of development (about three months) (Worley, Franz and Hendler 1977). In *L. groenlandica*, the embryos are held in the cardiac stomach throughout development (Figure 4.8) (Fisher 1930).

Brooding of embryos within the stomach or under the mouth has been assumed to interfere with the ability of females to feed. *Leptasterias hexactis* does not feed during brooding (Menge 1974) but *Anasterias rupicola* feeds while holding a brood in the mouth (Figure 4.9) (Blankley and Branch 1984). This would decrease the cost of brooding.

The young in the mass are attached individually to a long slender strand from the actinal interradial area just external to the mouth (Figure 4.9). Hendler and Franz (1982) suggested that the central mass to which embryos are attached is found in broods held in the stomach, but not in those held outside the stomach.

Within a species, the number of eggs brooded is variable but in general

Figure 4.8: A female *Leptasterias groenlandica* with upper body wall removed and viewed looking downwards through the stomach. Juveniles are brooded in the ventral stomach. ds: Dorsal stomach; g: gonads; m: mouth. (From Fisher 1930)

dependent upon the size of the parent. An individual *Leptasterias hexactis* of *c.* 2 g wet body weight had *c.* 250 eggs whereas one of *c.* 15 g wet body weight had *c.* 1250 eggs (Menge 1975). *L. polaris* reaches a much larger size. Individuals of *c.* 30 g wet body weight had 150 to 200 young whereas individuals of 100 to 200 g wet body weight had between 900 and 3000 young (Himmelman *et al.* 1982).

Brooding does not ensure survival, as Menge estimated 45% mortality of the brood of *Leptasterias hexactis.* Hendler and Franz (1982) found evidence for a 74% embryo mortality during brooding in *L. tenera.* The small (1.75 g wet body weight) simultaneous hermaphroditic asteroid *Asterina phylactica* also shows progressive mortality of the brood (Strathmann, Strathmann and Emson 1984). Larger individuals are less able to brood all the young they produce, mortality of the brood reaching 75–80%. This relationship between capacity to brood and body size was

Figure 4.9: (a) The undersurface of a female *Anasterias rupicola* showing a six-month-old brood. The individual is feeding on an isopod. (b) A juvenile attached to the brood mass. (From Blankley and Branch 1984)

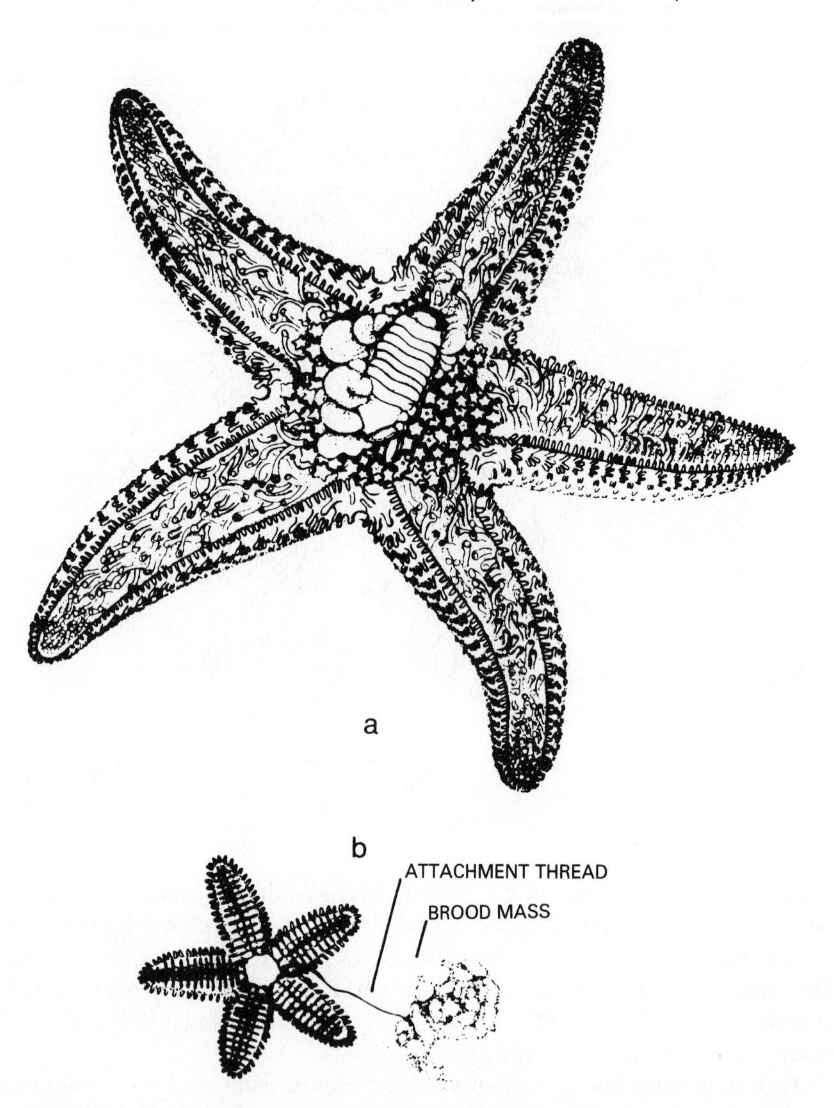

interpreted to support the allometry hypothesis for the association of brooding with small body size. As with ophiuroids, small size seems conducive to the evolution of brooding but not restrictive.

Fecundity in terms of the number of eggs produced by *Asterina phylactica* is similar to that of the similarly sized, non-brooding *Asterina minor* which is a simultaneous hermaphrodite. Strathmann *et al.* (1984) concluded that brooding in *A. phylactica* does not limit allocation to reproduction by a female, but does limit production of metamorphosing individuals and is more limiting for larger individuals. As Hendler (1979) found for a simultaneous hermaphroditic ophiuroid, Strathmann *et al.* reported that the allocation to the ovary was greater than that to the testes in *A. phylactica*, and that the difference in allocation increased with an increase in body size. In a sense, there is 'protandric simultaneous hermaphroditism', and perhaps the same bases as for protandric sequential hermaphroditism apply.

Asterina gibbosa and *A. phylactica* are closely related, occur together, partition the habitat, and show distinctly different life-history strategies (Crump and Emson 1983). *A. phylactica* is restricted to tide pools and is exposed to a very stressful environment. It reaches 15 mm in arm radius. *A. gibbosa* is found intertidally but in less exposed and stressful conditions. It reaches 50 mm in arm radius. Their reproduction characteristics differ greatly. Although both produce large eggs, *A. phylactica* is a simultaneous hermaphrodite, first reproduces at two years of age, reproduces one to three times, and lives up to four years. It broods its young. In contrast, *A. gibbosa* is a protandric hermaphrodite, first reproduces at four years of age, reproduces three to seven times, and lives for seven years or more. It does not brood its young. A greater proportion of acquired nutrients are allocated to reproductive output in *A. phylactica.* Thus, the reproductive characteristics of *A. phylactica* are associated with a habitat in which gonochoric reproduction and survival may be difficult. The lower longevity of *A. phylactica* might be associated with the lower allocation of nutrients to maintenance and higher allocation to reproduction or the higher predictability of pre-reproductive survival.

Hyman (1955) concluded that the body wall of phanerozonians (Paxillosida, Valvatida) was too stiff and the arms too short and broad to assume the brooding posture. However, the paxillosid *Luidia* can form the typical 'humped' feeding posture that involves flexion of the ray. In the paxillosid astropectinid *Leptychaster kerguelensis* (radius 66 mm), the upper surface is covered with a tesselated pavement composed of the heads of paxillae. The embryos develop in the space between the roof formed by these heads and the underlying epidermis. Juveniles in varying degrees of development occur simultaneously. The juveniles grow to a considerable size (radius 3.5 mm) before leaving the chamber.

In the spinulosid family Pterasteridae, a further refinement results in an aboral brood chamber formed by the paxillae. There is a supradorsal membrane supported by the crowns of the paxillae (Figure 3.11). *Diplopteraster verrucosus* has five or six arms and a radius up to *c.* 75 mm. A five-armed individual (radius 50 mm) had 25 young in the interradial chambers, four to six per interradius with a maximal radius of 9 mm. In *Hymenaster nobilis* (radius 300 mm) young develop beneath the membrane in the interradial areas and in the floor of the chamber formed by the osculum at the apical pole (Thomson 1876). The use of the chamber for brooding by some species of pterasterids seems to be an exploitation of a structure originally adapted for defence (Nance and Braithwaite 1981).

The genus *Odinella* is the only known member of its family which has a highly specialised brood chamber in each interradius (Figure 4.10) (Fisher 1940). *O. nutrix* has 11 to 14 rays and reaches a radius of 100 mm. The basal part of the ray is inflated and has numerous spines, some of which grow into the opposite body wall and lock together into a sort of circular secondary disc all around the genital regions of both males and females. Fisher concluded that this was necessary to prevent the loss of the rays which are very weakly attached to the disc. Once again, there has been exploitation of a structure developed for one purpose to serve as a brooding chamber. In the female, these spines form a basket in which the young develop until the radius reaches at least 6 mm. Five to nine young per chamber have varying degrees of development. Water circulates into the chamber by ciliary action.

A third exploitation of a body structure for brooding occurs in the

Figure 4.10: The upper surface of a female *Odinella nutrix* showing the interlocking dorso-lateral spines which form the brood chamber next to the disc (from Fisher 1930)

asterinid genus *Tremaster.* Members of this genus have an internal calcified duct in each interradius running from the lower to the upper body surfaces (Figure 4.11). These ducts presumably increase the flexibility of these armless asteroids, but the gonoducts also empty into them and embyros are brooded there (Jangoux 1982).

Brooding in echinoids is always external. The young simply occur among the spines or in marsupia (Table 4.7). In the extant regular echinoids, brooding seems confined to the cidaroids except for one euechinoid *Hypsiechinus coronatus.* This carries young in an elevated apical system (Mortensen 1903). Philip and Foster (1971) listed four species of the extinct regular euechinoid genus *Paradoxechinus* in which there was an annular depression around the raised apical system. They also reported lateral interambulacral marsupia in three other extinct regular echinoids. In cidaroids, the brood is always found around the peristome except for *Australis canaliculata* (Philip and Foster 1971). Here the young are found around the apical system (Thomson 1878). Whether the position of the young on the test is primarily for feeding or for protection is not known. Typically, there are no structural modifications associated with brooding in cidaroids, the young simply being held by the primary spines

Figure 4.11: The upper surface of *Tremaster mirabilis* showing openings of the interradial calcified ducts (from M. Jangoux, unpubl.)

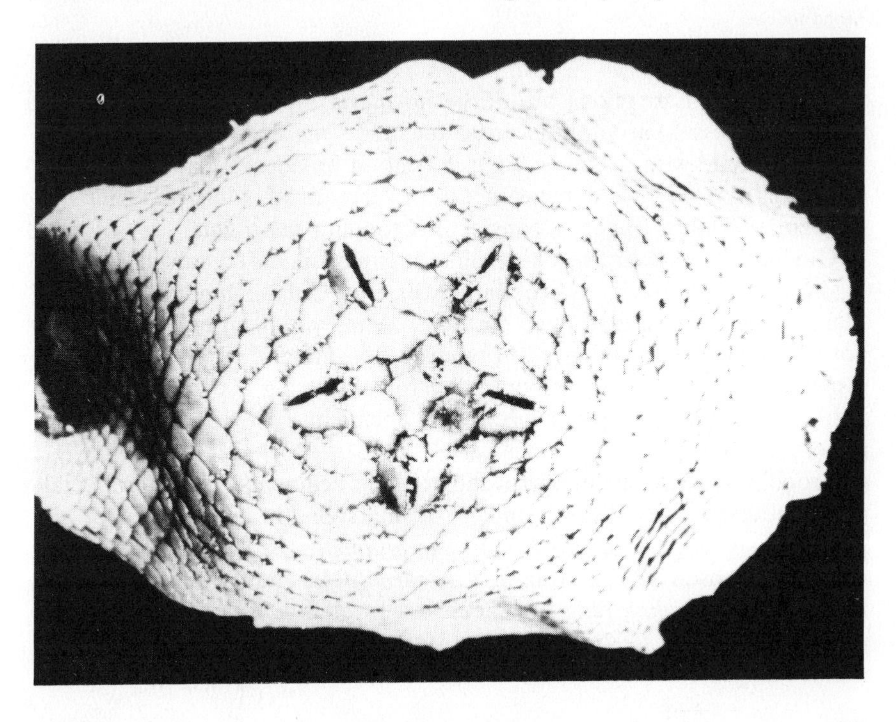

Table 4.7: Types of marsupia in echinoids (from Philip and Foster 1971)

Marsupium adapical

A. *Central*

- 1. Apical system raised
  - a. without annular depression
    - Camarodonta
    - *Hypsiechinus coronatus*[a]
  - b. with annular depression
    - Camarodonta
    - *Paradoxechinus granulosus*
    - *P. novus*
    - *P. profundus*
    - *P. stellatus*
- 2. Apical system sunken
  - Clypeasteroida
  - *Fibularia nutriens*[a]
  - *Pentedium curator*
  - Neolampadina
  - *Anochanus sinensis*[a]
  - *Tropholampas loveni*[a]
  - Spatangoida
  - *Plexechinus nordenskjoldi*[a]
  - *Peraspatangus brevis*
  - *P. depressus*
  - *Brissopneustes danicus*

B. *Lateral*

- 1. Ambulacral
  - Spatangoida
  - *Abatus*[a] 9 spp.
  - *Amphipneustes*[a] 8 spp.
  - *Tripylus*[a] 4 spp.
  - '*Tripylus*' *pseudoviviparus*
- 2. Interambulacral
  - Camarodonta
  - *Pentechinus mirabilis*
  - Stirodonta
  - *Goniopygus royoi*
  - *G. minor*
  - *Thylechinus said*

Marsupium adoral

A. *Circumoral*

- Cidaroida
- *Ctenocidaris geliberti*[a]
- *Rhynchocidaris triplopora*[a]

B. Anterior

- Clypeasteroida
- *Fossulaster halli*
- *F. exiguus*
- *Willungaster scutellaris*

[a]living species.

(Figure 4.12). Exceptions are *Rhynchocidaris triplopora*, which has an annular depression around the peristome, and *Ctenocidaris geliberti*, which has a peristone with a sunken edge. Young of broadcast-spawning regular echinoids may be found beneath the oral spines of adults (Tegner and Dayton 1977), and might be considered brooding *sensu latu.*

Obviously, the test of regular echinoids can be modified morphologically for brooding. The rarity of brooding by extant regular echinoids is curious. Marsupia might be costly, but a cost to brooding young among the spines is not obvious. The advantages of broadcast spawning in terms of increased dispersal and transfer of responsibility for producing a postmetamorphic individual of the appropriate size for independent existence must be more important.

Brooding also occurs in burrowing irregular echinoids. The lateral petalloid ambulacra are deeply sunken as marsupia in all species of the Antarctic spatangoid genera *Abatus*, *Amphipneustes*, and *Tripylus* (Figure 4.13) (Mortensen 1951). Although *A. cordatus* has only three gonads and three genital pores, young are incubated in all four marsupia in similar numbers (Magniez 1980). The apical system is depressed in the females of the neolampid cassiduloids *Anochanus sinensis* and *Tropholampas loveni*;

Figure 4.12: A female *Ctenocidaris nutrix* with juveniles among the spines (from Thomson 1876)

Figure 4.13: The upper surface of a female *Abatus philippi* showing brood chambers with eggs and juveniles (from Thomson 1876)

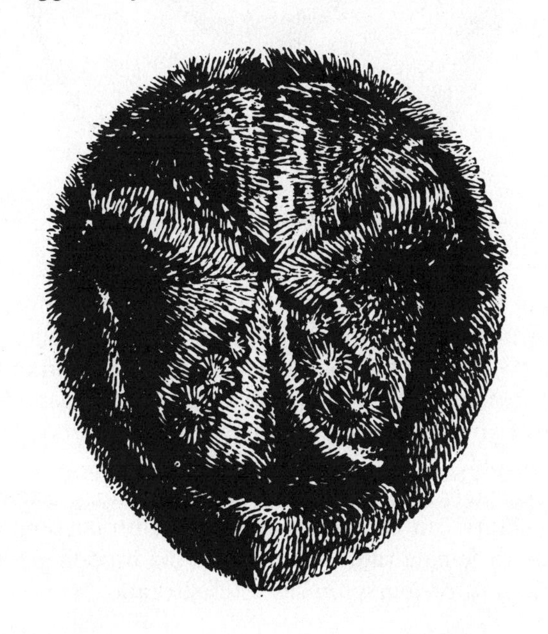

the clypeasteroid *Fibularia nutriens* (Figure 4.14), and the meridosternous spatangoid *Plexechinus nordenskjoldi.* Mortensen (1948b) reported that the tiny (3.3 mm maximal body length) *F. nutriens* was sexually dimorphic, with males being smaller than females. Thus a male *F. nutriens* produces less gametes than a female because it is smaller. In contrast, male and female *A. cordatus* are of the same size. The male produces more gametes because the depression of the petals is less, but invests less energy in them because of the difference in organic composition.

Figure 4.14: The upper surface of a female *Fibularia nutriens* showing the marsupium (from Mortensen 1948b)

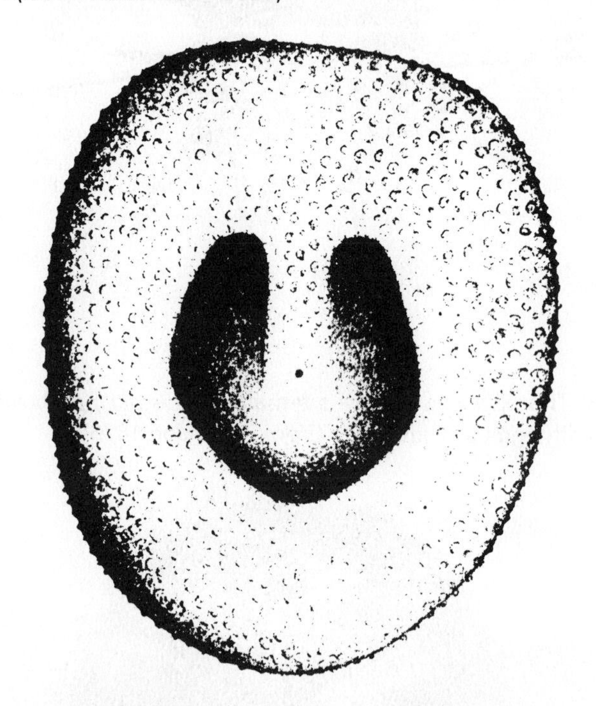

Brooding in the cassiduloid *Cassidulus caribaearum* does not involve a marsupium, with the young simply held among the spines of the upper body surface (Gladfelter 1978). A continuous succession of broods occurs so that most females carry young in two different stages of development. The number of young brooded increases with female body size. Mortality of premetamorphic young was greater than that of postmetamorphic young. The mortality indicates, as found for brooding asteroids, that pre-spawning fecundity can exceed actual juvenile production. In contrast to the high male to female ratio of 1:0.41 in the brooding *Ophiolepis kieri* (Hendler 1979), *C. caribaearum* has a 1:5 sex ratio.

The basic body form of both regular and irregular echinoids should not provide constraints on the development of marsupia associated with the apical system. The necessity for water circulation might be responsible for the restriction of the lateral, aboral marsupia to the ambulacra of spatangoids. The thinness of the body might contrain the development of marsupia in clypeasteroids. The association with particulate substrata might make marsupia on the lower body surface difficult for irregular echinoids.

Hyman (1955) listed the variety of types of brooding by holothuroids. The brood can be simply held on the tube feet of the bivium as in *Cladodactyla crocea* (Figure 4.15) (Thomson 1876). Tessellated dorsal scales form the brood chambers in *Psolus ephippifer* (Figure 4.16) (Thomson 1876). Depressions in the body wall near the tentacles of *Cucumaria joubini* and in the anterior interradii of the body wall of *Psolus koehleri* (Figure 4.17) are brood chambers (Vaney 1914). The deep-sea *Ocnus sacculus* is unusual in that the three dorsal internal pouches have no outlet (Figure 4.18) (Pawson 1983). Both dendrochirotid and apodid species brood their young in the coelom (e.g. *Synaptula hydriformis*) (H.L. Clark 1898).

As with asteroids and ophiuroids, brooding species of holothuroids have a small body size. The number of eggs produced similarly varies with the size of the individual. The allometric relationship between the number of eggs produced and body size in the intertidal *Cucumaria curata* has a slope ranging from *c.* two to more than four (Rutherford 1977). *C. curata*

Figure 4.15: A female *Cladodactyla crocea* with young on the body surface (from Thomson 1876)

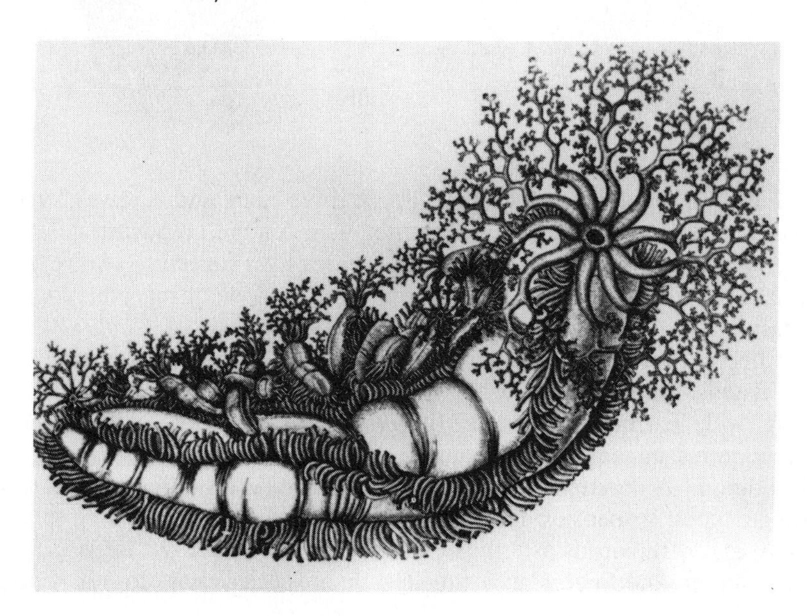

Figure 4.16: The upper surface of *Psolus ephippifer* (a) intact and (b) with several of the large tesselated plates removed. The exposed part of the these plates is flat, supported by a central column. (From Thomson 1876)

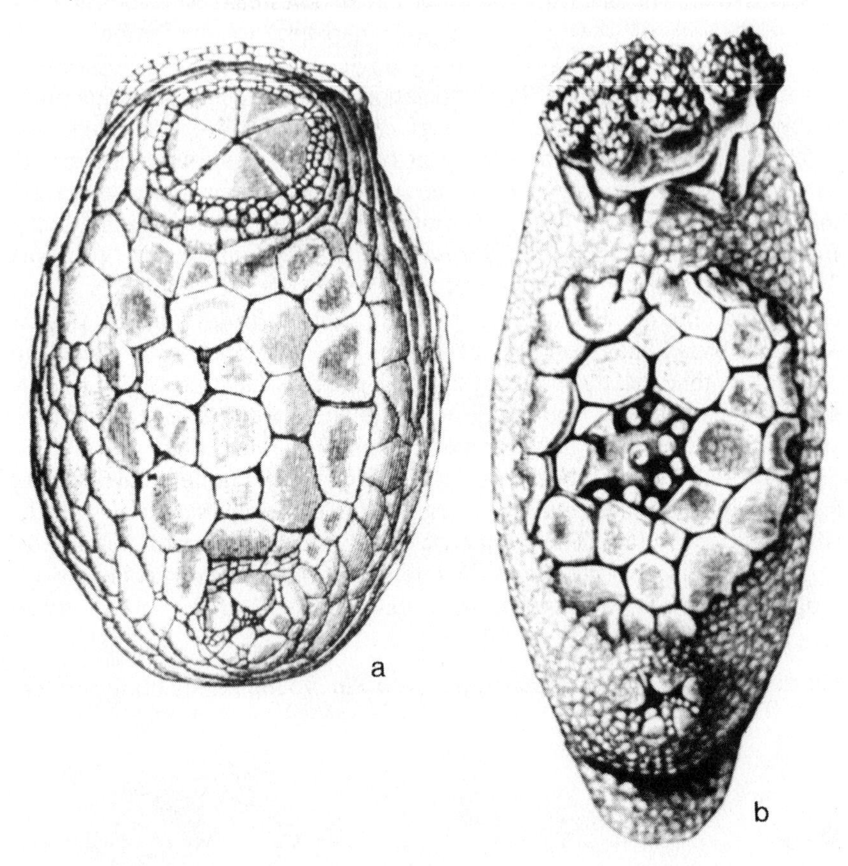

becomes reproductive at an age of about three years and survives for five years or more. This longevity is shorter than for the broadcast-spawning *Leptopentacta* (= *Cucumaria*) *elongata* which lives for ten to twelve years (Buchanan 1967), indicating a relation between mode of reproduction and longevity. Space is a limiting resource for populations of *C. curata*, as it may become for many brooding species. Survival of the young is negatively correlated with the density of the population.

As with marine invertebrates in general, the incidence of brooding in echinoderms increases with latitude (Thorson 1950; Mileikovsky 1971). Brooding in Antarctic and sub-Antarctic echinoderms is found in various trophic types: suspension-feeding holothuroids, sediment-ingesting spatangoids, and carnivorous asteroids. Species that brood are not large in body size, but are not necessarily smaller than species that do not brood.

Figure 4.17: (a) A female *Cucumaria joubini* with brood pouches filled with embryos. (b) A female *Psolus koehleri* from a lateral view showing the openings to the brood pouches behind the tentacles. (From Vaney 1914)

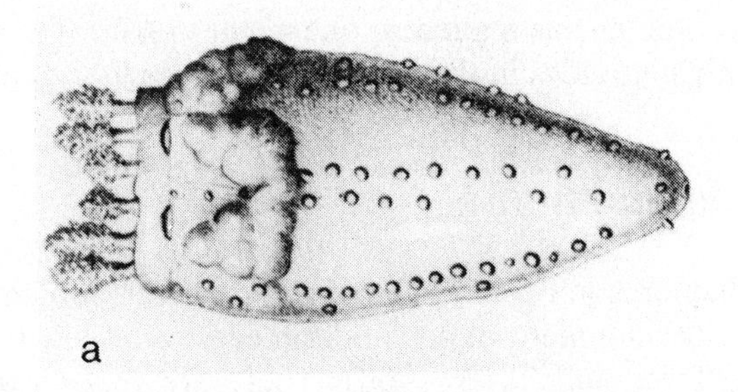

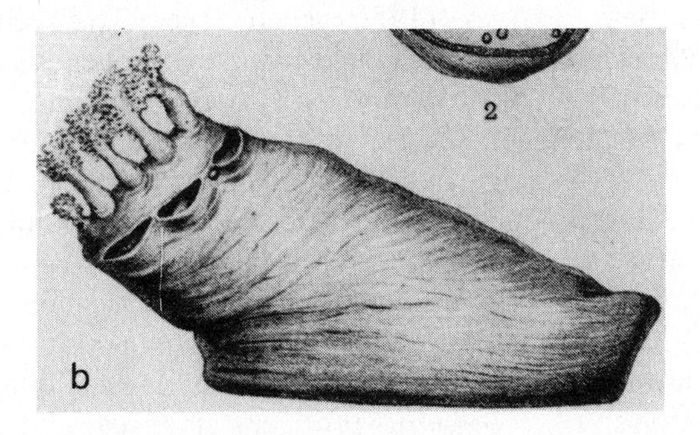

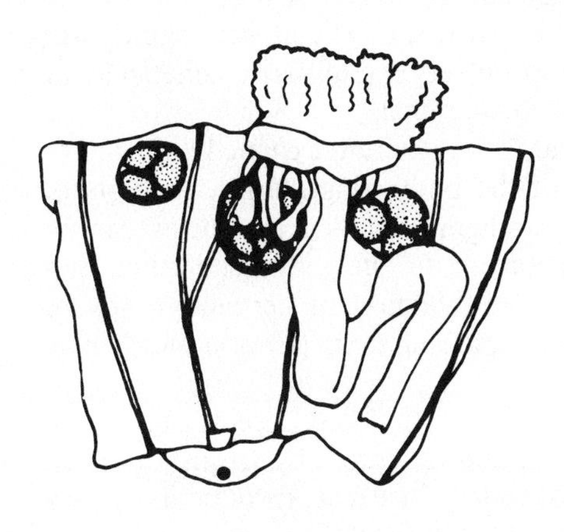

Figure 4.18: A female *Ocnus sacculus* dissected from the undersurface showing the three brood pouches with eggs (from Pawson 1983)

Although the deeper-water habitats may be relatively stable, some brooding species are found intertidally on the shores of sub-Antarctic islands where considerable physical variation occurs. If a small body size is conducive to brooding because of the inherent limited fecundity, and if low availability of food promotes this tendency, one might look for the adaptiveness of brooding in high latitudes in the trophic conditions found there.

4.4. ASEXUAL REPRODUCTION

Asexual reproduction can occur by parthenogenesis, fission and autotomy. Parthenogenetic development of echinoderm eggs can easily be induced in the laboratory (Tyler and Tyler 1966). There are reports of parthenogenesis in the field, Harvey (1956) cited several for echinoids and asteroids, and Mortensen (1936) thought that parthenogenesis might occur in several species of Antarctic ophiuroids. However, the only documented case of natural parthenogenesis is that for the small coral-reef asteroid *Ophidiaster granifer* (Yamaguchi and Lucas 1984).

Parthenogenetic reproduction by *O. granifer* leading to both stationary and dispersed larvae was interpreted to be related to the low fecundity and low population densities of this small asteroid. Two congeneric species of similar size and habit exist sympatrically with *O. granifer* on the coral reefs at Guam, but are rare. One, *O. squameus,* produces small planktotrophic larvae. The second, *O. robilliardi,* produces small eggs that would be assumed to have planktotrophic development. A population of *O. robilliardi* consisted only of females and included individuals which had been produced by fission, indicating maintenance of the population by asexual reproduction. The three species differ in reproductive strategies, despite similarities in size, morphology and behaviour. Unless many more unknown species with parthenogenesis exist, the cellular constraints preventing it under *in situ* conditions must be very conservative genetically or the advantages associated with it must be minimal.

A.M. Clark (1967) pointed out that the regenerative capacity of echinoderms is generally related to parts of the body, single arms, or parts of arms. In terms of Goss's (1969) paradigm, essential structures cannot regenerate and non-essential structures do not. Regeneration has developed into a means of asexual reproduction in certain groups of echinoderms by the processes of *fission* and *autotomy* (Emson and Wilkie 1980).

Asexual reproduction by fission (division of the body) occurs only in the asteroids, ophiuroids and holothuroids, and by autotomy (from a portion of an arm) only in asteroids. These two modes of asexual reproduction have a greater cost to the parent than reproduction involving gametes because of the greater transfer of mass to the 'offspring'. The autotomised parts or fission products must have sufficient mass and organisation to survive into a functioning individual. Reproduction by either fission or autotomy would have the consequences of low fecundity and a long period of time between reproductive events. Lack of dispersal ability, lack of the ability of a population to increase rapidly in size, and lack of outbreeding would be costs. The benefits should involve a high probability of survival of the 'offspring' and stability of the population. Like brooding, these characteristics may be particularly beneficial to species of echinoderms with a small body size. However, one must note that fission or autotomy may be difficult for large individuals. Reproduction by fission or autotomy does not seem to be an option for crinoids or echinoids because of the constraints of the body form.

Despite the potential for asexual reproduction by these groups, its extent is limited in terms of number of genera or even of congeneric species. These types of asexual reproduction may have disadvantages for most asteroids, ophiuroids and holothuroids.

4.4.1. Fission

Fission occurs in asteroids, ophiuroids and holothuroids. The occurrence of fission in these groups is limited to only a few species. Only 18 of *c.* 1600 species in only two of the 30 families of asteroids, and 34 of the *c.* 2000 species in nine of the 17 families of ophiuroids are fissiparous (A.M. Clark 1967; Mladenov and Emson 1984). Only two species of cucumarid dendrochirotids (*Ocnus* species) and four species of holothurid aspidochirotids (*Holothuria* species) are fissiparous (Emson and Wilkie 1980).

Fission in asteroids, ophiuroids and holothuroids thus seems correlated with a body wall that can be torn. All of these groups have members whose body wall is essentially a test. Any increase in rigidity should decrease the potential for fission. The presence of a rigid body is probably a factor preventing fission in echinoids and crinoids. The ability to close the wound resulting from fission should be important. The flexibility of the body wall and the great development of the circular muscles would seem to give great potential for sealing the wound in holothuroids. The ability to close the wound may be difficult in large asteroids and ophiuroids. Certainly the large wound which results from the division of a crinoid or echinoid would be difficult to repair.

Fission in asteroids involves the co-ordinated pull of two groups of arms in opposition causing the individual to split (Figure 4.19). Emson and Wilkie (1980) concluded that the structure of the body wall is very important in fission, occurring more rapidly in species that have loosely constructed body walls (e.g. *Coscinasterias calamaria*). It would seem that such a body wall is a prerequisite for the potential to occur, but not all

Figure 4.19: Fission in *Allostichaster polyplax* showing (a) the characteristic posture of an individual initiating fission, (b) an individual in the processes of fission, and (d) an individual near the end of fission (from Emson 1978)

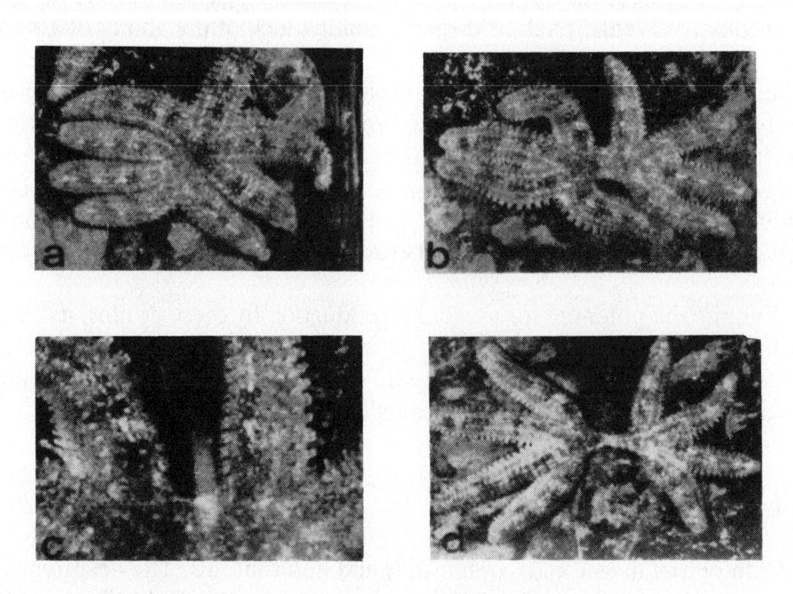

species which have such a body wall undergo fission.

The fission plane in *Coscinasterias acutispina* is indicated by a shallow furrow in which the muscle and tissue layers are relatively reduced (Yamazi 1950). The first indication of fission is the development of this bilateral furrow in the body wall on opposite sides of the disc. The development of the furrow must involve some change in state of the body wall involving an alteration in the strength or rigidity of the connective tissues. The capacity to soften the body wall and connective tissues is widespread in echinoderms (Motokawa 1984; Wilkie 1984). The furrow gradually deepens, and subsequent division seems to have a physical basis. The tube feet of an asteroid undergoing fission move in opposite directions, a process which Emson and Wilkie (1980) related to the contemporary dominance of either of two large arms or of one or more of a set of opposing arms. Repetitive fission by an individual is frequently along the same division line formed previously, indicating a weakening of the body wall or some other process leading to a disposition for division here, but this is not necessarily the case. Five-armed *Stephanasterias albula* also undergo fission by the furrow process, but six-armed individuals divide into two halves by a stretching process (Mladenov, Carson and Walker 1986).

The process of fission in the ophiuroid *Ophiocomella ophiactoides* contrasts with that described for asteroids as there is only a unilateral furrow developing in an interradial plane between two jaws (Mladenov, Emson,

Colpit and Wilkie 1983; Wilkie, Emson and Mladenov 1984). No special furrow plane is apparent. As in the asteroids, the tissue softens, furrows, and tears irregularly across the disc. The tube feet do not seem to pull the two halves apart as in asteroids, but the arms may be involved. The interradial rupture could minimise damage to the gonads, circumoral nerve ring and water vascular ring.

In holothuroids, initial constriction of the body wall is not common, twisting of the body wall is more frequent, but stretching of the body also occurs (Figure 4.20). The circular muscles contract and the tube feet pull in opposite directions until the tissues are severed. The structural change in the body wall is probably similar to the decrease in tensility found in fission in asterozoans and in evisceration in holothuroids (Byrne 1985).

Fission obviously does not result immediately in two individuals with the usual body form. The bodies of asteroids and ophiuroids that have undergone fission are usually distinctly asymmetrical with arms of different lengths (Figures 4.21. and 4.22). The number of arms is usually more than five. In ophiuroids the number is often even, and is usually six. Typically more than one madreporite and often more than one anus exist in asteroids. Fission may begin again before regeneration of new arms is complete in the asteroid *Stephanasterias albula* (Mladenov *et al.* 1986) and the ophiuroid *Ophiocomella ophiactoides* (Mladenov *et al.* 1983). This may maximise the production of new individuals by decreasing the time between successive fissions (Mladenov *et al.* 1983).

In holothuroids, the regenerating posterior end is dependent upon the body wall and dissolved organic material for regeneration. Despite this dependency, and despite the greater degree of reorganisation necessary, the posterior end regenerates more rapidly than does the anterior end

Figure 4.20: Fission in *Ocnus planci* by (a) stretching, and (b) twisting (Monticelli, from Emson and Wilkie 1980)

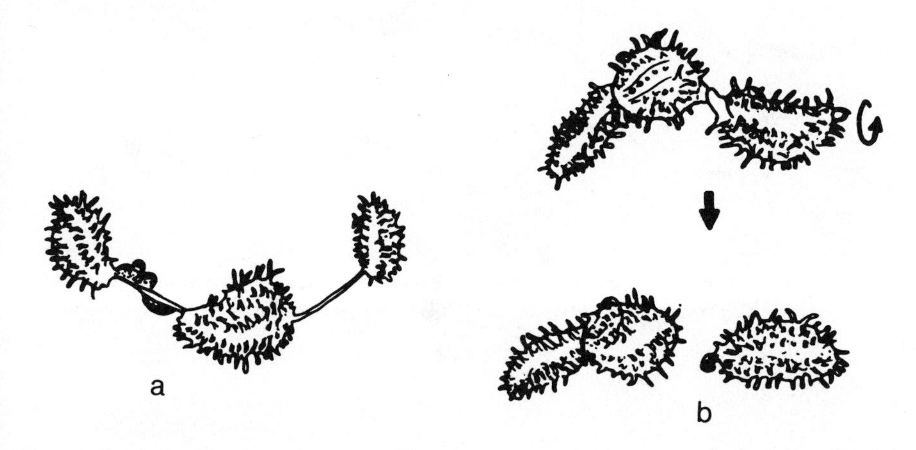

Figure 4.21: Body forms of *Coscinasterias acutispina* showing the results of fission through the disc (from Edmonson 1935)

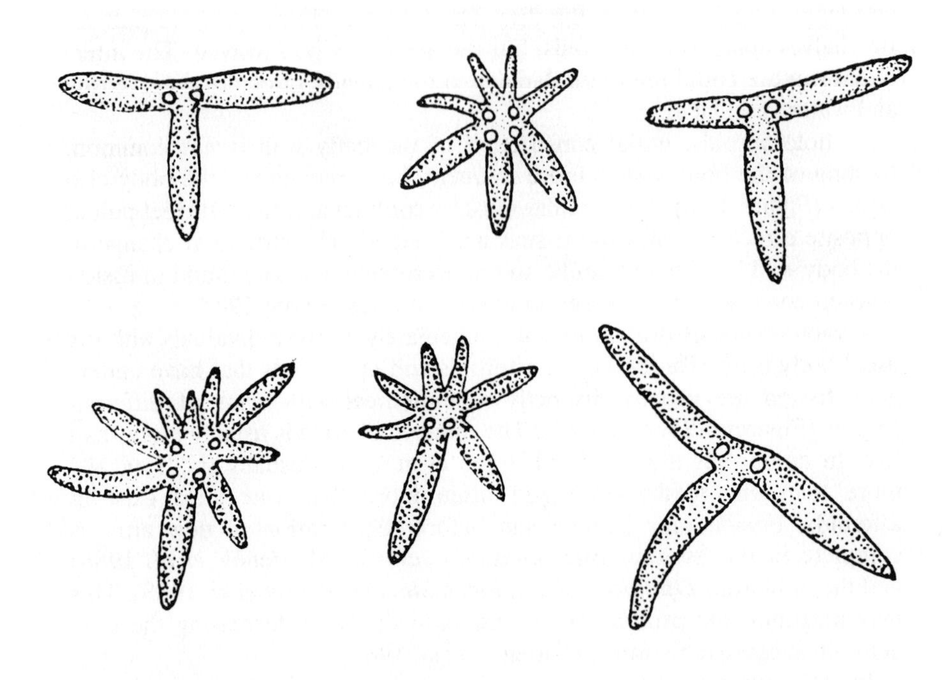

Figure 4.22: Body forms of *Ophiocomella ophiactoides* showing the results of fission through the disc (from Mladenov *et al.* 1983)

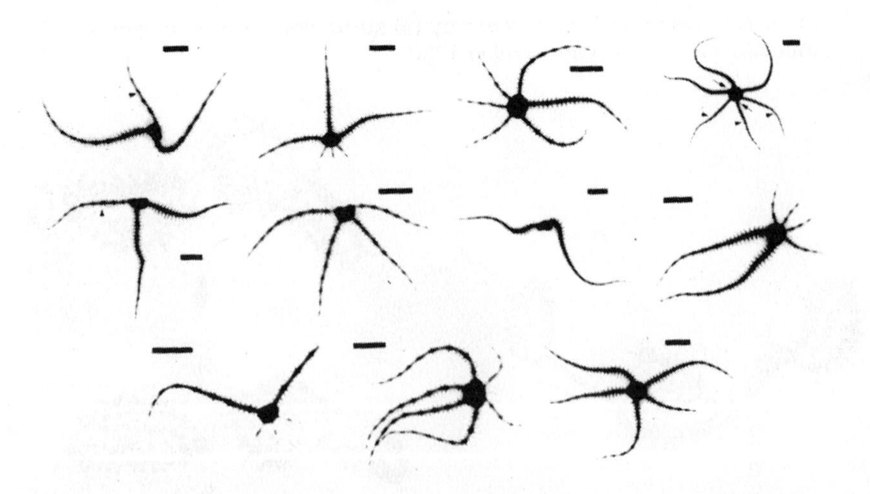

(Doty 1977). This might be correlated with occurrence of the gonads at the anterior end of the body in holothuroids. Until the regeneration of the anterior end the potential for sexual reproduction would be absent in the posterior fission product.

Energetic and population considerations have been postulated for the occurrence of fission primarily in species with small body sizes. In some species, populations differ as to whether fission occurs primarily in small individuals. Emson and Wilkie (1980) suggested that small individuals may be poorly nourished, as there is a good correlation between body size and food availability. Species that are closely related to those that reproduce by fission, and that are similar in size and structure, do not undergo fission. It is obvious that many factors are involved in asexual reproduction by fission. An appropriate body structure and size are prerequisites, but other controls must be present.

In the ophiuroids most closely studied (e.g. *Ophiactis virens*, *O. savignyi*, *Astroceras annulatum*, *Ophiothela danae*), fission occurs at any body size but is more prevalent in small individuals (Emson and Wilkie 1980). A small size and flexibility of arms are adaptive to the intertidal area (Emson and Wilkie 1980) and coralline algal and turtle-grass beds in shallow water (Emson *et al.* 1985).

Emson and Wilkie (1980) concluded that fissiparity in asteroids cannot be related to individual body size in some species (e.g. *Asterina burtoni*, *Coscinasterias tenuispina*, *Sclerasterias richardi*), but seems most common in small individuals of others (e.g. *Coscinasterias acutispina*, *Sclerasterias heteropes*), or is population dependent (e.g. *Nepanthia belcheri*, *Coscinasterias calamaria*). It is likely that all fissiparous species belong to the third category.

Fission seems to occur primarily in small *Holothuria difficilis* and *H. parvula*, but in all sizes of *H. surinamensis* and *Ocnus planci* (Emson and Wilkie 1980).

Of equal interest to the maximal size at which fission is possible is the restriction of fission to individuals above a minimal size, usually a radius of 6–8 mm in asteroids and a disc diameter of *c.* 4 mm in ophiuroids. This can perhaps be taken as the minimal size that can subdivide into functional individuals or those that have sufficient reserves to support regeneration until a functional state is achieved.

Sexual reproduction also occurs in fissiparous species. Emson and Wilkie (1980) reported that regeneration and sexual reproduction occur simultaneously in almost all fissiparous species of all three classes. They stated that it is the extent to which sexual development proceeds to completion which is variable. With fission, the number of gametes produced may be reduced both as a result of a decrease in size of the individual and potentially from a diversion of energy to regeneration (Crozier 1920). Emson and Wilkie pointed out that energy which otherwise would go for gametogenesis must be used for regeneration in individuals that had undergone fission. Even in non-fissiparous species, regeneration may take priority over reproduction (e.g. *Pisaster giganteus*, Harrold and Pearse 1980). Deposition of nutrient reserves even takes priority over arm regeneration in *Luidia clathrata* (Lawrence *et al.* 1986).

The fissiparous ophiuroid *Ophiocomella ophiactoides* has sexual reproduction throughout the year (Mladenov and Emson 1984). Gonads often occur in recently divided individuals, indicating the possibility of simultaneous sexual and asexual reproduction. Males are smaller than females, suggesting that the species is a protandrous hermaphrodite. The protandrous state would be an appropriate energetic response to produce a functional number of gametes at the smaller body size. Emson and Wilkie (1980) pointed out that the small size of fissiparous ophiuroids results in a limited capacity for sexual reproduction, but described (1984) successful reproduction in the fissiparous *Ophiactis savignyi* whose adult disc diameter was only *c.* 4 mm. Because of the limited fecundity resulting from the small body size, asexual reproduction of these ophiuroids seems to be the major form of reproduction and to be responsible for population maintenance (Emson *et al.* 1985). Sexual reproduction would provide for dispersal of small numbers of offspring. This could be important in shallow-water habitats where mass mortalities can occur unpredictably.

In *Nepanthia belcheri*, only fully regenerated individuals contain oocytes; if they undergo fission, the gonads regress and the individual develops through protandrous hermaphroditism (Ottesen and Lucas 1982). As with *Ophiocomella ophiactoides*, this phenomenon implies an energetic response related to body size, the ability to feed, and the ability to produce a functional number of gametes. Ottesen and Lucas proposed that reproduction by fission involves the greatest expenditure of energy per reproductive unit but gave the highest probability for entry of the unit into the population.

A population of the shallow-water (1–3 m) asteroid *Stephanasterias albula* that seems to be obligately fissiparous occurs off the coast of Maine (Mladenov *et al.* 1986). No allocation to sexual reproduction occurs as a differentiated reproductive system is completely lacking. Individuals in this population have a restricted diet and food supply, and their body size is much less than is possible for the species. Thus *S. albula* has exploited the potential for asexual reproduction by fission in asteroids as a means of maintaining a population with inadequate resources for sexual reproduction.

Populations of the fissiparous asteroids *Coscinasterias calamaria* differ in the degree to which they reproduce sexually and asexually (Crump and Barker 1985). Populations of large individuals had large gonads and a low incidence of fissiparity whereas those of small individuals had small gonads and a high incidence of fissiparity. The characteristics of body size and degree of gonadal development indicate that the former populations had a more abundant or better food availability.

It thus seems that fissiparity is a reponse developed in some asteroid species as a means of reproduction when nutrient resources are too low to support a large body size and gonadal growth. This implies that the better strategy is to grow large and reproduce sexually if possible. One should note that the fissiparous *Coscinasterias*, *Allostichaster* and *Sclerasterias* belong to the same subfamily, Asteriinae, as the non-fissiparous *Asterias*, *Leptasterias* and *Pisaster*. This indicates that constraints of body form have not prevented the development of fissiparity as an alternative mode of reproduction under conditions of inadequate food resources in the latter group.

Various kinds of physical stress initiate fission, and correlations of the degree of fission in asteroid species and physical stress in the field occur (Emson and Wilkie 1980). Although low availability of food has been indicted as a stress-inducing fission, this is probably incorrect. Low availability of food leads to conditions that are conducive to fission.

Holothuria atra on coral-reef flats shows asexual reproduction by fission as well as sexual reproduction. At Heron Island, Australia, fission is greater in individuals living in the gutter and mid-reef habitats than on the reef crest and lagoon (Harriott 1982). A direct relation of the incidence of fission to physical stress is supported by the low incidence found in the lagoon but not by the low incidence found on the reef crest. Although fission of echinoderms may be correlated with physical stress (Emson and Wilkie 1980), it is difficult to interpret fission as a response to physical stress *per se* as a means to increase fitness by increasing reproduction. Ebert (1978) suggested that fission by *H. atra* was an adaptation to the difficulty of recruitment from sexual reproduction. The observations of the differences in incidence of fission found at Heron Island may be indicative of food availability if this is a factor in incidence of fission as it seems to be for ophiuroids and asteroids.

Fission should lead to an increased density of individuals in a population, but the densities of those species that undergo fission (Emson and Wilkie 1980) are not necessarily greater than those of non-fissiparous species. The question, though, is whether the population of fissiparous species could persist or be at the densities that occur without fission (Ebert 1983). Although the density may not be any greater, the predictability of recruitment may be of considerable significance. Emson and Wilkie concluded that the importance of fissiparity in echinoderms is no doubt variable, depending on the relative frequency of successful sexual reproduction, but should be at least considerable and possibly essential. The latter would be true when fission is so frequent that sexual maturity is rare. Emson *et al.* (1985) concluded that fission can be the dominant form of reproduction in small ophiuroids and responsible for reliable local recruitment. In this case, sexual reproduction would result in dispersal of small numbers of larvae which could found new populations. This potential for dispersion provides an additional dimension that does not occur in other adaptations to low fecundity, abbreviated development or direct development which can be associated with brooding.

4.4.2. Autotomous asexual reproduction

Emson and Wilkie (1980) defined autotomous asexual reproduction in echinoderms as that resulting from the autotomy of an arm. Asexual reproduction by autotomy occurs in asteroids but not in ophiuroids or crinoids. Only seven species in the Ophidiasteridae (three species of *Linckia* and two species of *Ophidiaster*), Echinasteridae (*Echinaster luzonica*) and Asterinidae (*Nepanthia belcheri*) are known to have autotomous asexual reproduction.

The lack of autotomous asexual reproduction in crinoids and ophiuroids may be due in part to the lack of sufficient tissue to support the production of a functional calyx or disc. Thus, the fleshy asteroid arm and the presence of the plyoric caeca may be necessary for autotomous asexual reproduction. The presence of adhesive tube feet which enable an autotomised arm of an asteroid to attach to the substratum and to locomote could also be a factor. However, the rarity of their reproductive mode indicates that the developmental reorganisation that is involved does not occur easily. Certainly, the loss of arms by many species of asteroids seems to occur with sufficient frequency for the capacity to regenerate a new individual from the autotomised arm to be adaptive .

The incidence of autotomous asexual reproduction shows variation among populations. An upper limit to the size of an arm which would regenerate would not be expected, except possibly if the ability to close the wound was limited. A lower limit to the size of an arm which would regenerate could be possible. Rideout (1978) found no 'comets' of *Linckia multifora* with an arm of less than 8 mm in length. The growth rate of the regenerating arm was independent of the length of the principal arm, indicating that if a principal arm has a minimal size, the regeneration process has an inherent rate of development.

The autotomy of an arm in these asteroids is similar to the process of fission (Figure 4.23). A proximal part of the arm softens, followed by the pulling of the tube feet in opposition probably as co-ordination of their activity is lost through incapacitation of that portion of the nerve cord. This process is much different from that of autotomy of the arms of *Luidia* in which the suckerless tube feet are not involved and the arm autotomises rapidly. This process is more similar to the autotomy that Monks (1904) described for *Linckia columbiae*, which shows autotomous asexual repro-

Figure 4.23: Arm autotomy in *Linckia multifora* over a 3-min period (from Rideout 1978)

duction. Rideout (1978) suggested that the principal arms of 'comets' were autotomised to produce 'counter-comets' resulting in cycles promoting the production of new individuals (Figure 4.24). The basis for this increased susceptibility of the longest arm is unknown.

As with asexual reproduction by fission, the asteroid species that do possess autotomous asexual reproduction are small, but not more so than congeneric species that do not. This suggests that the phenomenon may be related to the habitat. The species with autotomous asexual reproduction are all tropical, shallow-water or reef-flat dwellers. *Linckia, Ophidiaster, Echinaster* and *Nepanthia* are microphagous surface-film feeders (see Jangoux 1982). The two species of *Linckia* that reproduce by autotomy are cryptic on the coral reef at Guam (Yamaguchi 1975). Although *Echinaster luzonica* is said to live exposed on the coral reef at Guam (Yamaguchi 1975), it is cryptic except when feeding nocturnally at

Figure 4.24: Cycle of autotomous asexual reproduction in *Linckia multifora*. C: Autotomised arm; D: young comet; E: comet; F: counter-comet formed by autotomisation of the principal arm of the comet; G to I: post counter-comet; J: disc; K: disc regenerating all arms. (From Rideout 1978)

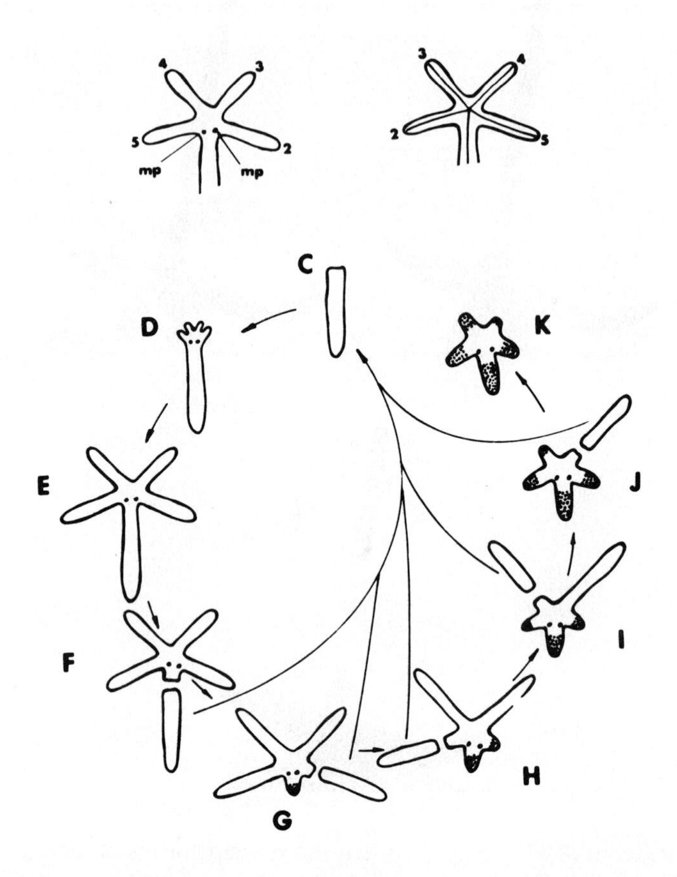

Eniwetak (pers. obs.). The three species of *Ophidiaster* on the coral reef at Guam are all cryptic (Yamaguchi 1975).

The three species of *Ophidiaster* show a complex of reproductive characters (Yamaguchi and Lucas 1984). *O. squameus* has only sexual reproduction with planktotrophic larvae, *O. granifer* reproduces asexually parthenogenetically (see above), and *O. robillardi* reproduces asexually by autotomy. *O. squameus* is very rare at Guam, whereas *O. granifer* is common and *O. robillardi* is rare (Yamaguchi 1975). *Linckia guildingi* and *L. multifora* reproduce asexually by autotomy, whereas *L. laevigata* reproduces sexually with planktotrophic larvae. *L. guildingi* is very rare at Guam, but both *L. multifora* and *L. laevigata* are very common (Yamaguchi 1975). *Echinaster luzonica* with asexual reproduction by autotomy is rare at Guam (Yamaguchi 1975) but common at Eniwetak (pers obs.).

It thus seems that some species of small, microphagous asteroids living in areas where food availability may be low and recruitment poor may have exploited the potential of arm autotomy as a means of asexual reproduction. They would thus have a means for facilitating recruitment into the population by asexual reproduction along with a means of dispersal by sexual reproduction (Ottesen and Lucas 1982). The success of autotomous asexual reproduction in the sense of producing a dense population seems to be variable. The maintenance of a population, whether dense or not, may be more important in the overall life history.

Appendix: Classification of Extant Echinoderms

(Modified from the *Treatise on Invertebrate Paleontology* and the *Synopsis and Classification of Living Organisms*)

PHYLUM ECHINODERMATA

A water vascular system of coelomic origin with radial water canals; a mesordermally derived endoskeleton composed of crystalline calcite in stereom microstructure.

1. Class Crinoidea

Ambulacra, mouth and anus on the upper body surface, stalked or secondarily stalkless with a reduced theca; radial arms, which are generally branched to form additional arms supported by internal skeletal elements.

1.1. Subclass Articulata

Calyx dicyclic or usually cryptodicyclic with five infrabasal, five basal and five radial plates, with the infrabasals reduced in size or absent in adults. Arms uniserial with muscular articulations between radials and arms and between some or all brachial plates. Brachials not part of the calyx, which is small. Non-muscular articulations present. Pinnules always present.

1.1.1. Order Cyrtocrinida. Calyx composed of large radials with no basals apparent. Permanently attached by short stalk or simple expansion of an attachment disc. Example: *Holopus.*

1.1.2. Order Bourgueticrinida. Calyx small, composed of five basals and five radials. Slender stalk attached to hard substratum by an irregular plate or to soft substratum by branched rootlike cirri. Examples: *Bathycrinus, Rhizocrinus.*

1.1.3. Order Isocrinida. Calyx small, radials with large muscular articulations. Stalked with true cirri at column internodes. Juveniles attached by terminal disc; post-juveniles with column broken at distal end. Examples: *Cenocrinus, Neocrinus, Endoxocrinus.*

1.1.4. Order Comatulida. Stalk present only in early postmetamorphic juveniles, autotomised after formation of first cirri-bearing columnal. The columnal or fused uppermost columnals forming a centrodorsal attached to the calyx. Examples: *Capillaster, Tropiometra, Neometra, Astrometra, Thalassometra, Crinometra, Notocrinus, Aporometra, Antedon.*

2. Class Asteroidea

Body usually star-shaped, formed of arms which merge gradually with the disc or are sharply set off from it, but may be pentagonal or circular without pronounced arms. Arms supported by skeletal elements in the body wall. Usually five arms but may be multiarmed as a result of additional radial water canals. Ambulacra and mouth on under body surface. Anus on upper body surface if present.

2.1. Order Paxillosida

Varied body form; papulae widely distributed on the upper body surface; conspicuous marginal plates define the ambitus; tube feet in two rows and taper to blunt points or small knobs; inconspicuous mouth plates; pedicellariae, when present, are spiniform or sessile; may lack an anus. Examples: *Luidia, Astropecten, Ctenodiscus.*

2.2. Order Notomyotida

Stellate body form; papulae restricted to proximal part of upper body surface, usually in an oval or bifurcate area near the base of the arm; conspicuous marginal plates which are frequently alternate rather than opposite; tube feet in two rows and with suckers; inconspicuous mouth plates; pedicellariae are fasciculate or pectinate. Examples: *Benthopecten, Cheiraster, Pectinaster.*

2.3. Order Valvatida

Varied body form; papulae widely distributed on the upper surface and sometimes on the undersurfaces; marginal plates may be conspicuous; tube feet in two rows and with suckers; mouth plates may be conspicuous; pedicellariae, when present, are valvate and in pits or depressions in the ossicles. Examples: *Odontaster, Hippasteria, Goniaster, Podosphaeraster, Ophidiaster, Asterina, Porania, Oreaster, Acanthaster.*

2.4. Order Velatida

Varied body form; papulae widely distributed on the upper surface and sometimes on the undersurface; marginal plates are usually inconspicuous; tube feet in two rows (except in *Diplopteraster*) and with suckers; prominent mouth plates; pedicellariae, when present, are spiniform or sessile. Examples: *Solaster, Hymenaster, Pteraster, Diplopteraster.*

2.5. Order Spinulosida

Rounded, blunt or tapering arms; widely distributed papulae; marginal plates are usually inconspicuous; tube feet in two rows and with suckers; prominent mouth plates; no pedicellariae. Examples: *Echinaster, Henricia.*

2.6. Order Forcipulatida

Usually a stellate body form; papulae widely distributed, often on both the upper surface and under surface; marginal plates are inconspicuous; tube feet usually in four rows and with suckers; dense ambulacral plates; inconspicuous mouth plates; pedicellariae, usually present, are crossed and/or straight with two valves and a basal piece. Examples: *Asterias, Pisaster, Marthasterias, Pycnopodia, Heliaster, Coscinasterias, Leptasterias, Evasterias.*

2.7. Order Brisingida

Small, flat, circular disc which is well marked off from long, attenuate arms which always number more than five; papulae widely distributed; inconspicuous marginal plates and usually in only one incomplete series; tube feet in two rows and with suckers; inconspicuous mouth plates; pedicellariae, always present, are abundant and crossed. Examples: *Brisinga, Freyella, Colpaster.*

3. Class Ophiuroidea

Flat disc almost always sharply distinct from slender, elongate arms. Arms supported by internal skeletal elements. Usually five arms but may be multiarmed. Arms may branch. Ambulacra and mouth on lower body surface. No anus.

3.1. Order Oegophiurida

Ambulacral groove covered by soft skin forming open canal; disc covered by skin with or without imbricating scales; no oral or radial shields, dorsal or ventral arm plates, genital plates or bursae; paired serial gonads and gastric caeca in proximal part of arms. Example: *Ophiocanops.*

3.2. Order Phrynophiurida

Disc and arms covered with skin which may have embedded plates or granules; ambulacral groove covered by ventral arm plates; radial shields; gonads usually completely in disc but may have extension in proximal part of arm; bursae. Examples: *Ophiomyxa, Gorgonocephalus, Asteroporpa, Astroboa.*

3.3. Order Ophiurida

Arms completely covered by dorsal, ventral and lateral plates, arranged segmentally with vertebrae; ambulacral grooves closed by lateral plates; usually radial and oral shields, genital plates. Examples: *Ophiura, Ophiolepis, Ophiocoma, Ophionereis, Ophioderma, Amphiura, Amphioplus, Ophiothrix.*

4. Class Echinoidea

Test composed of plates which have movable spines. Plates arranged in 20 meridional columns, two columns in each ambulacrum and interambulacrum, alternating in 10 pairs; mouth on lower surface; Aristotle's lantern which may be secondarily lost; internal radial water canals which extend from the oral ring to the apex of the body at the apical system.

4.1. Subclass Cidaroidea, Order Cidaroida

Regular echinoids with mouth on lower surface and anus opposite on upper surface; spherical or subspherical shape; ambulacra much narrower than interambulacra; two interambulacral columns in extant forms; perignathic girdle incomplete, composed of apophyses; grooved teeth (aulodont dentition); no buccal slits or extensions; each interambulacral plate with a single large primary tubercle with a large spine with a cortex. Examples: *Cidaris, Eucidaris, Phyllacanthus, Prionocidaris.*

4.2. Subclass Euechinoidea

Regular or irregular echinoids; interambulacra and ambulacra each with two columns; ambulacral plates compound in regular forms, simple in irregular forms; perignathic girdle complete (ambulacral and interambulacral elements).

4.2.1. Superorder Echinacea (abridged list of orders). Regular echinoids with mouth on lower surface with anus opposite on upper surface; keeled teeth; compound ambulacral plates; perignathic girdle with well developed auricles; well developed buccal slits and extensions.

4.2.1.1. *Order Echinoida.* Rounded (may be domed or subspherical); echinoid ambulacral plates; echinoid–camarodont lantern; large peristome with shallow buccal slits, Examples: *Echinus, Loxechinus, Paracentrotus, Psammechinus, Sterechinus, Lytechinus, Toxopneustes, Tripneustes, Echinometra, Colobocentrotus, Heterocentrotus, Evechinus, Strongylocentrotus, Allocentrotus, Hemicentrotus.*

4.2.1.2. *Order Arbacioida.* Rounded (domed); compound arbacioid ambulacral plates; arbacioid–stirodont lantern, arbacioid-type teeth. Examples: *Arbacia, Coelopleurus.*

4.2.2. Superorder Echinothuriacea. Regular echinoids with mouth on lower surface and anus opposite on upper surface; subspherical; echinothuriid-type ambulacral plates, echinothuriid aulodont and phormosomatid–aulodont lantern, grooved teeth, test low, flexible; buccal slits and extensions. Examples: *Echinothuria, Asthenosoma, Phormosoma.*

4.2.3. Superorder Diadematacea. Regular echinoids with oral mouth on lower surface and anus opposite on upper surface; subspherical, diadematoid or micropygid-type compound ambulacral plates; diadematoid–aulodont-type lantern, grooved teeth, hollow spines, imbricating plates; gills with deep gill slits, perignathic girdle with auricles. Examples: *Diadema, Echinothrix, Micropyga.*

4.2.4. Superorder Atelostomacea. Irregular echinoids with pronounced secondary bilateral symmetry; posterior anus; no gills; no Aristotle's lantern in adult; simple ambulacral plates; interambulacra wider than ambulacra adorally.

4.2.4.1. *Order Holasteroida.* Flattened round or oval shape; apical system elongate or disjunct; usually with petals; apical system and mouth may be opposite. Examples: *Sternopatagus, Urechinus, Stereopneustes.*

4.2.4.2. *Order Pourtalesioida.* Bottle-shaped; meridosternous plastron; ambulacra usually with single pores; test fragile. Examples: *Pourtalesia, Echinosigra*

4.2.4.3. *Order Spatangoida.* Mouth usually slightly anterior, compact apical system; plastron amphisternal; posterior periproct; phyllodes and fascioles present; no bourrelets. Examples: *Schizaster, Abatus, Moira, Brissopsis, Eupatagus, Meoma, Metalia, Spatangus, Lovenia.*

4.2.5. Superorder Gnathostomacea. Irregular echinoids with pronounced secondary bilateral symmetry; mouth central or subcentral on lower surface; anus posterior; lantern present but may be secondarily lost; teeth of triangular/diamond- or wedge-shaped types.

4.2.5.1. *Order Echinoneoida.* Ambulacral plates simple, or simple and pseudocompound, lantern may be absent in adults, echinoneoid-type lantern, non-petaloid ambulacra. Example: *Echinoneus.*

4.2.5.2. *Order Cassiduloida.* Fibulariid-type lantern present in juveniles, cassiduloid-type lantern support; compact apical system, adapical petaloid ambulacra. Example: *Cassidulus.*

4.2.5.3. *Order Clypeasteroida.* Petaloid ambulacra with pseudo-compound clypeasterid-type plates within petals. Clypeasterid-type lantern, arachnoidid- or clypeasterid-type tooth plates, Examples: *Clypeaster, Arachnoides.*

4.2.5.4. *Order Laganoida.* Petaloid ambulacra with simple ambulacral plates within petals. Fibulariid- or astriclypeid-type lantern; fibulariid-, rotulid-, astriclypeid- or mellitid-type tooth plates. Examples: *Mellita, Fibularia, Rotula, Echinarachnius.*

5. Class Holothuroidea

Ring of tentacles around the mouth; ossicles in the body wall may be absent; when present, the ossicles are usually microscopic and in a soft body wall, but are large and form a test in some species.

5.1. Order Dendrochirotida

Introvert and respiratory trees present; madreporite free in the body cavity; two gonadal tufts; no free tentacle ampullae; extremely branched tentacles; simple to complex calcareous ring; soft or firm body wall. Examples: *Placothuria, Psolus, Thyone, Sclerodactyla, Cucumaria.*

5.2. Order Dactylochirotida

Introvert and respiratory trees present; madreporite free in the body cavity; two gonadal tufts; no free tentacle ampullae; digitiform or digitate tentacles; simple calcareous ring; firm test. Examples: *Ypsilothuria, Rhopalodina.*

5.3. Order Aspidochirotida

Respiratory trees present; conspicuous bilateral body symmetry; shield-shaped tentacles; generally thick body wall. Examples: *Holothuria, Actinopyga, Stichopus.*

5.4. Order Elasipodida

Respiratory trees absent; conspicuous bilateral body symmetry; shield-shaped tentacles; generally gelatinous body wall, but occasionally plated. Examples: *Deima, Benthogone, Benthodytes, Pelagothuria.*

5.5. Order Apodida

Respiratory trees, tube feet, and anal papillae absent; simple, digitate, or pinnate tentacles; very thin body wall. Examples: *Synaptula, Chiridota, Leptosynapta, Euapta.*

5.6. Order Molpadida

Respiratory trees and anal papillae present; tube feet reduced and usually absent; simple or digitate tentacles. Examples: *Molpadia, Caudina.*

References and Further Reading

Agassiz, A (1869) On the habits of a few echinoderms, *Proc. Boston Soc. Hist.*, *13*, 104–7

Agassiz, A. (1873) *Revision of the Echini.* Illustrated Catalogue of the Museum of Comparative Zoology at Harvard College. University Press, Cambridge, Mass.

Agassiz, A. (1877) North American Starfishes, *Mem. Mus. Comp. Zool., Harvard College, 5* (1)

Agassiz, A. (1878) Report on the dredging operations of the United States Survey Steamer 'Blake'. II. Report on the Echini, *Bull. Mus. Comp. Zool. Harvard, 5*, 185–94

Agassiz, A. (1881) Report on the Echinoidea dredged by H.M.S. Challenger during the years 1873–1876. *Report on the scientific results of the voyage of H.M.S. Challenger during the years 1873–76. Zoology. 3*, part 9

Agassiz, A. (1908) The genus *Colobocentrotus, Mem. Mus. Comp. Zool., Harvard College, 39* (1)

Agassiz, A. and H.L. Clark. (1907) Hawaiian and other Pacific Echini. The Cidaridae, *Mem. Mus. Comp. Zool., Harvard College, 34* (1)

Agassiz, A. and H.L. Clark. (1908) Hawaiian and other Pacific Echini. The Salenidae, Arbaciadae, Aspidodiadematidae, and Diadematidae, *Mem. Mus. Comp. Zool., Harvard College, 34* (2)

Agassiz, A. and H.L. Clark. (1909) Hawaiian and other Pacific Echini. The Echinothuridae, *Mem. Mus. Comp. Zool., Harvard College, 34* (3)

Agassiz, L. (1841) *Monographies d'Échinodermes Vivant et Fossiles. Échinities. Famille des Clypéastroides. Secconde Monographie, Des Scutelles*, Neuchatel

Anderson, J.M. (1954) Studies on the cardiac stomach of the starfish, *Asterias rubens, Biol. Bull.*, *107*, 157–73

Andrew, N.L. and L.D. Stocker. (1986) Dispersion and phagokinesis in the echinoid *Evechinus chloroticus* (Val.), *J. Exp. Mar. Biol. Ecol.*, *100*, 11–23

Ausich, W.I. (1980) A model for niche differentiation in Lower Mississippian crinoid communities, *J. Paleontol.*, *54*, 273–88

Austin, W.C. (1966) *Feeding mechanisms, digestive tracts and circulatory systems in the ophiuroids* Ophiothrix spiculata *Le Conte, 1851 and* Ophiura luetkeni (*Lyman, 1860*), Unpublished Ph.D. thesis, Stanford University, Stanford

Baker, A.N. (1980) Euryalinid Ophiuroidea (Echinodermata) from Australia, New Zealand, and the south-west pacific Ocean, *New Zealand J. Zool.*, *7*, 11–83

Baker, S.L. (1973) *Growth of the red sea urchin* Strongylocentrotus franciscanus (*Agassiz*) *in two natural habitats.* Unpublished M.S. thesis, San Diego State University

Bakus, G.J. (1968) Defensive mechanisms and ecology of some trophical holothurians, *Mar. Biol.*, *2*, 23–32

Bakus, G.J. (1973) The biology and ecology of trophical holothurians, in O.A. Jones and R. Endean (eds), *Biology and geology of coral reefs. II: Biology*, Academic Press, New York, pp. 325-67

Bakus, G.J. (1974) Toxicity in holothurians; a geographical pattern, *Biotropica, 6*, 229-36

Bamford, D. (1982) Epithelial absorption, in M. Jangoux and J.M. Lawrence (eds), *Echinoderm nutrition*, A.A. Balkema, Rotterdam, pp. 317–30.

Barker, M.F. (1985) Reproduction and development in *Goniocidaris umbraculum* a brooding echinoid, in B.F. Keegan and B.D.S. O'Connor (eds), *Echinodermata*, A.A. Balkema, Rotterdam, pp. 207–14

Barnes, A.T., L.B. Quetin, J.J. Childress and D.L. Pawson. (1976) Deep-sea macroplanktonic sea cucumbers: suspended sediment feeders captured from deep submergence vehicle, *Science*, *194*, 1083-5

Barrington, E.J.W. (1967) *Invertebrate structure and function*, Houghton Mifflin, Boston

Bather, F.A. (1900) The Echinoderma, in E.R. Lankester (ed.), *A treatise on zoology*, Adam & Charles Black, London

Bedford, A.P. and P.G. Moore. (1985) Macrofaunal involvement in the sublittoral decay of kelp debris: the sea urchin *Psammechinus miliaris* (Gmelin) (Echinodermata: Echinoidea), *Estuarine, Coastal and Shelf Sci.*, *20*, 19-40

Beijnink, F.B., I. Van der Sluis and P.A. Voogt. (1984) Turnover rates of fatty acid and amino acid in the coelomic fluid of the sea star *Asterias rubens*: implications for the route of nutrient translocation during vitellogenesis, *Comp. Biochem. Physiol.*, *78B*, 761-7.

Beijnink, F.B. and P.A. Voogt. (1984) Nutrient translocation in the sea star: whole-body and microautoradiography after ingestion of radiolabeled leucine and palmitic acid, *Biol. Bull.*, *167*, 669-82

Beijnink, F.B. and P.A. Voogt. (1985) Pathways of nutrient translocation during vitellogenesis in the sea star, *Asterias rubens*, in B.F. Keegan and B.D.S. O'Connor (eds), *Echinodermata*, A.A. Balkema, Rotterdam, pp. 477-80

Beklemishev, W.N. (1969) *Principles of comparative anatomy of invertebrates. 1. Promorphology*, University of Chicago Press, Chicago

Bell, B.M. and R.W. Frey. (1969) Observations on ecology and the feeding and burrowing mechanisms of *Mellita quinquiesperforata* (Leske), *J. Paleontol.*, *43*, 553-60

Berrill, M. (1966) The ethology of the synaptid holothurian, *Opheodesoma spectabilis*, *Can. J. Zool.*, *44*, 457-82

Billett, D.S.M., B. Hansen and Q.J. Huggett. (1985) Pelagic Holothurioidea (Echinodermata) of the northeast Atlantic, in B.F. Keegan and B.D.S. O'Connor (eds), *Echinodermata*, A.A. Balkema, Rotterdam, pp. 399-411

Binyon, J. (1972) *Physiology of echinoderms*, Pergamon Press, Oxford

Binyon, J. (1980) Osmotic and hydrostatic permeability of the integument of the starfish *Asterias rubens* L., *J. Mar. Biol. Ass. UK*, *60*, 627-30

Birkeland, C. (1974) Interactions between a sea pen and seven of its predators, *Ecol. Monogr.*, *44*, 211-32

Bismarck, O. von. (1959) Vërsuch einer Analyse die Stoffwechselintensität ('Ruheumsatz') von *Asterias rubens* L., beeinflussenden Faktoren, *Kieler Meeresforsch.*, *15*, 164-86

Black, R., C. Codd, D.Hebbert, S. Vink and J. Burt. (1984) The functional significance of the relative size of Aristotle's lantern in the sea urchin *Echinometra mathaei* (de Blainville), *J. Exp. Mar. Biol. Ecol.*, *77*, 81–97

Black, R., M.S. Johnson and J.T. Trendall. (1982) Relative size of Aristotle's lantern in *Echinometra mathaei* occuring at different densities, *Mar. Biol.*, *71*, 101-6

Blake, D.B. (1983) Some biological controls of the distribution of shallow water sea stars (Asteroidea; Echinodermata), *Bull. Mar. Sci.*, *33*, 703-12

Blankley, W.O. and G.M. Branch. (1984) Co-operative prey capture and unusual brooding habits of *Anasterias rupicola* (Verrill) (Asteroidea) at sub-Antarctic Marion Island, *Mar. Ecol. Prog. Ser.*, *20*, 171-6

Bookbinder, L.H. and J.M. Shick. (1986) Anaerobic and aerobic energy metabolism in ovaries of the sea urchin *Strongylocentrotus droebachiensis*, *Mar. Biol.* *93*, 103-10

Boolootian, R.A. and J.L. Campbell. (1964) A primitive heart in the echinoid *Strongylocentrotus purpuratus*, *Science*, *145*, 173-5

Boolootian, R.A. and A.R. Moore. (1956) Hermaphroditism in echinoids, *Biol. Bull.*, *111*, 328-35

Bourseau, J.P. and M. Roux. (1985) Bathymétrie et variabilité morphologique chez les Pentacrinidae (Echinodermes–Crinoïdes pédonculés) du Pacifique occidentale, in B.F. Keegan and B.D.S. O'Connor (eds), *Echinodermata*, A.A. Balkema, Rotterdam, pp. 175-80

Bowmer, T. and B.F. Keegan. (1983) Field survey of the occurrence and significance of regeneration in *Amphiura filiformis* (Echinodermata: Ophiuroidea) from Galway Bay, west coast of Ireland, *Mar. Biol.*, *74*, 65-71

Breimer, A. (1978) Recent crinoids, in R.C. Moore and C. Teichert (eds), *Treatise on invertebrate paleontology. Part T, Echinodermata 2, 1*, University of Kansas Press, Lawrence, and the Geological Society of America, Inc., Boulder, pp. 9–58

Breimer, A. and N.G. Lane. (1978) Ecology and paleoecology, in R.C. Moore and C. Teichert (eds), *Treatise on invertebrate paleontology, Part T. Echinodermata 2, 1*, University of Kansas Press, Lawrence, and the Geological Society of America, Boulder, pp. 316-47

Breimer, A. and G.D. Webster. (1975) A further contribution to the paleoecology of fossil stalked crinoids, *Proc. Kon. Ned. Akad. Wet.*, *B*, *78*, 149-67

Broertjes, J.J.S. and G. Posthuma. (1978) Direct visualization of the haemal system in starfish by a staining procedure, *Experientia*, *34*, 1243-5

Broertjes, J.J.S., G. Posthuma, P. Den Breejen and P.A. Voogt. (1980) Evidence for an alternative transport route for use of vitellogenesis in the sea-star *Asterias rubens* (L.), *J. Mar. Biol. Ass UK*, *60*, 157-62

Brookbank, J.W. (1968) Spawning season and sex ratio of echinoids, *Quart. J. Fla. Acad. Sci.*, *30*, 177-83

Browmer, J.C. (1976) Evolution of Melocrinitie, *Thalassia Jugosl.*, *12*, 41-9

Brown, W.I. and J.M. Shick. (1979) Biomodal gas exchange and the regulation of oxygen uptake in holothurians, *Biol. Bull.*, *156*, 272-88

Brumbaugh, J.H. (1985) *The anatomy, diet and tentacular feeding mechanism of the dendrochirote holothurian* Cucumaria curata *Cowles 1907.* Unpublished Ph. D. thesis, Stanford University, Stanford

Brun, E. (1969) Aggregation of *Ophiothrix fragilis* (Abildgaard) (Echinodermata; Ophiuroidea), *Nytt Mag. Zool.*, *17*, 153-60

Buchanan, J.B. (1964) A comparative study of some features of the biology of *Amphiura filiformis* and *Amphiura chiajei* (Ophiuroidea) considered in relation to their distribution, *J. Mar. Biol. Ass. UK*, *44*, 565-76

Buchanan, J.B. (1967) Dispersion and demography of some infaunal echinoderm populations, in N. Millott (ed.), *Echinoderm Biology, Symp. Zool. Soc. Lond.* (20), pp. 1-11

Buchanan, J.B. (1969) Feeding and the control of volume within the tests of regular sea-urchins, *J. Zool., Lond.*, *159*, 51–64.

Budington, R.A. (1942) The ciliary transport-system of *Asterias forbesi*, *Biol. Bull.*, *83*, 438-50

Burkhardt, A., W. Hansmann, K. Märkel and H.-J. Niemann. (1983) Mechanical design in spines of diadematoid echinoids (Echinodermata, Echinoidea), *Zoomorphology*, *102*, 189-203

Burla, H., V. Ferlin, B. Pabst and G. Ribi. (1972) Notes on the ecology of *Astropecten aranciacus*, *Mar. Biol.*, *14*, 235-41

Burton, M.P. (1964) Haemal system of regular echinoids, *Nature*, *204*, 1218

Byrne, M. (1985) The mechanical properties of the autotomy tissues of the holothurian *Eupentacta quinquesemita* (Selenka) and the effects of certain

physico-chemical agents, *J. Exp. Biol., 117,* 69-86
Byrne, M. and A.R. Fontaine. (1981) The feeding behaviour of *Florometra serratissima* (Echinodermata: Crinoidea), *Can. J. Zool., 59,* 11-18
Byrne, M. and A.R. Fontaine. (1983) Morphology and function of the tube feet of *Florometra serratissima* (Echinodermata: Crinoidea), *Zoomorphology, 102,* 175-87
Calow, P. (1984a) Economics of ontogeny — adaptational aspects, in B. Shorrocks (ed.), *Evolutionary ecology,* Blackwell Scientific Publications, Oxford, pp. 81-104
Calow, P. (1984b) Exploring the adaptive landscapes of invertebrate life cycles, *Adv. Invert. Reprod., 3,* 329-42
Calow, P. and C.R. Townsend. (1981) Energetics, ecology and evolution, in C.R. Townsend and P. Calow (eds), *Physiological ecology: an evolutionary approach to resource use.* Sinauer Associates, Sunderland, Mass., pp. 3-19
Cameron, J.L. and P.V. Fankboner. (1984) Tentacle structure and feeding processes in life stages of the commercial sea cucumber *Parastichopus californicus* (Stimpson), *J. Exp. Mar. Biol. Ecol., 81,* 193-209
Cameron, J.L. and P.V. Fankboner. (1986) Reproductive biology of the commercial sea cucumber *Parastichopus californicus* (Stimpson) (Echinodermata: Holothuroidea). I. Reproductive periodicity and spawning behaviour, *Can. J. Zool., 64,* 168-75
Campbell, A.C. (1972) The form and function of the skeleton in pedicellariae from *Echinus esculentus* L., *Tissue and Cell, 4,* 647-61.
Campbell, A.C. (1973) Observations on echinoid pedicellariae. I. Stem responses and their significance, *Mar. Behav. Physiol., 2,* 33-61.
Campbell, A.C. (1976) Observations on the activity of echinoid pedicellariae: III. Jaw responses of globiferous pedicellariae and their significance, *Mar. Behav. Physiol., 4,* 25-39
Campbell, A.C. (1983) Form and function of pedicellariae, *Echinoderm Studies, 1,* 139-67
Campbell, A.C., J.K.G. Dart, S.M. Head and R.F.G. Ormond. (1973) The feeding activity of *Echinostrephus molaris* (de Blainville) in the central Red Sea, *Mar. Behav. Physiol., 2,* 155-69
Campbell, A.C., and M.S. Laverack. (1968) The responses of pedicellariae from *Echinus esculentus* (L.), *J. Exp. Mar. Biol. Ecol., 2,* 191-214
Campbell, D.B. (1984) Foraging movements of the sea star *Asterias forbesi* (Desor) (Echinodermata: Asteroidea) in Narragansett Bay, Rhode Island USA, *Mar. Behav. Physiol., 11,* 185-98
Carpenter, P.H. (1884) Report on the Crinoidea collected during the voyage of H.M.S. 'Challenger' during the years 1873–76. Part I. General morphology with description of the stalked crinoids, *Report of the scientific results of the voyage of H.M.S. Challenger during the years 1873–76. Zoology, 11,* part 11
Carpenter, P.H. (1888) Report on the Crinoidea collected during the voyage of H.M.S. 'Challenger' during the years 1873–1876. Part II. The Comatulae. *Report of the scientific results of the voyage of H.M.S. Challenger during the years 1873–76. Zoology, 26,* part 26
Carpenter, W.B. (1876) Supplemental note to a paper 'On the structure, physiology, and development of *Antedon* (*Comatula* Lamk.) *rosaceus*', *Proc. Roy. Soc. Lond, 169,* 1-4
Caster, K.E. (1967) Homoiostelea, in R.C. Moore (ed.), *Treatise on invertebrate paleontology,* Part S. Echinodermata 1, *2,* University of Kansas Press, Lawrence, and the Geological Society of America, New York, pp. 611-27.
Castilla, J.C. (1972) Responses of *Asterias rubens* to bivalve prey in a Y-maze, *Mar.*

Biol., *12*, 222-8
Chadwick, H.C. (1907) *Antedon*, *Proc. Trans. Liverpool Biol. Soc.*, *21*, 371-417
Chesher, R.H. (1963) The morphology and function of the frontal ambulacrum of *Moira atropos* (Echinoidea: Spatangoida), *Bull. Mar. Sci.*, *13*, 549-73
Chesher, R.H. (1969) Contribution to the biology of *Meoma ventricosa* (Echinoidea: Spatangoida), *Bull. Mar. Sci.*, *19*, 72-110
Chia, F.-S. (1969) Histology of the pedicellariae of the sand dollar, *Dendraster excentricus* (Echinodermata), *J. Zool.*, *Lond.*, *157*, 503-7
Chia, F.-S. (1977) Structure and function of the genital papillae in a tropical sand dollar, *Arachnoides placenta* (L.) with a discussion on the adaptive significance of genital papillae in echinoids, *J. Exp. Mar. Biol. Ecol.*, *27*, 187-94
Chia, F.-S. (1985) Selection, storage and elimination of heavy sand particles by the juvenile sand dollar, *Dendraster excentricus* (Eschscholtz), in B.F. Keegan and B.D.S. O'Connor (eds), *Echinodermata*, A.A. Balkema, Rotterdam, pp. 215-21
Chia, F.-S. and H. Amerongen. (1975) On the prey-catching pedicellariae of a starfish, *Stylasterias forreri* (de Loriol), *Can. J. Zool.*, *53*, 748-55
Chia, F.-S., and J.B. Buchanan. (1969) Larval development of *Cucumaria elongata* (Echinodermata: Holothuroidea), *J. Mar. Biol. Ass. UK*, *49*, 151-9
Christensen, A.A. (1957) The feeding behavior of the seastar *Evasterias troschelii* Stimpson, *Limnol. Oceanogr.*, *2*, 180-97.
Christensen, A.M. (1970) Feeding biology of the sea-star *Astropecten irregularis* Pennant, *Ophelia*, *8*, 1-134
Clark, A.H. (1915) A monograph of the existing crinoids. Vol. 1. The comatulids. Part 1, *Bull. US Nat. Mus.*, *82*
Clark, A.H. (1921) A monograph of the existing crinoids. Vol. 1. The comatulids. Part 2, *Bull. US Nat. Mus.*, *82*
Clark, A.H. (1931) A monograph of the existing crinoids. Vol. 1. The comatulids. Part 3. Superfamily Comasterida, *Bull. US Nat. Mus.*, *82*
Clark, A.H. (1941) A monograph of the existing crinoids. Vol. 1. The comatulids. Part 4a. Superfamily Mariametrida (except the family Colobometridae), *Bull. US Nat. Mus.*, *82*
Clark, A.M. (1967) Variable symmetry in fissiparous Asterozoa, *Symp. Zool. Soc. Lond.* (20), 143–57
Clark, A.M. (1977) Notes on deep-water Atlantic Crinoidea, *Bull. Brit. Mus. Nat. Hist.* (*Zool.*), *31*, 159-86
Clark, A.M. (1982) Notes on Atlantic Asteroidea. 2. Luidiidae, *Bull. Brit. Mus. Nat. Hist.* (*Zool.*), *42*, 157-84
Clark, A.M. (1984) Notes on Atlantic and other Asteroidea. 4. Families Poraniidae and Asteropseidae, *Bull. Brit. Mus. Nat. Hist.* (*Zool.*), *47*, 19-51
Clark, H.L. (1898) *Synapta vivipara.* A contribution to the morphology of the echinoderms, *Mem. Biol. Lab., Johns Hopkins Univ.*, *4*, 53-88
Clark, H.L. (1899) The synaptas of the New England coast,'*Bull. US Fish. Comm.*, *19*, 21-31
Clark, H.L. (1912) Hawaiian and other Pacific Echini. The Pedinidae, Phymosomatidae, Stomopneustidae, Echinidae, Temnopleuridae, Strongylocentrotidae, and Echinometridae. *Mem. Mus. Comp. Zool., Harvard College. 34* (4)
Clark, H.L. (1914) Hawaiian and other Pacific Echini. The Clypeasteridae, Arachnoididae, Laganidae, Fibulariidae, and Scutellidae. *Mem. Mus. Comp. Zool., Harvard College*, *46* (1)
Clark, H.L. (1915) The comatulids of Torres Strait: with special reference to their habits and reactions, *Carnegie Inst. Wash., Publ. 212*, 97-125
Clark, H.L. (1917) Hawaiian and other Pacific Echini. The Echinoneidae, Nucleolitidae, Urechinidae, Echinocorythidae, Calylmnidae, Pourtalesiidae, Palaeosto-

matidae, Aeropsidae, Palaeopneustidae, Hemiasteridae, and Spatangidae. *Mem. Mus. Comp. Zool., Harvard College, 46* (2)

Clark, H.L. (1921) The echinoderm fauna of Torres Strait: its composition and its origin, *Carnegie Inst. Wash., Publ. 214*

Clark, R.B. (1971) Structural gadgetry and functional adaptation, in J.E. Smith (ed.), *The invertebrate panorama*, Universe Books, New York, pp. 110-30

Clements, L.A.J. and S.E. Stancyk. (1984) Particle selection by the burrowing brittlestar *Micropholis gracillima* (Stimpson) (Echinodermata: Ophiuroidea), *J. Exp. Mar. Biol. Ecol., 84*, 1-13

Cognetti, G. and R. Delavault. (1960) Recherches sur la sexualité d'*Echinaster sepositus* (Echinoderme, Astéride). Etude des glandes génitales chez les animaux des côtes de Livourne, *Cah. Biol. Mar., 1*, 421-32

Colacino, J.M. (1973) *A mathematical model of oxygen transport in the tube foot-ampullar system of the sea-cucumber,* Thyone briareus *L.* Unpublished. Ph.D. thesis. State University of New York, Buffalo

Conand, C. (1981) Sexual cycle of three commercially important holothurian species (Echinodermata) from the lagoon of New Caledonia, *Bull. Mar. Sci., 31*, 532–43

Conand, C. (1982) Reproductive cycle and biometric relations in a population of *Actinopyga echinites* (Echinodermata: Holothuroidea) from the lagoon of New Caledonia, western tropical Pacific, in J.M. Lawrence (ed.), *Echinoderms, Proceedings of the International Conference, Tampa Bay,* A.A. Balkema, Rotterdam, pp. 437-42

Conand, C. (1985) Distribution, reproductive cycle and morphometric relationships of *Acanthaster planci* (Echinodermata: Asteroidea) in New Caledonia, western trophical Pacific, in B.F. Keegan and B.D.S. O'Connor (eds), *Echinodermata,* A.A. Balkema, Rotterdam, pp. 499-506

Costello, D.P. (1946) The swimming of *Leptosynapta, Biol. Bull., 90,* 93-6

Costelloe, J. and B.F. Keegan. (1984) Feeding and related morphological structures in the dendrochirote *Aslia lefevrei* (Holothuroidea: Echinodermata), *Mar. Biol., 84,* 135-42

Cowen, R. (1981) Crinoid arms and banana plantations: an economic harvesting analogy, *Paleobiology, 7,* 332-43

Crozier, W.J. (1916) The rhythmic pulsation of the cloaca of holothurians, *J. Exp. Zool., 20,* 297-356

Crozier, W.J. (1920) Notes on some problems of adaptation. 2. On the temporal relations of asexual propagation and gametic reproduction in *Coscinasterias tenuispina*: with a note on the direction of progression and on the significance of the madreporites, *Biol. Bull., 39,* 116-29

Crump, R.G., and M.F. Barker. (1985) Sexual and asexual reproduction in geographically separated populations of the fissiparous asteroid *Coscinasterias calamaria* (Gray), *J. Exp. Mar. Ecol., 88,* 109-27

Crump, R.G. and R.H. Emson (1983) The natural history, life history and ecology of the two British species of *Asterina, Field Studies, 5,* 867-82

Cuénot, L. (1948) Anatomie, éthologie et systématique des Échinodermes, in P.P. Grassé (ed.), *Traité de Zoologie, 11,* Masson, Paris, pp. 3-363

Currey, J.D. (1975) A comparison of the strength of echinoderm spines and mollusc shells, *J. Mar. Biol. Ass. UK, 55,* 419-24

Dafni, J. (1983) Aboral depressions in the tests of the sea urchin *Tripneustes* cf. *gratilla* (L.) in the Gulf of Eilat, Red Sea, *J. Exp. Mar. Biol. Ecol., 67,* 1-15

Dafni, J. (1985) Effect of mechanical stress on the calcification pattern in regular echinoid skeletal plates, in B.F. Keegan and B.D.S. O'Connor (eds), *Echinodermata,* A.A. Balkema, Rotterdam, pp. 233-6

Da Silva, J., J.L. Cameron and P.V. Fankboner. (1986) Movement and orientation patterns in the commercial sea cucumber *Parastichopus californicus* (Stimpson) (Holothuroidea: Aspidochirotida), *Mar. Behav. Physiol., 12,* 133-47.

Dayton, P.K. (1975) Experimental evaluation of ecological dominance in a rocky intertidal algal community, *Ecol. Monogr., 45,* 137-59

Dayton, P.K., G.A. Robilliard and R.T. Paine. (1970) Benthic faunal zonation as a result of anchor ice at McMurdo Sound, Antarctica, in M.W. Holdgate (ed.), *Antarctic Ecology,* Vol. *1,* Academic Press, New York, pp. 244-58

Dayton, P.K., R.J. Rosenthal, L.C. Mahen and T. Antezana. (1977) Population structure and foraging biology of the predaceous Chilean asteroid *Meyenaster gelatinosus* and the escape biology of its prey, *Mar. Biol., 39,* 361-70

Dearborn, J.H. (1977) Foods and feeding characteristics of Antarctic asteroids and ophiuroids, in *Adaptations within Antarctic ecosystems,* Smithsonian Institution, Washington, D.C, pp. 293-326

De Ridder, C. (1982) Feeding and some aspects of the gut structure in the spatangoid echinoid, *Echinocardium cordatum,* in J.M. Lawrence (ed.), *Echinoderms: Proceedings of the International Conference, Tampa Bay,* A.A. Balkema, Rotterdam, pp. 5-10

De Ridder C. and M. Jangoux. (1982) Digestive systems: Echinoidea, in M. Jangoux and J.M. Lawrence (eds), *Echinoderm nutrition,* A.A. Balkema, Rotterdam, pp. 213-34

De Ridder, C. and M. Jangoux. (1985) Origine des sédiments ingeres et durée du transit digestif chez l'oursin spatangide, *Echinocardium cordatum* (Pennant) (Échinodermata), *Ann. Inst. Océanogr., 61,* 51-8

De Ridder, C. and J.M. Lawrence. (1982) Food and feeding mechanisms: Echinoidea, in M. Jangoux and J.M. Lawrence (eds), *Echinoderm nutrition,* A.A. Balkema, Rotterdam, pp. 57-115

Delavault, R. (1960) La sexualité chez *Echinaster sepositus* Gray du Golfe de Naples, *Pubbl. Staz. Napoli, 32,* 41-7

Dix, T.G. (1970) Biology of *Evechinus chloroticus* (Echinoidea: Echinometridae) from different localities, *N.Z. J. Mar. Freshwater Res., 4,* 385-405

Dix, T.G. (1977) Reproduction in Tasmanian populations of *Heliocidaris erythrogramma* (Echinodermata: Echinometridae). *Aust. J. Mar. Freshwat. Res., 28,* 509-20

Döderlein, L. (1906) Die Echinoiden der deutschen Tiefsee-Expedition. *Wissenschaftliche Ergebnisse der Deutschen Tiefsee-Expedition auf dem Dampfer 'Valdivia' 1898–1899, V* (2)

Döderlein, L. (1912) Die Arme der Gorgonocephalinae, *Zool. Jahrb., Suppl. 15* (2) 257-74

Dotan, A. and L. Fischelson. (1985) Morphology of spines of *Heterocentrotus mammillatus* (Echinodermata, Echinoidae) and its ecological significance, in B.F. Keegan and B.D.S. O'Connor (eds), *Echinodermata,* A.A. Balkema, Rotterdam, pp. 253-60

Doty, J.E. (1977) *Fission in* Holothuria atra *and holothurian population growth.* Unpublished M.S. thesis, University of Guam, Agana

Downey, M.E. (1972) *Midgardia xandaros,* new genus, new species, a large brisingid starfish from the Gulf of Mexico, *Proc. Biol. Soc. Wash., 84,* 421-6

Downey, M.E. and G.M. Wellington. (1978) Rediscovery of the giant seastar *Luidia superba* A.H. Clark in the Galapagos Islands, *Bull. Mar. Sci., 28,* 375-6

Duncan, P.M. (1883) On some points in the morphology of the test of the Temnopleuridae, *J. Linn. Soc. Zool., 16,* 343-58

Duncan, P.M. and W.P. Sladen. (1885) On the family Arbaciadae, Gray. Part I. The morphology of the test in the genera *Coelopleurus* and *Arbacia*, *J. Linn. Soc.*, *19*, 25-57

Durham, J.W. (1966) Clypeasteroids, in R.C. Moore (ed.), *Treatise on invertebrate paleontology, Part U. Echinodermata 3, 2*, University of Kansas Press, Lawrence, and the Geological Society of America, New York, pp. 450-91

Durham, J.W. and K.E. Caster. (1966) *Helicoplacoids*, in R.C. Moore (ed.) *Treatise on invertebrate paleontology, Part U.* Echinodermata 3. *1*, University of Kansas Press, Lawrence, and the Geological Society of America, New York, pp. 131-6

Durham, J.W., H.B. Fell, A.G. Fischer, P.M. Kier, R.V. Melville, D.L. Pawson and C.D. Wagner. (1966) Echinoidea, in R.C. Moore (ed.), *Treatise on invertebrate paleontology, Part U, Echinodermata 3*, University of Kansas Press, Lawrence, and the Geological Society of America, New York, pp. 212–640.

Ebert. T.A. (1967) Negative growth and longevity in the purple sea urchin *Strongylocentrotus purpuratus* (Stimpson), *Science*, *157*, 557-8

Ebert, T.A. (1968) Growth rates of the sea urchin *Strongylocentrotus purpuratus* related to food availability and spine abrasion, *Ecology*, *49*, 1075-81

Ebert, T.A. (1971) A preliminary quantitive survey of the echinoid fauna of Kealakekua and Honauunau Bays, Hawaii, *Pac. Sci.*, *25*, 112-31

Ebert, T.A. (1975) Growth and mortality of post-larval echinoids, *Amer. Zool.*, *15*, 755-75

Ebert, T.A. (1978) Growth and size of the tropical sea cucumber *Holothuria* (*Halodeima*) *atra* Jäger at Enewetak, Marshall Islands, *Pac. Sci.*, *32*, 183-91

Ebert, T.A. (1980) Relative growth of sea urchin jaws: an example of plastic resource allocation, *Bull. Mar. Sci. 30*, 467-74

Ebert, T.A. (1982) Longevity, life history, and relative body wall size in sea urchins, *Ecol. Monogr.*, *52*, 353-94

Ebert, T.A. (1983) Recruitment in echinoderms, *Echinoderm Studies 1*, 169-203

Edmonson, C.H. (1935) Autotomy and regeneration in Hawaiian starfishes, *Bernice Bishop Mus.*, *Occasional Paper 11*, 3-19

Ellers, O. and M. Telford. (1984) Collection of food by oral surface podia in the sand dollar, *Echinarachnius parma* (Lamarck), *Biol. Bull.*, *166*, 574-82

Ellington, W.R. (1982) Intermediary metabolism, in M. Jangoux and J.M. Lawrence (eds), *Echinoderm nutrition*, A.A. Balkema, Rotterdam, pp. 395–415

Emlet, R., L. McEdward and R.R. Strathmann. (1987) Echinoderm larval ecology viewed from the egg, *Echinoderm Studies*, *2*, in press

Emson, R. (1978) Some aspects of fission in *Allostichaster polyplax*, in D.S. McLusky and A.J. Berry (eds), *Physiology and behaviour of marine organisms*, Pergamon Press, Oxford, pp. 321-9

Emson, R. (1985) Bone idle — A recipe for success, in B.F. Keegan and B.D.S. O'Connor (eds), *Echinodermata*, A.A. Balkema, Rotterdam, pp. 25-30

Emson, R.H. and R.G. Crump. (1976) Brooding in *Asterina gibbosa* (Pennant), *Thalassia Jugoslav.*, *12*, 99-108

Emson, R.H. and R.G. Crump. (1979) Description of a new species of *Asterina* (Asteroidea), with an account of its ecology, *J. Mar. Biol. Ass. UK*, *59*, 77-94

Emson, R.H., P.V. Mladenov and I.C. Wilkie. (1985) Patterns of reproduction in small Jamaican brittle stars: fission and brooding predominate, in M.L. Reaka (ed.), *The ecology of coral reefs*, NOAA Symp. Ser. Undersea Res., *3*, 87-100

Emson, R. and I.C. Wilkie. (1980) Fission and autotomy in echinoderms, *Oceanogr. Mar. Biol. Ann. Rev.*, *18*, 155-250

Emson, R. and I.C. Wilkie. (1982) The arm-coiling response of *Amphipholis squamata* (Delle Chiaje) in J.M. Lawrence (ed.), *Echinoderms: Proceedings of the International Conference, Tampa Bay*, A.A. Balkema, Rotterdam, pp. 11-18

Emson, R.H., and I.C. Wilkie. (1984) An apparent instance of recruitment following sexual reproduction in the fissiparous brittlestar *Ophiactis savignyi*, *J. Exp. Mar. Biol. Ecol.*, *77*, 23-8

Engstrom, N.A. (1980) Reproductive cycles of *Holothuria* (*Halodeima*) *floridana*, *H.* (*H.*) *mexicana* and their hybrids (Echinodermata: Holothuroidea) in southern Florida, USA, *Int. J. Invert. Reprod.*, *2*, 237-44

Engstrom, N.A. (1982) Immigration as a factor in maintaining populations of the sea urchin *Lytechinus variegatus* (Echinodermata: Echinoidea) in seagrass beds on the southwest coast of Puerto Rico, *Stud. Neotrop. Fauna Env. 17*, 51-60.

Estes, J.A. (1974) *Population numbers, feeding behaviour and the ecological importance of sea otters in the western Aleutian Islands, Alaska.* Unpublished Ph.D. thesis, University of Arizona, Tucson

Eylers, J.P. (1976) Aspects of skeletal mechanics of the starfish *Asterias forbesi*, *J. Morphol.*, *149*, 353-68

Fankboner, P.V. (1978) Suspension-feeding mechanisms of the armoured sea cucumber *Psolus chitinoides* Clark, *J. Exp. Mar. Biol. Ecol.*, *31*, 11-25

Fankboner, P.V. (1979) Self-suturing by a synaptid sea cucumber (Holothuroidea: Echinodermata), *Experientia*, *34*, 729

Fankboner, P.V. (1981) A re-examination of mucus feeding by the sea cucumber *Leptopentacta* (= *Cucumaria*) *elongata*, *J. Mar. Biol. Ass. UK*, *61*, 679-83

Fankboner, P.V. and J.L. Cameron. (1985) Seasonal atrophy of the visceral organs in a sea cucumber, *Can J. Zool.*, *63*, 2888-92

Farmanfarmaian, A. (1966) The respiratory physiology of echinoderms, in R.A. Boolootian (ed.), *Physiology of Echinodermata*, Interscience Publishers, New York, pp. 245-65

Farmanfarmaian, A. (1969) Intestinal absorption and transport in *Thyone.* I. Biological aspects, *Biol. Bull. 137*, 118-31

Farmanfarmaian, A., A.C. Giese, R.A. Boolootian and J. Bennett. (1958) Annual reproductive cycles in four species of west coast starfishes, *J. Exp. Zool.*, *138*, 355-67

Farmanfarmaian, A. and J.H. Phillips. (1962) Digestion, storage and translocation of nutrients in the purple sea urchin (*Strongylocentrotus purpuratus*), *Biol. Bull.*, *123*, 105-20

Feder, H.M. (1955) On the methods used by the starfish *Pisaster ochraceus* in opening three types of bivalve molluscs, *Ecology*, *36*, 764-7

Fell, H.B. (1946) The embryology of the viviparous ophiuroid *Amphipholis squamata* Delle Chiaje, *Trans. Roy. Soc. N.Z.*, *75*, 419-64

Fell, H.B. (1963) The evolution of the echinoderms, *Smithsonian Inst.*, *Annual Rept 1962*, 457-90

Fell, H.B. (1966) Morphology, in R.C. Moore (ed.), *Treatise on Invertebrate Paleontology, Part U. Echinodermata 3, 2*, University of Kansas Press, Lawrence, and the Geological Society of America, New York, pp. 368-72

Fell, H.B. and R.C. Moore. (1966) General features and relationships of echinozoans, in R.C. Moore (ed.), *Treatise on invertebrate paleontology, Part U. Echinodermata 3, 1*, University of Kansas Press, Lawrence, and the Geological Society of America, New York, pp. 108–18

Fell, H.B. and D.L. Pawson. (1966) Echinacea, in R.C. Moore (ed.), *Treatise on invertebrate paleontology, Part U*, Echinodermata 3, *2*, University of Kansas Press, Lawrence, and the Geological Society of America, New York, pp. 367-72

Fenaux, L., G. Malara, C. Cellario, R. Charra and I. Palazzoli. (1977) Évolution des constituants biochimiques des principaux compartiments de l'oursin *Arbacia lixula* (L.) au cours d'un cycle sexual et effets d'un jeune de courte durée au cours de la maturation sexuelle, *J. Exp. Mar. Biol. Ecol.*, *28*, 17-30

Féral, J.-P., and C. Massin. (1982) Digestive systems: Holothuroidea, in M. Jangoux and J.M. Lawrence (eds), *Echinoderm nutrition*, A.A. Balkema, Rotterdam, pp. 191-212

Ferber, I. (1976) Functional morphology of *Lovenia elongata* (Gray) (Echinoidea: Spatangoida). *Thalass. Jugoslav.*, *12*, 123-8

Ferber, I. and J.M. Lawrence. (1976) Distribution, substratum preference and burrowing behaviour of *Lovenia elongata* (Echinoidea: Spatangoida) in the Gulf of Eilat ('Aqaba), Red Sea, *J. Exp. Mar. Biol. Ecol.*, *22*, 207-25

Ferguson, J.C. (1964) Nutrient transport in starfish. II. Uptake of nutrients by isolated organs, *Biol. Bull.*, *126*, 391-406

Ferguson, J.C. (1970) An autoradiographic study of the translocation and utilization of amino acids by starfish, *Biol. Bull.*, *138*, 14-25

Ferguson, J.C. (1982a) A comparative study of the net metabolic benefits derived from the uptake and release of free amino acids by marine invertebrates, *Biol. Bull. 162*, 1-17

Ferguson, J.C. (1982b) Nutrient translocation, in M. Jangoux and J.M. Lawrence (eds), *Echinoderm nutrition*, A.A. Balkema, Rotterdam, pp. 373-93

Ferguson, J.C. (1982c) Support of metabolism of superficial structures through direct net uptake of dissolved primary amines in echinoderms, in J.M. Lawrence (ed.), *Echinoderms, Proceedings of the International Conference, Tampa Bay*, A.A. Balkema, Rotterdam, pp. 345-51

Ferguson, J.C. (1984) Translocative functions of the enigmatic organs of starfish — the axial organ, hemal vessels, Tiedemann's bodies, and rectal caeca: an autoradiographic study, *Biol. Bull.*, *166*, 140-55

Ferguson, J.C. (1985) Hemal transport of ingested nutrients by the ophiuroid, *Ophioderma brevispinum*, in B.F. Keegan and B.D.S. O'Connor (eds) *Echinodermata*, A.A. Balkema, Rotterdam, pp. 623-6

Fischer, A.G. (1966) Spatangoids, in R.C. Moore (ed.), *Treatise on invertebrate paleontology, Part U. Echinodermata 3, 1.* University of Kansas Press, Lawrence, and the Geological Society of America, New York, pp. 543-628

Fish, J.D. (1967) The biology of *Cucumaria elongata* (Echinodermata: Holothuroidea), *J. Mar. Biol. Ass. UK*, *47*, 129-43

Fisher, W.K. (1911) Asteroidea of the north Pacific and adjacent waters. Part 1. Phanerozonia and Spinulosa, *Bull. US Nat. Mus.*, *76*

Fisher, W.K. (1919) Contributions to the biology of the Philippine Archipelago and adjacent waters. Starfishes of the Philippine Seas and adjacent waters, *Bull. US Nat. Mus.*, *100, 3*

Fisher, W.K. (1928) Asteroidea of the north Pacific and adjacent waters. Part 2. Forcipulata (part), *Bull. US Nat. Mus.*, *76*

Fisher, W.K. (1930) Asteroidea of the north Pacific and adjacent waters. Part 3. Forcipulata (concluded), *Bull. US Nat. Mus. 76*

Fisher, W.K. (1940) Asteroidea, *Discovery Reports. 20*, 69-306

Fong, W. and K.H. Mann (1980) Role of gut flora in the transfer of amino acids through a marine food chain, *Can. J. Fish. Aquat. Sci.*, *37*, 88-96

Fontaine, A.R. (1965) The feeding mechanisms of the ophiuroid *Ophiocomina nigra*, *J. Mar. Biol. Ass. UK*, *45*, 373-85

Foster-Smith, R.L. (1978) An analysis of water flow in tube-living animals, *J. Exp. Mar. Biol. Ecol.*, *34*, 73-95

Fricke, H.W. (1971) Fische als Feinde trophischer Seeigel, *Mar. Biol.*, *9*, 328-38

Fuji, A. (1960) Studies on the biology of the sea urchin. I. Superficial and histological gonadal changes in gametogenic process of two sea urchins, *Strongylocentrotus nudus* and *S. intermedius*, *Bull. Fac. Fish., Hokkaido Univ.*, *11*, 1-14

Fuji, A. (1967) Ecological studies on the growth and food consumption of Japanese common littoral sea urchin, *Strongylocentrotus intermedius* (A. Agassiz), *Mem.*

Fac. Fish., Hokkaido Univ., 15, 83-160
Gemmill, J.F. (1914) The development and certain points in the adult structure of *Asterias rubens* L., *Phil. Trans. Roy. Soc. Lond. B., 205,* 213-94
Gemmill, J.F. (1915) On the ciliation of asteroids and on the question of ciliary nutrition in certain species, *Proc. Zool. Soc. Lond., 1915,* 1-19
Ghiold, J. (1979) Spine morphology and its significance in feeding and burrowing in the sand dollar *Mellita quinquiesperforata* (Echinodermata: Echinoidea), *Bull. Mar. Sci., 29,* 481-90
Ghiold, J. (1982) Observations on the clypeasteroid *Echinocyamus pusillus* (O.F. Müller), *J. Exp. Mar. Biol. Ecol., 61,* 57-74
Ghiold J. (1983) The role of external appendages in the distribution and life habits of the sand dollar *Echinarachnius parma* (Echinodermata: Echinoidea), *J. Zool., Lond, 200,* 405-19.
Ghiold, J. (1984) Adaptive shifts in clypeasteroid evolution — feeding strategies in the soft-bottom realm, *N. Jb. Geol. Paläont. Abh., 169,* 41-73
Ghiold, G. and A. Seilacher. (1982) Burrowing and feeding in sand dollars as reflected in the functional differentiation of external structures, *N. Jb. Geol. Paläont. Abh., 164,* 221-8
Giese, A.C. (1961) Further studies on *Allocentrotus fragilis,* a deep sea echinoid, *Biol. Bull., 121,* 141-50
Giese, A.C. (1966) On the biochemical constitution of some echinoderms, in R.A. Boolootian (ed.), *Physiology of Echinodermata,* Interscience Publishers, New York, pp. 757-96
Giese, A.C., Krishnaswamy, B.S. Vasu and J.M. Lawrence. (1964) Reproductive and biochemical studies on a sea urchin, *Stomopneustes variolaris,* from Madras Harbor, *Comp. Biochem. Physiol., 13,* 367-80
Gislén, T. (1924) Echinoderm studies, *Zool. Bidr. Uppsala, 9,* 1–316
Gladfelter, W.B. (1978) General ecology of the cassiduloid urchin *Cassidulus caribaearum, Mar. Biol., 47,* 149-60
Glynn, P.W. (1965) Active movements and other aspects of the biology of *Astichopus* and *Leptosynapta* (Holothuroidea), *Biol. Bull., 129,* 106-23
Gonor, J.J. (1972) Gonad growth in the sea urchin, *Strongylocentrotus purpuratus* (Stimpson) (Echinodermata: Echinoidea) and the assumptions of the gonad index methods, *J. Exp. Mar Biol. Ecol., 10,* 89-103
Gonor, J.J. (1973) Sex ratio and hermaphroditism in Oregon intertidal populations of the echinoid *Strongylocentrotus purpuratus, Mar. Biol., 19,* 278-80
Goodbody, I. (1960) The feeding mechanism in the sand dollar *Mellita sexiesperforata* (Leske), *Biol. Bull., 119,* 80-6
Goss, R. (1969) *Principles of regeneration,* Academic Press, New York
Greenberg, M.J. (1985) Ex bouillabaisse lux: the charm of comparative physiology and biochemistry, *Amer. Zool., 25,* 737-49
Greenway, M. (1977) *The production and utilization of* Thalassia testudinum *König in Kingston Harbour, Jamaica.* Unpublished Ph.D. thesis, University of the West Indies, Kingston
Grimmer, J.C. and N.D. Holland. (1979) Haemal and coelomic circulatory systems in the arms and pinnules of *Florometra serratissima* (Echinodermata: Crinoidea), *Zoomorphology, 94,* 93–109
Grimmer, J.C., N.D. Holland and C.G. Messing. (1984) Fine structure of the stalk of the bourgueticrinid sea lily *Democrinus conifer* (Echinodermata: Crinoidea), *Mar. Biol., 81,* 163-76
Hagström, B.E. and S. Lönning. (1964) Morphological variation in *Echinus esculentus* from the Norwegian west coast, *Sarsia, 17,* 39-46
Halpern, J.A. (1970) Growth rate of the tropical sea star *Luidia senegalensis*

(Lamarck), *Bull. Mar. Sci.*, *20*, 626-33

Hamann, O. (1887) Beiträge zur Histologie der Echinodermen. 3. Die Echiniden, *Z. Naturw., Jena*, *21*, 87-266

Hamann, O. (1889) Anatomie der Ophiuren und Crinoiden, *Z. Naturw., Jena*, *23*, 233-388

Hammond, L.S. (1982) Analysis of grain-size selection by deposit-feeding holothurians and echinoids (Echinodermata) from a shallow reef lagoon, Discovery Bay, Jamaica, *Mar. Ecol. Prog. Ser.*, *8*, 25-36

Hancock, D.A. (1955) The feeding behaviour of starfish on Essex oyster beds, *J. Mar. Biol. Ass. UK*, *34*, 313-31

Hancock, D.A. (1974) Some aspects of the biology of the sunstar *Crossaster papposus* (L.), *Ophelia*, *13*, 1-30

Hansen, B. (1972) Photographic evidence of a unique type of walking in deep-sea holothurians, *Deep-Sea Res.*, *19*, 461-2

Hansen, B. (1975) Systematics and biology of the deep-sea holothurians, *Galathea Report*, *13*

Hanson, J. and G. Gust. (1986) Circulation of perivisceral fluid in the sea-urchin *Lytechinus variegatus*, *Mar. Biol.*, *92*, 125-34

Harriott, V. (1982) Sexual and asexual reproduction of *Holothuria atra* Jaeger at Heron Island Reef, Great Barrier Reef, *Australian Mus. Mem.* (16), 53-66

Harrold, C. and J.S. Pearse. (1980) Allocation of pyloric caecum reserves in fed and starved sea stars, *Pisaster giganteus* (Stimpson): somatic maintenance comes before reproduction, *J. Exp. Mar. Biol. Ecol.*, *48*, 169-83

Harrold, C. and D.C. Reed. (1985) Food availability, sea urchin grazing, and kelp forest community structure, *Ecology*, *66*, 1160-9

Harvey, E.B. (1956) *The American* Arbacia *and other sea urchins*, Princeton University Press, Princeton

Haude, R. and F. Langenstrassen. (1976) *Rotasaccus dentifer* n.g., n. sp., ein devonischer Ophiocistioide (Echinodermata) mit 'holothuroiden' Wandskleriten und 'echinoidem' Kauapparat, *Paläont. Z.*, *50*, 130-50

Haugh, B.N. and B.M. Bell. (1980) Classification schemes, in T.W. Broadhead and J.A. Waters, *Echinoderms, notes for a short course. Studies in geology*, *3*, University of Tennessee, Knoxville, pp. 94-9

Heatwole, D.W. and S.E. Stancyk. (1985) Spawning and functional morphology of the reproductive system in the ophiuroid, *Hemipholis elongata* (Say), in B.F. Keegan and B.D.S. O'Connor (eds) *Echinodermata*, A.A. Balkema, Rotterdam, pp. 469–74

Heddle, D. (1967) Versatility of movement and the origin of the asteroids, *Symp. Zool. Soc. Lond.* (20), 125-41

Heffernan, J.M., and S.A. Wainwright. (1974) Locomotion of the holothurian *Euapta lappa* and redefinition of peristalsis, *Biol. Bull.*, *147*, 95-104

Hendler, G. (1975) Adaptational significance of the patterns of ophiuroid development, *Amer. Zool.*, *15*, 691-715

Hendler, G. (1979) Sex-reversal and viviparity in *Ophiolepsis kieri*, n.sp., with notes on viviparous brittlestars from the Caribbean (Echinodermata: Ophiuroidea), *Proc. Biol. Soc. Wash.*, *92*, 783–95

Hendler, G. (1982) Slow flicks show star tricks: elapsed-time analysis of basketstar (*Astrophyton muricatum*) feeding behavior, *Bull. Mar. Sci.*, *32*, 909–18

Hendler, G. (1984) The association of *Ophiothrix lineata* and *Callyspongia vaginalis*: A brittlestar-sponge cleaning symbiosis?', *PSZNI: Mar. Ecol.*, *5*, 9-27

Hendler, G. and D.R. Franz. (1982) The biology of a brooding seastar, *Leptasterias tenera*, in Block Island Sound, *Biol. Bull.*, *162*, 273-89

Hendler, G. and J.E. Miller. (1984) *Ophioderma devaneyi* and *Ophioderma ensiferum*, new brittlestar species from the western Atlantic (Echinodermata: Ophiuroidea), *Proc. Biol. Soc. Wash.*, *97*, 442-61

Hermans, C.O. (1983) The duo-gland adhesive system, *Oceanogr. Mar. Biol. Ann. Rev.*, *21*, 283-339

Herold, R.C. (1969) Hermaphrodite specimen of the sand dollar, *Echinarachnius parma*, *J. Fish. Res. Bd Can.*, *26*, 1965-6

Hérouard, E. (1906) Sur *Pelagothuria bouvieri* (Holothurie pélagique nouvelle) recueillie pendant la Campagne du yacht 'Princess-Alice' en 1905, *Bull. Mus. Océanogr. Monaco*, (60), 1-6

Herreid, C.F., II, C.R. DeFesi and V.F. LaRussa. (1977) Vascular follicle system of the sea cucumber, *Stichopus moebii*, *J. Morphol.*, *154*, 19-38

Herreid, C.F., II, V.F. LaRussa and C.R. DeFesi. (1976) Blood vascular system of the sea cucumber, *Stichopus moebii*, *J. Morphol.*, *150*, 423-52

Hilgers, H. and H. Splechtna. (1976) Struktur- und Funktionsanalyse ophiocephaler Pedizellarien von *Sphaerechinus granularis* Lam., *Echinus acutus* Lam. und *Paracentrotus lividus* Lam. (Echinodermata, Echinoidea), *Zoomorphologie*, *86*, 61-80

Himmelman, J.H. and T.H. Carefoot. (1975) Seasonal changes in calorific values of three Pacific coast seaweeds, and their significance to some marine invertebrate herbivores, *J. Exp. Mar. Biol. Ecol.*, *18*, 139-51

Himmelman, J.H. and D.H. Steele. (1971) Foods and predators of the green sea urchin *Strongylocentrotus droebachiensis* in Newfoundland waters, *Mar. Biol.*, *9*, 315-22

Himmelman, J.H., Y. Lavergne, A. Cardinal, G. Martel and P. Jalbert. (1982) Brooding behaviour of the northern sea star *Leptasterias polaris*, *Mar. Biol.*, *68*, 235-40

Hoffman, C.K. (1873) Zur Anatomie der Asteriden, *Niederl. Arch. Zool.*, *2*, 1-32

Holland, N.D. (1970) The fine structure of the axial organ of the feather star *Nemaster rubiginosa* (Echinodermata: Crinoidea), *Tissue and Cell*, *2*, 625-36

Holland, N.D. and M.T. Ghiselin. (1970) A comparative study of the gut mucous cells in thirty-seven species of the class Echinoidea (Echinodermata). *Biol. Bull.*, *138*, 286-305

Hoshiai, T. (1963) Some observations on the swimming of *Labidoplax dubia* (Semper), *Bull. Mar. Biol. Sta. Asamushi*, *11*, 167-70

Hotchkiss, F.H.C. (1979) Case studies in the teratology of starfish, *Proc. Acad. Nat. Sci. Philadephia*, *131*, 139-57

Hunter, R.D. and H.Y. Elder. (1967) Analysis of burrowing mechanism in *Leptosynapta tenuis* and *Golfingia gouldi*, *Biol. Bull.*, *133*, 471

Hyman, L.H. (1955) *The invertebrates: Echinodermata, the coelomate Bilateria*. McGraw-Hill, New York

Ikeda, H. (1931) A biometric study of the sexual dimorphism and sex ratio in *Temnopleurus toreumaticus* (Klein), *Annot. Zool., Jap.*, *13*, 233–42

Jackson, L.E. and A.R. Fontaine. (1984) Nutrient and nitrogenous waste concentrations in the fluid transport systems of a holothurian, *Comp. Biochem. Physiol.*, *74A*, 123-38

Jangoux, M. (1982) Digestive systems: Asteroidea, in M. Jangoux and J.M. Lawrence (eds), *Echinoderm nutrition*, A.A. Balkema, Rotterdam, pp. 235-72

Jangoux, M. and M. Vloebergh. (1973) Contribution a l'étude du cycle annuel de réproduction d'une population d'*Asterias rubens* (Échinodermata, Astéroidea) du littoral Belge, *Neth. J. Sea Res.*, *16*, 389-403

Jennings, H.S. (1907) Behaviour of the starfish, *Asterias forreri* de Loriol, *Univ. Calif. Pub. Zool.*, *4*, 53-185

Jensen, M. (1966) The response of two sea-urchins to the sea-star *Marthasterias glacialis* (L.) and other stimuli, *Ophelia*, *3*, 209-19

Jensen, M. (1981) Morphology and classification of Euechinoidea Bronn, 1860, a cladistic analysis, *Vidensk. Meddr. Dansk Naturh. Foren.*, *143*, 7-99

Jensen, M. (1982) Pedicellariae in the classification of echinoids, in J.M. Lawrence (ed.), *Echinoderms*; *Proceedings of the International Conference, Tampa Bay*, A.A. Balkema, Rotterdam, pp. 111-15

Jensen, M. (1985) Functional morphology of test, lantern and tube feet ampullae system in flexible and rigid sea urchins (Echinoidea) in B.F. Keegan and B.D.S. O'Connor, *Echinodermata*, A.A. Balkema, Rotterdam, pp. 281-8

Johansen, K. and J.A. Petersen. (1971) Gas exchange and active ventilation in a starfish, *Pteraster tesselatus*, *Z. Vergl. Physiol.*, *71*, 365-81

John, D.D. (1938) Crinoidea, *Discovery Repts*, *18*, 121-222

Jost, P. (1982) 'Optimal foraging', un test de la théorie avec l'astérie *Astropecten aranciacus*, *Symbioses*, *15*, 227-9

Kawamoto, N. (1927) The anatomy of *Paracaudina chilensis* (J. Müller) with especial reference to the perivisceral cavity, the blood and the water vascular systems in their relation to blood circulation, *Sci. Rep. Tohoku Imp. Univ.*, *2*, 239-64

Kerkut, G.A. (1953) The forces exerted by the tube feet of the starfish during locomotion, *J. Exp. Biol.*, *30*, 575-83

Kholodov, V.I. (1975) Assimilation of different forms of food by *Strongylocentrotus droebachiensis* (O.F. Müller), *Hydrobiol. J.*, *11*, 41-6

Kidron, J., L. Fishelson and B. Moau. (1972) Cytology of an unusual case of hermaphroditic gonads in the tropical sea urchin *Tripneustes gratilla* from Eilat (Red Sea), *Mar. Biol.*, *14*, 260-3

Kier, P.M. (1962) Revision of the cassiduloid echinoids, *Smithsonian Misc. Coll.*, *144*

Kier, P.M. (1966) Cassiduloids, in R.C. Moore (ed.) *Treatise on invertebrate paleontology, Part U. Echinodermata 3, 2*, University of Kansas Press, Lawrence, and the Geological Society of America, New York, pp. 492–523

Kier, P.M. (1974) Evolutionary trends and their functional significance in the post-Paleozoic echinoids, *J. Paleontology*, *48* (3), suppl.

Kier, P.M. (1982) Rapid evolution in echinoids, *Paleontology*, *25*, 1-9

Kim, Y.S. (1968) Histological observation of the annual change in the gonad of the starfish, *Asterias amurensis* Lutken, *Bull. Fac. Fish., Hokkaido Univ.*, *19*, 97-108

King, H.D. (1898) Regeneration in *Asterias vulgaris*, *Arch. Entwicklungsmech. Org.*, *7*, 351-63

Klinger, T.S. (1982) Feeding rates of *Lytechinus variegatus* Lamarck (Echinodermata: Echinoidea) on differing physiognomies of an artifical food of uniform composition, in J.M. Lawrence (ed.), *Echinoderms*: *Proceedings of the International Conference, Tampa Bay*, A.A. Balkema, Rotterdam, pp. 29-32

Klinger, T.S. (1984) *Feeding of a marine generalist grazer*: Lytechinus variegatus (*Lamarck*) (*Echinodermata*: *Echinoidea*). Unpublished PhD thesis, University of South Florida, Tampa

Klinger, T.S., H.L. Hsieh, R.A. Pangallo, C.P. Chang and J.M. Lawrence. (1986) The effect of temperature on feeding, digestion and absorption of *Lytechinus variegatus* (Lamarck) (Echinodermata: Echinoidea), *Physiol. Zool.*, *59*, 332-6

Klinger, T.S. and J.M. Lawrence. (1985) The hardness of the teeth of five species of echinoids (Echinodermata), *J. Nat. Hist.*, *19*, 917-20

Koehler, R. (1883) Recherches sur les Echinides du côtes de Provence, *Ann. Mus. Hist. Nat. Marseille. Zool. 1, Mem. 3*

Komatsu, M. (1983) Development of the sea-star, *Archaster typicus* with a note on male-on-female superposition, *Annot. Zool. Japon.*, *56*, 187-95

Könnecker, G. and B.F. Keegan. (1973) In situ behavioural studies on echinoderm aggregations. Part I. *Pseudocucumis mixta, Helgolander Wiss. Meeresunters.*, *24*, 157-62

Lahaye, M.-C., and M. Jangoux. (1985a) Functional morphology of the podia and ambulacral grooves of the comatulid crinoid *Antedon bifida* (Echinodermata), *Mar. Biol.*, *86*, 307-18

Lahaye, M.-C. and M. Jangoux. (1985b) Post-spawning behavior and early development of the comatulid crinoid, *Antedon bifida*, in B.F. Keegan and B.D.S. O'Connor (eds), *Echinodermata*, A.A. Balkema, Rotterdam, pp. 181-4

Lambert, A., L. De Vos and M. Jangoux.(1984) Functional morphology of the pedicellariae of the asteroid *Marthasterias glacialis* (Echinodermata), *Zoomorphology*, *104*, 122-30

Landenberger, D.E. (1968) Studies on selective feeding in the Pacific starfish *Pisaster* in Southern California, *Ecology*, *49*, 1062-75

Lane, J.M. (1977) *Bioenergetics of the sand dollar,* Mellita quinquiesperforata (*Leske, 1778*). Unpublished PhD thesis University of South Florida, Tampa

Lane, N.G. (1984) Predation and survival among inadunate crinoids, *Paleobiology*, *10*, 453-8

Lane, N.G. and G.D. Webster. (1980) Crinoidea, in T.W. Broadhead and J.A. Waters (eds), *Echinoderms. Notes for a short course, Studies in geology*, *3*, University of Tennessee, Knoxville, pp. 144-57

Larson, B.R., R.L. Vadas and M. Keser. (1980) Feeding and nutritional ecology of the sea urchin *Strongylocentrotus droebachiensis* in Maine, USA, *Mar. Biol.*, *59*, 49-62

Lawrence, J.M. (1966) *Lipid levels in the body fluid, blood, and tissues of some marine molluscs and echinoderms in relation to the nutritional and reproductive state.* Unpublished PhD thesis, Stanford University, Stanford

Lawrence, J.M. (1973) Level, content, and caloric equivalents of the lipid, carbohydrate, and protein in the body components of *Luidia clathrata* (Echinodermata: Asteroidea: Platyasterida) in Tampa Bay, *J. Exp. Mar. Biol. Ecol.*, *11*, 263-74

Lawrence, J.M. (1975) On the relationships between marine plants and sea urchins, *Oceanogr. Mar. Biol. Ann. Rev.*, *13*, 213-86

Lawrence, J.M. (1976a) On the role of the tube feet and spines in the righting response of sea urchins (Echinodermata: Echinoidea), *Amer. Zool.*, *16*, 228

Lawrence, J.M. (1976b) Patterns of lipid storage in postmetamorphic marine invertebrates, *Amer. Zool.*, *16*, 747-62

Lawrence, J.M. (1983) Alternate states of populations of *Echinometra mathaei* (de Blainville) (Echinodermata: Echinoidea) in the Gulf of Suez and the Gulf of Aqaba, *Bull. Inst. Oceanogr. Fish.*, *9*, 141-7

Lawrence, J.M. (1985) The energetic echinoderm, in B.F. Keegan and B.D.S. O'Connor (eds), *Echinodermata*, A.A. Balkema, Rotterdam, pp. 47-67

Lawrence, J.M. (1987a) Bioenergetics and echinoderms, in T.J. Pandian and F.J. Vernberg (eds), Academic Press, New York

Lawrence, J.M. (1987b) Une histoire de deux etoiles: l'effet du nombre des bras sur la biologie, *Bull. Mus. Nat. Nantes*, in press

Lawrence, J.M. and A. Guille. (1982) Numerical and caloric densities of ophiuroids and holothuroids (Echinoidermata) from Kerguelen (South Indian Ocean) and Banyuls (Mediterranean Sea), in J.M. Lawrence (ed.), *Echinoderms, Proceedings of the International Conference, Tampa Bay*, A.A. Balkema, Rotterdam, pp. 313-17

Lawrence, J.M. and L. Hughes-Games. (1972) The diurnal rhythm of feeding and passage of food through the gut of *Diadema setosum* (Echinodermata: Echinoidea), *Israel J. Zool., 21*, 13-16

Lawrence, J.M., T.S. Klinger, J.B. McClintock, S.A. Watts, C.-P. Chen, A. Marsh, and L. Smith. (1986) Allocation of nutrient resources to body components by regenerating *Luidia clathrata* (Say) (Echinodermata; Asteroidea). *J. Exp. Mar. Biol. Ecol., 102*, 47-53

Lawrence, J.M. and J.M. Lane. (1982) The utilization of nutrients by post-metamorphic echinoderms, in M. Jangoux and J.M. Lawrence (eds), *Echinoderm nutrition*, A.A. Balkema, Rotterdam, pp. 331-71

Lawrence J.M., J.B. McClintock and A. Guille. (1985) Organic level and caloric content of eggs of brooding asteroids and an echinoid (Echinodermata) from Kerguelen (South Indian Ocean), *Int. J. Invert. Reprod. Dev., 7*, 249-57

Lawrence, J.M. and P.W. Sammarco. (1982) Effects of feeding on the environment: Echinoidea, in M. Jangoux and J.M. Lawrence (eds), *Echinoderm nutrition*, A.A. Balkema, Rotterdam, pp. 499-519

Leighton, D.L. (1968) *A comparative study of food selection and nutrition in the abalone* Haliotis rufescens *Swainson and the sea urchin*, Strongylocentrotus purpuratus (*Stimpson*). Unpublished PhD thesis, University of California, San Diego

Levin, V.S. (1979) The composition of food particles of aspidochirote holothurians from the upper sublittoral zone of the Indo-west Pacific, *Biol. Morya* (6), 20-7

Lewis, J.B. (1958) The biology of the tropical sea urchin *Tripneustes esculentus* Leske in Barbados, British West Indies, *Can. J. Zool., 36*, 607-21

Lewis, J.B. and G.S. Storey. (1984) Differences in morphology and life history traits of the echinoid *Echinometra lucunter* from different habitats, *Mar. Ecol. Prog. Ser., 15*, 207-11

Liddell, W.D. (1982) Suspension feeding by Caribbean comatulid crinoids, in J.M. Lawrence (ed.), *Echinoderms, Proceedings of the International Conference, Tampa Bay*, A.A. Balkema, Rotterdam, pp. 33-9

Littler, M.M. and D.S. Littler. (1980) The evolution of thallus form and survival strategies in benthic marine macroalgae: field and laboratory tests of a functional form model, *Amer. Nat., 116*, 25-44

Littler, M.M. and D.S. Littler. (1984) Relationships between macroalgal functional form and substrata stability in a subtropical rocky-intertidal system, *J. Exp. Mar. Biol. Ecol., 74*, 13-34

Lovén, S. (1874) Études sur les Échinoïdées, *Kong. Svenska Vetensk.-Akad. Handl. 11* (7)

Lovén, S. (1883) On *Pourtalesia.* A genus of Echinoidea, *Kong. Svenska Vedensk.-Akad. Handl. 19* (7)

Lowe, E.F. (1978) *Relationships between biochemical and caloric composition and reproductive cycle in* Asterias vulgaris (*Echinodermata*: *Asteroidea*) *from the Gulf of Maine.* Unpublished PhD thesis., University of Maine, Orono

Lowe, E.F. and J.M. Lawrence. (1976) Absorportion efficiencies of *Lytechinus variegatus* (Lamarck) (Echinodermata: Echinoidea) on selected marine plants, *J. Exp. Mar. Biol. Ecol., 21*, 223-34

Lucas, J.S. (1984) Growth, maturation and effects of diet in *Acanthaster planci* (L.) (Asteroidea) and hybrids reared in the laboratory, *J. Exp. Mar. Biol. Ecol., 79*, 129-47

Lyman, T. (1877) Mode of forking among astrophytons, *Proc. Boston Soc. Nat. Hist., 19*, 2-8

Lyman, T. (1882) Report on the Ophiuroidea dredged by H.M.S. Challenger during the years 1873–1876, *Report on the scientific results of the voyage of H.M.S. Challenger during the years 1873–76. Zoology. 5*, part 14

Macurda, D.B., Jr and D.L. Meyer. (1974) Feeding posture of the modern stalked crinoids, *Nature, 247,* 394-6

Macurda, D.B., Jr and D.L. Meyer. (1976) The identification and interpretation of stalked crinoids (Echinodermata) from deep-water photographs, *Bull. Mar. Sci., 26,* 205-15

Macurda, D.B., Jr, and M. Roux. (1981) The skeletal morphology of the isocrinid crinoids *Annacrinus wyvillethomsoni* and *Diplocrinus maclearanus, Contr. Mus. Paleontology, Univ. Mich., 25,* 169-219

Madsen, F.J. (1950) Echinoderms collected by the Atlantide Expedition 1945–46. I. Asteroidea. With remarks on other sea-stars from tropical and northern West Africa, *Atlantide Report,* (1), 169-222

Madsen, F.J. (1961) The Porcellanasteridae, *Galathea Rep., 4,* 33-176

Madsen, F.J. (1981) Records of a porcellanasterid, *Styracaster elongatus* (Echinodermata, Asteroidea), from the Caribbean, with remarks on growth and notes on some other species of the genus, *Steenstrupia, 7,* 309-19

Magniez, P. (1980) *Le cycle sexuel d'*Abatus cordatus (*Echinoidea*: *Spatangoida*): *modalités d'incubation et évolution histologique et biochimique des gonades.* Unpublished thesis, Doctrat 3⁰ Cycle, University of Paris

Magniez, P. (1983) Reproductive cycle of the brooding echinoid *Abatus cordatus* (Echinodermata) in Kerguelen (Antarctic Ocean): changes in the organ indices, biochemical composition and caloric content of the gonads, *Mar. Biol., 74,* 55-64

Magnus, D.B. (1963) Der Federstern *Heterometra savignyi* im Roten Meer, *Natur. Mus. Frankf., 93,* 355-68

Mangum, C. and W. Van Winkle. (1973) Response of aquatic invertebrates to declining oxygen conditions, *Amer. Zool., 13,* 529-41

Märkel, K. (1970) Morphologie der Seeigelzähne. III. Die Zähne der Diadematoida und Echinothuroida (Echinodermata, Echinoidea), *Z. Morph. Tiere, 66,* 189-211

Märkel, K. (1981) Experimental morphology of coronal growth in regular echinoids, *Zoomorphology, 97,* 31-52

Märkel, K., P. Gorny and K. Abraham. (1977) Microarchitecture of sea urchin teeth, *Fortschr. Zool., 24,* 103-14

Märkel, K. and U. Röser. (1983) Calcite-resorption in the spine of the echinoid *Eucidaris tribuloides, Zoomorphology, 103,* 43-58

Märkel, K. and H. Titschack. (1969) Morphologie der Seeigelzähne. I. Der Zahn von *Stylocidaris affinis* (Phil.) (Echinodermata, Echinoidea), *Z. Morph. Tiere, 64,* 179-200

Marx, W. (1929) Über sekundäre Geschlechtsmerkmale bein *Psammechinus miliaris* und *Echinocyamus pusillus, Zool. Anz., 80,* 331-5

Massin, C. (1982) Food and feeding mechanisms: Holothuroidea, in M. Jangoux and J.M. Lawrence (eds), *Echinoderm nutrition,* A.A. Balkema, Rotterdam, pp. 43-55

Matsumoto, H. (1929) Outline of a classification of Echinodermata, *Sci. Rept. Tôhoku Imp. Univ., Sendai, Second Ser.* (*Geol.*), *13,* 27-33

Mauzey, K.P., C. Birkeland and P.K. Dayton. (1968) Feeding behavior of asteroids and escape responses of their prey in the Puget Sound region, *Ecology, 49,* 603–19

May, R.M. (1925) Les réactions sensorielles d'une ophiure (*Ophionereis reticulata,* Say), *Biol. Bull., 53,* 372-402

McClintock, J.B. (1984) *An optimization study on the feeding behavior of* Luidia clathrata (*Say*) (*Echinodermata*: *Asteroidea*). Unpublished PhD thesis, University of South Florida, Tampa

McClintock, J.B., T.S. Klinger and J.M. Lawrence. (1983) Extraoral feeding in *Luidia clathrata* (Say) (Echinodermata: Asteroidea), *Bull. Mar. Sci., 33,* 171-2

McClintock, J.B., T.S. Klinger and J.M. Lawrence. (1984) Chemoreception in *Luidia clathrata* (Echinodermata: Asteroidea): qualitative and quantitative aspects of chemotactic responses to low molecular weight compounds, *Mar. Biol., 84,* 47-52

McClintock, J.B. and J.M. Lawrence. (1985) Size selectivity of prey by *Luidia clathrata* (Say) (Echinodermata: Asteroidea): effect of nutritive condition and age, in B.F. Keegan and B.D.S. O'Connor (eds), *Echinodermata,* A.A. Balkema, Rotterdam, pp. 533-9

McClintock, J.B. and T.J. Robnett, Jr. (1986) Size selective predation by the asteroid *Pisaster ochraceus* on the bivalve *Mytilus californianus*: a cost-benefit analysis, *PSZNI Mar. Ecol., 7,* 321-32

McNamara, K.J. and G.M. Philip. (1984) A revision of the spatangoid echinoid *Pericosmus* from the Tertiary of Australia, *Rec. West. Aust. Mus., 11,* 319-56

McPherson, B.F. (1965) Contributions to the biology of the sea urchin *Tripneustes ventricosus, Bull. Mar. Sci., 15,* 228-44

McPherson, B.F. (1968) Contributions to the biology of the sea urchin *Eucidaris tribuloides* (Lamarck), *Bull. Mar. Sci., 18,* 400-43

Melville, R.V. and J.W. Durham. (1966) Skeletal morphology, in R.C. Moore (ed.), *Treatise on invertebrate paleontology, Part U. Echinodermata 3, 1,* University of Kansas Press, Lawrence, and the Geological Society of America, New York, pp. 220-57

Menge, B.A. (1970) *The population ecology and community role of the predaceous asteroid,* Leptasterias hexactis (*Stimpson*). Unpublished PhD thesis, University of Washington, Seattle

Menge, B.A. (1972) Competition for food between two intertidal starfish species and its effect on body size and feeding, *Ecology, 53,* 635-44

Menge, B.A. (1974) Effect of wave action and competition on brooding and reproductive effort in the seastar, *Leptasterias hexactis, Ecology, 55,* 84-93

Menge, B.A. (1975) Brood or broadcast? The adaptive significance of different reproductive strategies in the two intertidal sea stars *Leptasterias hexactis* and *Pisaster ochraceus, Mar. Biol., 131,* 87-100

Menge, B.A. (1979) Coexistence between the seastars *Asterias vulgaris* and *A. forbesi* in a heterogeneous environment: a non-equilibrium explanation, *Oecologia, 41,* 245-72

Merrill, R.J. and E.S. Hobson. (1970) Field observations of *Dendraster excentricus,* a sand dollar of western North America, *Amer. Midland Nat., 83,* 595-624

Messing, C.G. (1985) Submersible observations of deep-water crinoid assemblages in the trophical western Atlantic Ocean, in B.F. Keegan and B.D.S. O'Connor (eds), *Echinodermata,* A.A. Balkema, Rotterdam, pp. 185-93

Meyer, D.L. (1971) The collagenous nature of problematical ligaments in crinoids (Echinodermata), *Mar. Biol., 9,* 235–41

Meyer, D.L. (1972) *Ctenantedon,* a new antedonid crinoid convergent with comasterids, *Bull. Mar. Sci., 22,* 53-66

Meyer, D.L. (1973) Feeding behavior and ecology of shallow-water unstalked crinoids (Echinodermata) in the Caribbean Sea, *Mar. Biol., 22,* 105-29

Meyer, D.L. (1979) Length and spacing of the tube feet in crinoids (Echinodermata) and their role in suspension-feeding, *Mar. Biol., 51,* 361-9

Meyer, D.L. (1980) Ecology and biogeography of living classes, in T.W. Broadhead and J.A. Waters (eds), *Echinoderms: Notes for a short course, Studies in Geology 3,* University of Tennessee, Knoxville, pp. 1-14

Meyer, D.L. (1982) Food and feeding mechanisms: Crinozoa, in M. Jangoux and

J.M. Lawrence (eds), *Echinoderm nutrition*, A.A. Balkema, Rotterdam, pp. 25-42
Meyer, D.L. (1985) Evolutionary implications of predation on Recent comatulid crinoids from the Great Barrier Reef, *Paleobiology, 11*, 154-65
Meyer. D.L. and N.G. Lane. (1976) The feeding behaviour of some Paleozoic crinoids and recent basketstars, *J. Paleontol., 50*, 472-80
Meyer, D.L. and D.B. Macurda, Jr. (1976) Distribution of shallow-water crinoids near Santa Marta, Colombia, *Mitt. Inst. Colombo-Alemán Invest. Cient., 8*, 141-56
Meyer, D.L. and D.B. Macurda, Jr. (1977) Adaptive radiation of the comatulid crinoids, *Paleobiology, 3*, 74-82
Meyer, D.L. and D.B. Macurda, Jr. (1980) Ecology and distribution of the shallow-water crinoids of Palau and Guam, *Micronesica, 16*, 59-99
Mileikovsky, S.A. (1971) Types of larval development in marine bottom invertebrates, their distribution and ecological significance: a reevaluation, *Mar. Biol., 10*, 193-213
Miller, J.S.(1821) *A natural history of the Crinoidea, lily-shaped animals with observations on the genera* Asteria, Euryale, Comatula and Marsupites, C. Frost, Bristol
Miller, P.J. (1979) Adaptiveness and implications of small size in teleosts, in P.J. Miller (ed.), *Fish phenology: anabolic adaptiveness in teleosts*, Academic Press, London, pp. 263-306
Miller, R.J., K.H. Mann and D.J. Scarratt. (1971) Production potential of a seaweed–lobster community in eastern Canada, *J. Fish. Res. Bd Canada, 28*, 1733-8
Millott, N. and H.G. Vevers. (1964) Axial organ and fluid circulation in echinoids, *Nature, 204*, 1216-17
Millott, N. and H.G. Vevers. (1968) The morphology and histochemistry of the echinoid axial organ, *Phil. Trans. Roy. Soc. Lond., B, 253*, 201-30
Mladenov, P.V., S.F. Carson and C.W. Walker. (1986) Reproductive ecology of an obligately fissiparous population of the sea star *Stephanasterias albula* (Stimpson), *J. Exp. Mar. Biol. Ecol., 96*, 155-75
Mladenov, P.V. and F.S. Chia. (1983) Development, settling behaviour, metamorphosis and pentacrinoid feeding and growth of the feather star *Florometra serratissima, Mar. Biol., 73*, 309-23
Mladenov, P.V. and R.H. Emson. (1984) Divide and broadcast: sexual reproduction in the West Indian brittle star *Ophiocomella ophiactoides* and its relationship to fissiparity, *Mar. Biol., 81*, 273-82
Mladenov, P.V., R.H. Emson, L.V. Colpit and I.C. Wilkie. (1983) Asexual reproduction in the West Indian brittle star *Ophiocomella ophiactoides* (H.L. Clark) (Echinodermata: Ophiuroidea), *J. Exp. Mar. Biol. Ecol., 72*, 1-23
Moitoza, D.J. and D.W. Phillips. (1979) Prey defense, predator preference, and nonrandom diet: the interactions between *Pycnopodia helianthoides* and two species of sea urchins, *Mar. Biol., 53*, 299-304
Monks, S.P. (1904) Variability and autotomy of *Phataria, Proc. Acad. Nat. Sci. Phila., 56*, 596-600
Mooi, R. and M. Telford. (1982) The feeding mechanism of the sand dollar *Echinarachnius parma* (Lamarck), in J.M. Lawrence (ed.), *Echinoderms: Proceedings of the International Conference, Tampa Bay,* A.A. Balkema, Rotterdam, pp. 51-7
Moore, H.B. (1932) A hermaphoditic sea urchin, *Nature, 130*, 59
Moore, H.B. (1935) A comparison of the biology of *Echinus esculentus* in different habitats. Part II, *J. Mar. Biol. Ass. UK, 20*, 109-28

Moore, H.B. (1936) The biology of *Echinocardium cordatum*, *J. Mar. Biol. Ass. UK*, *20*, 655-72

Moore, H.B., T. Jutare, J.C. Bauer and J.A. Jones. (1963) The biology of *Lytechinus variegatus*, *Bull. Mar. Sci.*, *132*, 23-53

Moore, H.B., T. Jutare, J.A. Jones, B.F. McPherson and C.F.E. Roper. (1963) A contribution to the biology of *Tripneustes esculentus*, *Bull. Mar. Sci.*, *13*, 267-81

Moore, H.B. and N.N. Lopez. (1966) The ecology and productivity of *Moira atropos* (Lamarck), *Bull. Mar. Sci.*, *16*, 648-67

Mortensen, T. (1903) Echinoidea (part 1), *The Danish Ingolf-Expedition*, *4* (1)

Mortensen, T. (1921) *Studies of the development and larval forms of echinoderms*, G.E.C. Gad, Cophenhagen

Mortensen T. (1923) *A monograph of the Echinoidea. I. Cidaroida*, C.A. Reitzel, Copenhagen

Mortensen, T. (1932) Papers from Dr. Th. Mortensen's Pacific Expedition 1914–16. LX. On an extraordinary ophiuroid, *Ophiocanops fugiens* Koehler. With remarks on *Astrogymnotes*, *Ophiopteron*, and on an albino *Ophiocoma*, *Vidensk. Medd. Dansk naturh. Foren.*, *93*, 1-21

Mortensen, T. (1933) Papers from Dr. Th. Mortensen's Pacific Expedition 1914–16. LXIII. Biological observations on ophiurids, with descriptions of two new genera and four new species, *Vidensk. Medd. Dansk naturh. Foren.*, *93*, 171-94

Mortensen, T. (1935) *A monograph of the Echinoidea. II. Bothriocidaroida, Melonechinoidea, Lepidocentroida, and Stirodonta*, C.A. Reitzel, Cophenhagen

Mortensen, T. (1936) Echinoidea and Ophiuroidea, *Discovery Reports*, *12*, 199-348

Mortensen, T. (1937) Contributions to the study of the development and larval forms of echinoderms. III, *D. Kgl. Danske Vidensk. Selsk. Skrifter, Naturv. og Math. Afd. 9.* *7* (1)

Mortensen, T. (1938) Contributions to the study of the development and larval forms of echinoderms. IV, *D. Kgl. Danske Vidensk. Selsk. Skrifter, Naturv. og Math. Afd. 9.* *7* (3)

Mortensen, T. (1940) *A monograph of the Echinoidea. III. 1. Aulodonta with additions to vol. II* (*Lepidocentroida and Stirodonta*), C.A. Reitzel, Copenhagen

Mortensen, T. (1943a) *A monograph of the Echinoidea. III. 2. Camarodonta. I. Orthopsidae, Glyphocyphidae, Temnopleuridae* and *Toxopneustidae*, C.A. Reitzel, Copenhagen

Mortensen, T. (1943b) *A monograph of the Echinoidea. III. 3. Camarodonta. II. Echinidae, Strongylocentrotidae, Parasalenidae, Echinometridae*, C.A. Reitzel, Copenhagen

Mortensen, T. (1948a) *A monograph of the Echinoidea. IV. 1. Holctypoida, Cassiduloida*, C.A. Reitzel, Copenhagen

Mortensen, T. (1948b) *A monograph of the Echinoidea, IV. 2. Clypeasteroida. Clypeasteridae, Arachnoididae, Fibulariidae, Laganidae and Scutellidae*, C.A. Reitzel, Copenhagen

Mortensen, T. (1950) *A monograph of the Echinodea. V. 1. Spatangoida. I. Protosternata, Meridosternata, Amphisternata I. Palaeopneustidae, Palaeostomatidae, Aeropsidae, Toxasteridae, Micrasteridae, Hemiasteridae*, C.A. Reitzel, Copenhagen

Mortensen, T. (1951) *A monograph of the Echinoidea. V.2. Spatangoida. II. Amphisternata. II. Spatangidae, Loveniidae, Pericosmidae, Schizasteridae, Brissidae*, C.A. Reitzel, Copenhagen

Moss, J.E. and J.M. Lawrence. (1972) Changes in carbohydrate, lipid, and protein levels with age and season in the sand dollar *Mellita quinquiesperforata* (Leske), *J. Exp. Mar. Biol. Ecol.*, *8*, 225-39

Moss, M.L. and M. Meehan. (1967) Sutural connective tissues in the test of an echinoid *Arbacia punctulata, Acta Anat.*, *66*, 279-304

Motokawa, T. (1984) Connective tissue catch in echinoderms, *Biol. Rev.*, *59*, 255-70

Myers, A.C. (1977) Sediment processing in a marine subtidal sandy bottom community: I. physical aspects, *J. Mar. Res.*, *35*, 609-32

Nance, J.M., and L.F. Braithwaite. (1979) The function of mucous secretions in the cushion star *Pteraster tesselatus* Ives. *J. Exp. Mar. Biol. Ecol.*, *40*, 259-66

Nance, J.M. and L.F. Braithwaite. (1981) Respiratory water flow and production of mucus in the cushion star, *Pteraster tesselatus* Ives (Echinodermata: Asteroidea), *J. Exp. Mar. Biol. Ecol.*, *50*, 21-31

Nauen, C.E. (1978) The growth of the sea star, *Asterias rubens*, and its role as benthic predator in Kiel Bay, *Kieler Meeresforsch.*, *4*, 68-81

Nichols, D. (1959) Changes in the chalk heart-urchin *Micraster* interpreted in relation to living forms. *Phil. Trans. R. Soc. Lond. B*, *242*, 347-437

Nichols, D. (1960) The histology and activities of the tube-feet of *Antedon bifida*, *Quart J. Microsc. Sci.*, *101*, 105-17

Nichols, D. (1966) Functional morphology of the water-vascular system, in R.A. Boolootian (ed.), *Physiology of Echinodermata*, Interscience Publishers, New York, pp. 219–44

Nichols, D. (1969) *Echinoderms*, Hutchinson University Library, London

Nichols, D. (1972) The water-vascular system in living and fossil echinoderms, *Palaeontology*, *15*, 519-38

Nichols, D. (1982) A biometrical study of populations of the European sea-urchin *Echinus esculentus* (Echinodermata: Echinoidea) from four areas of the British Isles, *Australian Mus. Mem.* (16), 147-63

Nigrelli, R.F. and S. Jakowska. (1960) Effect of holothurin, a steroid saponin from the Bahamian sea cucumber (*Actinopyga agassizi*), on various biological systems, *Ann. N. Y. Acad. Sci.*, *90*, 884-92

Nojima, S. and H. Mukai. (1985) A preliminary report on the distribution pattern, daily activity and moving pattern of a seagrass grazer, *Tripneustes gratilla* (L.) (Echinodermata: Echinoidea), in Papua New Guinea seagrass beds. *Spec. Pub. Mukashima Mar. Biol. Sta.*, pp. 173-83

Nutting, C. (1919) Barbados Antigua Expedition, *Univ. Iowa Stud. Nat. Hist.*, *8*, 1-274

Ogden, J.C., R.A. Brown and N. Salesky. (1973) Grazing by the echinoid *Diadema antillarum* Philippi. Formation of halos around West Indian patch reefs, *Science*, *182*, 715-16

Oguro, C., T. Shōsaku and M. Komatsu. (1982) Development of the brittle-star, *Amphipholis japonica* Matsumoto, in J.M. Lawrence (ed.), *Echinoderms, Proceedings of the International Conference, Tampa Bay*, A.A. Balkema, Rotterdam, pp. 491-6

Ohta. S. (1985) Photographic observations of the swimming behavior of the deep-sea pelagothuriid holothurian *Enypniastes* (Elasipoda, Holothuroidea), *J. Oceanograph. Soc. Jap.*, *41*, 121-33

Okada, M. and Y. Tahara. (1970) Normal development of secondary sexual characters in the sea urchin, *Hemicentrotus pulcherrimus*, *Zool. Mag.*, *79*, 46-52

Oldfield, S.C. (1976) The form of the globiferous pedicellarial ossicles of the regular echinoid, *Psammechinus miliaris* Gmelin, *Tissue and Cell*, *8*, 93-9

Olmsted, J.M.D. (1917) The comparative physiology of *Synaptula hydriformis* (Lesueur), *J. Exp. Zool.*, *24*, 333-79

O'Neill, P.L. (1978) Hydrodynamic analysis of feeding in sand dollars, *Oecologia*, *34*, 157-74

Ormond, R.F.G., N.J. Hanscomb and D.H. Beach. (1976) Food selection and learning in the crown-of-thorns starfish, *Acanthaster planci* (L.), *Mar. Behav. Physiol., 4,* 93-105

Ottesen, P.O. and J.S. Lucas. (1982) Divide or broadcast: interrelation of asexual and sexual reproduction in a population of the fissiparous hermaphroditic sea-star *Nepanthia belcheri* (Asteroidea: Asterinidae), *Mar. Biol., 69,* 223-33

Paine, V.L. (1926) Adhesion of the tube feet in starfishes, *J. Exp. Zool., 45,* 361-6

Paine, V.L. (1929) The tube feet of starfishes as autonomous organs, *Amer. Nat., 63,* 517

Parker, G.A. (1985) Population consequences of evolutionary stable strategies, in R.M. Sibly and R.H. Smith (eds), *Behavioural ecology: ecological consequences of adaptive behaviour.* Blackwell Scientific Publications, Oxford, pp. 33-58

Parker, G.H. (1921) The locomotion of the holothurian *Stichopus panimensis* Clark, *J. Exp. Zool., 33,* 205-8

Parker, G.H. (1927) Locomotion and righting movements in echinoderms, especially in *Echinarachnius, Amer. J. Psychol., 39,* 167-80

Parker, G.H. (1936) Direction and means of locomotion in the regular sea-urchin *Lytechinus, Mem. Mus. Roy. Hist. Nat. Belg., Deux. Ser., 3,* 197-208

Parker, G.H. and M.V. Alstyne. (1932) Locomotor organs of *Echinarachnius parma, Biol. Bull., 62,* 195-201

Patent, D.H. (1968) *The general and reproductive biology of the basket star,* Gorgonocephalus caryi (*Echinodermata, Ophiuroidea*). Unpublished PhD thesis, University of California, Berkeley

Paul, C.R.C. and A.B. Smith. (1984). The early radiation and phylogeny of echinoderms, *Biol. Rev., 59,* 443-81

Pawson, D.L. (1966) Phylogeny and evolution of the holothuroids, in R.C. Moore (ed.), *Treatise on invertebrate paleonotology, Part U. Echinodermata 3, 2,* University of Kansas Press, Lawrence, and the Geological Society of America, Inc., New York, pp. 641-6

Pawson, D.L. (1967) The psolid holothurian genus *Lissothuria, Proc. US Nat. Mus., 122,* 1-17

Pawson, D.L. (1970) The marine fauna of New Zealand: sea cucumbers (Echinodermata: Holothuroidea), *Bull. N. Z. Dept. Scient. Ind. Res., 201*

Pawson, D.L. (1976) Some aspects of the biology of deep-sea echinoderms, *Thalassia Jugosl., 12,* 287-93

Pawson, D.L. (1982a) Deep-sea echinoderms in the Tongue of the Ocean, Bahama Islands: a survey, using the research submersible *Alvin, Austral. Mus. Mem.* (16), 129-45

Pawson, D.L. (1982b) Holothuroidea, in S.P. Parker (ed.), *Synopsis and classification of living organisms,* McGraw-Hill, New York, pp. 791-818

Pawson, D.L. (1983) *Ocnus sacculus* new species (Echinodermata: Holothuroidea), a brood-protecting holothurian from southeastern New Zealand, *N. Z. J. Mar. Freshwat. Res., 17,* 227-30

Pawson, D.L. and H.B. Fell. (1965) A revised classification of the dendrochirote holothurians, *Breviora* (214), 1-7

Pawson, D.L. and J.E. Miller. (1979) Secondary sex characters in *Coelopleurus floridanus* A. Agassiz, 1872 (Echinodermata: Echinoidea), *Bull. Mar. Sci., 29,* 581-6

Pearse, A.S. (1908) Observations on the behavior of the holothurian, *Thyone briareus* (Lesueur), *Biol. Bull., 15,* 259-88

Pearse, J.S. (1967) Coelomic water volume control in the Antarctic sea-star *Odontaster validus, Nature, 216,* 1118-19

Pearse, J.S. (1969) Reproductive periodicities of Indo-Pacific invertebrates in the

Gulf of Suez. II. The echinoid *Echinometra mathaei* (de Blainville), *Bull. Mar. Sci., 19,* 580-613

Pearse, J.S. and B. Phillips. (1968) Continuous reproduction in the Indo-Pacific sea urchin *Echinometra mathaei* at Rottnest Island, western Australia, *Aust. J. Mar. Freshwat. Res., 19,* 161-72

Pelseneer, P. (1926) La proportion relative des sexes chez les animaux et particulierement chez les mollusques. *Mem. Acad. Roy. Bruxelles, 8,* 1-258

Penney, A.J. and G.L. Griffiths. (1984) Prey selection and the impact of the starfish *Marthasterias glacialis* (L.) and other predators on the mussel *Choromytilus meriodionalis* (Krauss), *J. Exp. Mar. Biol. Ecol., 75,* 19-36

Pennington, J.T. (1985) The ecology of fertilization of echinoid eggs: the consequences of sperm dilution, adult aggregation, and synchronous spawning, *Biol. Bull., 169,* 417-30

Pentreath, R.J. (1970) Feeding mechanisms and the functional morphology of podia and spines in some New Zealand ophiuroids (Echinodermata), *J. Zool., Lond., 161,* 395-429

Pentreath, R.J. (1971) Respiratory surfaces and respiration in three New Zealand intertidal ophiuroids, *J. Zool., Lond., 163,* 397-412

Péquignat, E. (1970) Biologie des *Echinocardium cordatum* (Pennant) de la Baie de Seine, *Forma Functio, 2,* 121-68

Péquignat, E. (1972) Some new data on skin digestion and absorption in sea urchins and sea stars (*Asterias* and *Henricia*), *12,* 28-41

Percy, J.A. (1971) *Thermal acclimatization and acclimation in the echinoid,* Strongylocentrotus droebachiensis (*O.F. Müller, 1776*). Unpublished PhD thesis, Memorial University of Newfoundland, St John's

Perrier, E. (1875) L'appareil circulatoire des oursins, *Arch. Zool. Exp. Gen., 4,* 604-43

Phelan, T.F. (1977) Comments on the water vascular system, food grooves, and ancestry of the clypeasteroid echinoids, *Bull. Mar. Sci., 27,* 400-22

Philip, G.M. and R.J. Foster. (1971) Marsupiate Tertiary echinoids from south-eastern Australia and their zoogeographic significance, *Palaeontology, 14,* 666-95

Pianka, E.R. (1976) Natural selection of optimal reproductive tactics, *Amer. Zool., 16,* 775-84

Powell, E.N. (1977) Particle size selection and sediment reworking in a funnel feeder, *Leptosynapta tenuis* (Holothuroidea, Synaptidae), *Int. Revue Ges. Hydrobiol., 62,* 385–408

Propp, N. (1977) Ecology of the sea urchin *Strongylocentrotus droebachiensis* of the Barents Sea: metabolism and regulation of abundance. *Biol. Mor.* (1), 39–51

Prouho, H. (1888) *Recherches sur le* Dorocidaris papillata *et quelques autres Échinides de la Méditerranée.* Unpublished Doc. Sci. Nat. thesis, University of Paris

Prouho, H. (1890) Du rôle des pédicellaires gemmiformes des oursins, *C. R. Acad. Sci. Paris, 111*

Quatrefages, A. de (1842) Memoire sur la Synapte de Duvernoy (*Synapta duvernea,* Nob), *Ann. Sci. Nat.* (*Zool.*), *17,* 19-93

Randall, J.E. (1967) Food habits of reef fishes of the West Indies, *Stud. Trop. Oceanogr. Miami, 5,* 665-847

Randall, J.E., R.E. Schroeder and W.A. Starck, II. (1964) Notes on the biology of the echinoid *Diadema antillarum, Carib. J. Sci., 4,* 421-33

Rasmussen, B.N. (1965) On taxonomy and biology of the North Atlantic species of the genus *Henricia* Gray, *Meddel. Danm. Fisk. Havunders., N. S., 4,* 157-213

Rasmussen, H.W. and H. Sieverts-Doreck. (1978) Classification, in R.C. Moore and

C. Teichert (eds), *Treatise on invertebrate paleontology, Part T. Echinodermata 2. Crinoidea. 3,* University of Kansas Press, Lawrence, and the Geological Society of America, Boulder, pp. 813-928

Raup D.M. (1966) Geometric analysis of shell coiling: general problems, *J. Paleontol., 40,* 1178-90

Régis, M.-B. (1978) *Croissance de deux Echinoïdes du Golfe de Marseille* (Paracentrotus lividus *(Lmk) et* Arbacia lixula *L.). Aspects écologiques de la Microstructure du Squelette et de l'Evolution des Indices physiologiques*: Thesis, Université d'Aix-Marseille, Marseille

Reese, E.S. (1966) The complex behavior of echinoderms, in R.A. Boolootian (ed.), *Physiology of Echinodermata,* Interscience Publishers, New York, pp. 157-218

Regnéll, G. (1966) Edrioasteroids, in R.C. Moore (ed.), *Treatise on invertebrate paleontology,* Part U. *Echinodermata 3, 1,* University of Kansas Press, Lawrence, and The Geological Society of America, Inc., New York, pp. 136-73

Reimer, R.D. and A.A. Reimer. (1975) Chemical control of feeding in four species of tropical ophiuroids of the genus *Ophioderma, Comp. Biochem. Physiol., 51A,* 915-27

Reyss, D. and J. Soyer. (1965) Étude de deux vallées sousmarines de la mer Catalane, *Bull. Inst. Océanogr. Monaco, 65,* 1-27

Rhoads, D.C. and D.K. Young. (1971) Animal–sediment relations in Cape Cod Bay, Massachusetts. II. Reworking by *Molpadia oolitica* (Holothuroidea), *Mar. Biol., 11,* 255-61

Ribi, G., R. Schrer and P. Ochsner. (1977) Stomach contents and size-frequency distributions of two coexisting sea star species, *Astropecten aranciacus* and *A. bispinosus,* with reference to competition, *Mar. Biol., 43,* 181-5

Rideout, R.S. (1978) Asexual reproduction as a means of population maintenance in the coral reef asteroid *Linckia multifora* on Guam, *Mar. Biol., 47,* 287-95

Roberts, D. (1979) Deposit-feeding mechanisms and resource partitioning in tropical holothurians, *J. Exp. Mar. Biol. Ecol., 37,* 43-56

Roberts, D. and C. Bryce. (1982) Further observations on tentacular feeding mechanisms in holothurians, *J. Exp. Mar. Biol. Ecol., 59,* 151-63

Robertson, D.A. (1972) Volume changes and oxygen extraction efficiency in the holothurian, *Stichopus mollis* (Hutton), *Comp. Biochem. Physiol., 43A,* 795-800

Robilliard, G.A. (1971) Feeding behaviour and prey capture in an asteroid, *Stylasterias forreri, Syesis, 4,* 191-5

Romanes, G.J. (1885) *Jelly-fish, star-fish and sea-urchins, being a research on primitive nervous systems,* Kegan Paul, Trench & Co., London

Rosenthal, R.J. and J.R. Chess. (1972) A predator–prey relationship between the leather star, *Dermasterias imbricata,* and the purple urchin, *Strongylocentrotus purpuratus, Fish. Bull. U. S., 70,* 205-16

Roux, M. (1975) Microstructural analysis of the crinoid stem, *Paleontol. Contrib., Univ. Kansas, 75,* 1-6

Roux, M. (1987) The ecology, evolution, and biogeography of stalked crinoids, *Echinoderm Studies II*

Rumrill, S.S. (1982) *Contrasting reproductive patterns among ophiuroids (Echinodermata) from southern Monterey Bay, USA.* Unpublished MS thesis, University of California, Santa Cruz

Rumrill, S.S. and J.S. Pearse. (1985) Contrasting reproductive periodicities among north-eastern Pacific ophiuroids, in B.F. Keegan and B.D.S. O'Connor (eds). *Echinodermata.* A.A. Balkema, Rotterdam, pp. 633-8

Rutherford, J.C. (1977) Variation in egg numbers between populations and between years in the holothurian *Cucumaria curata, Mar. Biol., 43,* 175-80

Rutman, J. and L. Fishelson. (1969) Food composition and feeding behavior of shallow-water crinoids at Eilat (Red Sea), *Mar. Biol.*, *3*, 46-57

Salsman, G.G. and W.H. Tolbert. (1965) Observations on the sand dollar, *Mellita quinquiesperforata*, *Limnol. Oceanogr.*, *10*, 152-5

Sars, M. (1867) Om Echinodermer og Coelenterater fundne ved Lofoten, *Vidensk..-Selsk. Forh.*, 19-23

Scheibling, R.E. (1979) *The ecology of* Oreaster reticulatus (*L.*) (*Echinodermata: Asteroidea*) *in the Carribean.* Unpublished PhD thesis, McGill University, Montreal

Scheibling, R.E. (1980) Dynamics and feeding activity of high-density aggregations of *Oreaster reticulatus* (Echinodermata: Asteroidea) in a sand patch habitat, *Mar. Ecol. Prog. Ser.*, *2*, 321-7.

Scheibling, R.E. (1981a) The annual reproductive cycle of *Oreaster reticulatus* (L.) (Echinodermata: Asteroidea) and interpopulation differences in reproductive capacity, *J. Exp. Mar. Biol. Ecol.*, *54*, 39-54

Scheibling, R.E. (1981b) Optimal foraging movements of *Oreaster reticulatus* (L.) (Echinodermata: Asteroidea), *J. Exp. Mar. Biol. Ecol.*, *51*, 173-85

Scheibling, R.E. (1982) Feeding habits of *Oreaster reticulatus* (Echinodermata: Asteroidea), *Bull. Mar. Sci.*, *33*, 504-10

Scheibling, R.E. and J.M. Lawrence. (1982) Differences in reproductive strategies of morphs of the genus *Echinaster* (Echinodermata: Asteroidea) from the eastern Gulf of Mexico, *Mar. Biol.*, *70*, 51-62

Seilacher, A. (1979) Constructional morphology of sand dollars, *Paleobiology*, *5*, 191-221

Semper, C. (1868) *Reisen im Archipel der Philippinen, Zweiter Theil, Wissenschaftliche Resultate. I. Holothurien*, Wilhelm Engelmann, Leipzig

Shapiro, H. (1935) A case of functional hermaphroditism in the sea urchin, *Arbacia punctulata*, and an estimate of the sex ratio, *Amer. Nat.*, *69*, 286–8

Sharp, D.T. and I.E. Gray. (1962) Studies on factors affecting the local distribution of two sea urchins, *Arbacia punctulata* and *Lytechinus variegatus*, *Ecology*, *43*, 309-13

Shick, J.M. (1976) Physiological and behavioral responses to hypoxia and hydrogen sulfide in the infaunal asteroid *Ctenodiscus crispatus*, *Mar. Biol.*, *37*, 279-89

Shick, J.M. (1983) Respiratory gas exchange in echinoderms, *Echinoderm Studies I*, 67-110

Shick, J.M., K.C. Edwards and J.H. Dearborn. (1981) Physiological ecology of the desposit-feeding sea star *Ctenodiscus crispatus*: ciliated surfaces and animal–sediment interactions, *Mar. Ecol. Prog. Ser.*, *5*, 165-84

Shirley, T.C. (1982) The importance of echinoderms in the diet of fishes of a sublittoral rock reef, in B.R. Chapman and J.W. Tunnell (eds), *South Texas Fauna*, Caesar Kelberg Wildlife Res. Inst., pp. 49-55

Sibly, R.M. (1981) Strategies of digestion and defecation, in C.R. Townsend and P. Calow (eds), *Physiological ecology*: *an evolutionary approach to resource use*, Sinauer Associates, Sunderland, pp. 109-39

Sides, E.M. (1982) Estimates of partial mortality for eight species of brittle-stars, in J.M. Lawrence (ed.), *Echinoderms, Proceedings of the International Conference, Tampa Bay*, A.A. Balkema, Rotterdam, p. 327

Sides, E.M. and J.D. Woodley. (1985) Niche separation in three species of *Ophiocoma* (Echinodermata: Ophiuroidea) in Jamaica, West Indies, *Bull. Mar. Sci.*, *36*, 701-15

Singletary, R.L. (1970) *The biology and ecology of* Amphipolis coniortoides, Ophionephthys limnicola, *and* Micropholis gracillima. Unpublished PhD thesis, University of Miami, Coral Gables

Sladen, W.P. (1889) Report on the Asteroidea collected by H.M.S. Challenger during the years 1873–76, *Report on the scientific results of the voyage of H.M.S. Challenger during the years 1873–76. Zoology. 30*, part 51

Sloan, N.A. and S.M.C. Robinson. (1983) Winter feeding by asteroids on a subtidal sandbed in British Columbia, *Ophelia, 22*, 125-40

Smith, A.B. (1978) A functional classification of the coronal pores of regular echinoids, *Palaeontology, 21*, 759-89

Smith, A.B. (1980a) The structure, function, and evolution of tube feet and ambulacral pores in irregular echinoids, *Palaentology, 23*, 39-83

Smith, A.B. (1980b) Stereom microstructure of the echinoid test, *Spec. Pap. Palaeontol.*, (25)

Smith, A.B. (1980c) The structure and arrangement of echinoid tubercles, *Phil. Trans. R. Soc. Lond.* B, *289*, 1-54

Smith, A.B. (1981) Implications of lantern morphology for the phylogeny of post-Palaeozoic echinoids, *Palaeontology, 24*, 779-801

Smith, A.B. (1984) Classification of the Echinodermata, *Palaeontology, 27*, 431–59

Smith, D.P.B. (1975) *Studies on sea urchins. I. Movement and behavior in the long-spined Caribbean sea-urchin* Diadema antillarum (*Echinodermata*: *Echinoidea*); *II. Straight-line conduction without straight-line connection*: *a model for the echinoid nerve net.* Unpublished PhD thesis, University of Wisconsin, Madison

Smith, D.S., S.A. Wainwright, J. Baker and M.L. Cayer. (1981) Structural features associated with movement and 'catch' of sea-urchin spines, *Tissue and Cell, 13*, 299-320

Smith, J.E. (1937) The structure and function of the tube feet in certain echinoderms, *J. Mar. Biol. Ass. UK, 22*, 345-57

Smith, J.E. (1941) The reproductive system and associated organs of the brittle-star *Ophiothrix fragilis, Quart. J. Micr. Sci., 82*, 267-309

Smith, J.E. (1947) The activities of the tube feet of *Asterias rubens* L. I. The mechanics of movement and of posture, *Quart. J. Micr. Sci., 88*, 1-14

Smith, J.E. (1950) Some observations on the nervous mechanisms underlying the behaviour of starfishes, *Symp. Soc. Exp. Biol.* (4), 196-220

Smith, J.E. (1965) Echinodermata, in T.H. Bullock and G.A. Horridge (eds), *Structure and function in the nervous systems of invertebrates. II*, W.H. Freeman, San Francisco, pp. 1519-58

Smith, R.H. (1971) *Reproductive biology of a brooding sea-star*, Leptasterias pusilla (*Fisher*), *in the Monterey Bay region.* Unpublished PhD thesis, Stanford University, Stanford

Smith, T.B. (1983) Tentacular ultrastructure and feeding behaviour of *Neopentadactyla mixta* (Holothuroidea: Dendrochirota), *J. Mar. Biol. Ass. UK, 63*, 301-11

Spencer, W.K. (1951) Early palaeozoic starfish, *Phil. Trans. R. Soc. Lond., B, 235*, 87-129

Spencer, W.K. and C.W. Wright. (1966) Asterozoans, in R.C. Moore (ed.), *Treatise on invertebrate paleontology, Part U. Echinodermata 3, 1*, University of Kansas Press, Lawrence, and the Geological Society of Amercia, New York, pp. 4-107

Sprinkle, J. (1983) Patterns and problems in echinoderm evolution, *Echinoderm Studies, 1*, 1-18

Stancyk, S.E. (1970) *Studies on the biology and ecology of ophiuroids at Cedar Key, Florida.* Unpublished MS thesis, University of Florida, Gainesville

Stephens, G.C., M.J. Volk S.H. Wright and P.S. Backlund. (1978) Transepidermal accumulation of naturally occurring amino acids in the sand dollar *Dendraster*

excentricus, *Biol. Bull.*, *154*, 335-47
Stephenson, D.G. (1980) Symmetry and suspension-feeding in pelmatozoan echinoderms, in M. Jangoux (ed.), *Echinoderms: present and past*, A.A. Balkema, Rotterdam, pp. 53-8
Stott, F.C. (1955) The food canal of the sea-urchin *Echinus esculentus* and its functions, *Proc. Zool. Soc. Lond.*, *125*, 63-86
Stott, R.S.H. H.R. Hepburn, I. Joffe and J.J.A. Heffron. (1974) The mechanical defensive mechanism of a sea cucumber, *S. Afr. J. Sci.*, *70*, 46-8
Strathmann, R.R. (1975) Limitations on diversity of forms: branching of ambulacral systems of echinoderms, *Amer. Nat.*, *109*, 177-90
Strathmann, R.R. (1981) The role of spines in preventing structural damage to echinoid tests, *Paleobiology*, *7*, 400–6
Strathmann, R.R. and C. Chaffee. (1984) Constraints on egg masses. 2. Effect of spacing, size and number of eggs on ventilation of embryos in jelly, adherent groups, or thin walled capsules, *J. Exp. Mar. Biol. Ecol.*, *84*, 85-93
Strathmann, R.R., M.F. Strathmann and R.H. Emson. (1984) Does limited brood capacity link adult size, brooding, and simultaneous hermaphroditism? A test with the starfish *Asterina phylactica*, *Amer. Nat.*, *123*, 796-818
Swan, E.F. (1961) Seasonal evisceration occurring in the gonad of *Parastichopus californicus* (Stimpson), *Science*, *133*, 1078-9
Swan, E.F. (1966) Growth, autotomy, and regeneration, in R.A. Boolootian (ed.), *Physiology of Echinodermata*, Interscience Publishers, New York, pp. 397-434
Swann, M.M. (1954) Secondary sex differences in five European species of sea-urchin, *Pubbl. Staz. Zool. Napoli*, *25*, 198-9
Taghon, G.L. (1982) Optimal foraging by deposit-feeding invertebrates: roles of particle size and organic coating, *Oecologia*, *52*, 295-304
Taghon, G.L., R.F.L. Self and P.A. Jumars. (1978) Predicting particle selection by deposit feeders: a model and its implications, *Limnol. Oceanogr.*, *23*, 752-9
Tahara, Y. and M. Okada. (1968) Normal development of secondary sexual characters in the sea urchin, *Echinometra mathaei*, *Publ. Seto. Mar. Biol. Lab.*, *16*, 41-50
Tahara, Y., M. Okada and N. Kobayashi. (1958) Secondary sexual characters in Japanese sea-urchins, *Pubs. Seto Mar. Biol. Lab.*, *7*, 165-72
Tahara, Y., M Okada and N. Kobayashi. (1960) Further notes on the sexual dimorphism in Japanese sea urchins, *Pubs. Seto Mar. Biol. Lab.*, *8*, 183-9
Tegner, M.J. (1980) Multispecies considerations of resource management in southern California kelp beds, *Can. Tech. Rep. Fish. Aquat. Sci.* (954), 125-43
Tegner, M.J. and P.K. Dayton. (1977) Sea urchin recruitment patterns and implications of commercial fishing, *Science*, *196*, 324-6
Tegner, M.J. and L.A. Levin. (1983) Spiny lobsters and sea urchins: analysis of a predator–prey interaction, *J. Exp. Mar. Biol. Ecol.*, *73*, 125-50
Telford, M. (1981) A hydrodynamic interpretation of sand dollar morphology, *Bull. Mar. Sci.*, *31*, 605-22
Telford, M. (1983) An experimental analysis of lunule function in the sand dollar *Mellita quinquiesperforata*, *Mar. Biol.*, *76*, 125-34
Telford, M. (1985a) Domes, arches and urchins: the skeletal architecture of echinoids (Echinodermata), *Zoomorphology*, *105*, 114-24
Telford, M. (1985b) Structural analysis of the test of *Echinocyamus pusillus* (O.F. Müller), in B.F. Keegan and B.D.S. O'Connor (eds), *Echinodermata*, A.A. Balkema, Rotterdam, pp. 353-9
Telford, M. and A. Harold. (1982) Lift, drag and camber in the northern sand dollar, *Echinarachnius parma*, in J.M. Lawrence (ed.), *Echinoderms, Pro-*

ceedings of the International Conference, Tampa Bay, A.A. Balkema, Rotterdam, pp. 235-41

Telford, M., A.S. Harold and R. Mooi. (1983) Feeding structures, behavior, and microhabitat of *Echinocyamus pusillus* (Echinoidea: Clypeasteroida), *Biol. Bull., 165,* 745-57

Telford, M., R. Mooi and O. Ellers. (1985) A new model of podial deposit feeding in the sand dollar, *Mellita quinquiesperforata* (Leske): the sieve hypothesis challenged, *Biol. Bull., 169,* 431-48

Théel, H. (1882) Report on the Holothuroidea dredged by H.M.S. Challenger during the years 1873–76. Part I., *Report on th scientific results of the voyage of H.M.S. Challenger during the years 1873–76. Zoology. 4,* part 13

Théel H. (1886) Report on the Holothuroidea dredged by H.M.S. Challenger during the years 1873–1876. Part II. *Reports on the scientific results of the voyage of H.M.S. Challenger during the years 1873--76. Zoology. 14,* part 39

Thompson, D'A.W. (1961) *On growth and form,* J.T. Bonner (ed.), Cambridge University Press, Cambridge

Thomson, C.W. (1876) Notice of some peculiarities in the mode of propagation of certain echinoderms of the Southern Sea, *J. Linn. Soc. Lond., Zool., 15,* 44-79

Thomson, C.W. (1878) *The voyage of the 'Challenger'. The Atlantic. II,* Harper, New York

Thorson, G. (1934) On the reproduction and larval stages of the brittle-stars *Ophiocten sericeum* (Forbes) and *Ophiura robusta* Ayres in east Greenland, *Medd. Grønland, 100* (4)

Thorson, G. (1950) Reproductive and larval ecology of marine bottom invertebrates, *Biol. Rev., 25,* 1-45

Thum, A.B. and J.C. Allen. (1976) Reproductive ecology of the lamp urchin *Echinolampas crassa* (Bell) 1880, from a subtidal biogenous ripple train, *Trans. Roy. Soc. S. Afr., 42,* 23-33

Timko, P.L. (1975) *High density aggregation in* Dendraster excentricus (*Eschscholtz*): *analysis of strategies and benefits concerning growth, age structure, feeding, hydrodynamics and reproduction.* Unpublished PhD thesis, University of California, Los Angeles

Timko, P.L. (1976) Sand dollars as suspension feeders: a new description of feeding in *Dendraster excentricus, Biol. Bull., 151,* 247-59

Town, J.C. (1979) *Aspects of the biology of* Astrostole scabra (*Hutton, 1872*). Unpublished PhD thesis, University of Canterbury, New Zealand

Town, J.C. (1980) Movement, morphology, reproductive periodicity, and some factors affecting gonad production in the sea star *Astrostole scabra* Hutton, *J. Exp. Mar. Biol. Ecol., 44,* 111-32

Townsend, C.R. and R.N. Hughes. (1981) Maximizing net energy returns from foraging, in C.R. Townsend and P. Calow (eds), *Physiological ecology; an evolutionary approach to resource use,* Blackwell Scientific Publications, Oxford, Sinauer Associates, Sunderland, Mass., pp. 86–108

Tsuchiya, M. and M. Nishihira. (1985) Agonistic behavior and its effect on the dispersion pattern in two types of the sea urchin, *Echinometra mathaei* (Blainville), *Galaxea, 4,* 37-48

Tsurnamal, M. and J. Marder. (1966) Observations on the basket star *Astroboa nuda* (Lyman) on coral reefs at Eilat (Gulf of Aqaba), *Isr. J. Zool., 15,* 9-17

Turner, R.L. (1974) Post-metamorphic growth of the arms in *Ophiophragmus filograneus* (Echinodermata: Ophiuroidea) from Tampa Bay, Florida (USA), *Mar. Biol., 24,* 273-7

Turner, R.L. and J.H. Dearborn. (1979) Organic and inorganic composition of post-metamorphic growth stages of *Ophionotus hexactis* (E.A. Smith) (Echinodermata: Ophiuroidea) during intraovarian incubation, *J. Exp. Mar. Biol. Ecol.*, *36*, 41-51

Turner, R.L., D.W. Heatwole and S.E. Stancyk. (1982) Ophiuroid discs in stingray stomachs: evasive autotomy or partial consumption of prey, in J.M. Lawrence (ed.), *Echinoderms: Proceedings of the International Conference, Tampa Bay*, A.A. Balkema, Rotterdam, pp. 331-5

Tyler, A. and B.S. Tyler. (1966) Physiology of fertilization and early development, in R.A. Boolootian (ed.), *Physiology of Echinodermata*, Interscience Publishers, New York, pp. 683–741

Ubaghs, G (1967) General characteristics, in R.C. Moore (ed.), *Treatise on invertebrate paleontology, Part S. Echinodermata 1, 1*, University of Kansas Press, Lawrence, and the Geological Society of America, New York, pp. 3-60

Ubaghs, G. (1978) Skeletal morphology of fossil crinoids, in R.C. Moore and C. Teichert (eds), *Treatise on invertebrate paleontology, Part T. Echinodermata 2, 1*, University of Kansas Press, Lawrence, and the Geological Society of America, Boulder, pp. 58-216

Uexkull, J. von (1899) Die Physiologie der Pedicellarien, *Zeitsch. Biol.*, *37*, 334-403

Vadas, R.L. (1977) Preferential feeding, an optimization strategy in sea urchins, *Ecol. Monogr.*, *47*, 337-71

Vadas, R.L., R.W. Elner, P.E. Garwood and I.G. Babb. (1986) Experimental evaluation of aggregation behavior in the sea urchin *Strongylocentrotus droebachiensis*, *Mar. Biol.*, *90*, 433-48

Van Veldhuizen, H.D. and P.W. Phillips. (1978) Prey capture by *Pisaster brevispinus* (Asteroidea: Echinodermata) on soft substrate, *Mar. Biol.*, *48*, 89-97

Vaney, C. (1914) Holothuries, *Deuxieme Expédition Antarctique Francaise (1908–1910) commandée par le Dr Jean Charcot. Sciences Naturelles: Documents Scientifiques*

Velimirov, B. (1985) Niche repartition in coexisting sea cucumber species in a kelp bed community, in B.F. Keegan and B.D.S. O'Connor (eds), *Echinodermata*, A.A. Balkema, Rotterdam, pp. 465-9

Vevers, H.G. (1949) The biology of *Asterias rubens* L.: growth and reproduction, *J. Mar. Biol. Ass. UK*, *28*, 165-87

Voogt, P.A. (1982) Steroid metabolism, in M. Jangoux and J.M. Lawrence (eds), *Echinoderm nutrition*, A.A. Balkema, Rotterdam, pp. 417-36

Walker, C.W. (1979) Ultrastructure of the somatic portion of the gonads in asteroids, with emphasis on flagellated-collar cells and nutrient transport, *J. Morph.*, *162*, 127-62

Walker, C.W. (1980) Spermatogenic columns, somatic cells and the structural microenvironment of germinal cells in the testes of sea-stars, *J. Morph.*, *166*, 81-107

Walker, C.W. (1982) Nutrition of development, in M. Jangoux and J.M. Lawrence (eds), *Echinoderm nutrition*, A.A. Balkema, Rotterdam, pp. 449-68

Walker, C.W. and J. Fineblit. (1982) Interactions between adult and embryonic tissues during brooding in *Axiognathus squamata*, in J.M. Lawrence (ed.), *Echinoderms: Proceedings of the International Conference, Tampa Bay*, A.A. Balkema, Rotterdam, p. 523

Warner, G.F. (1971) On the ecology of a dense bed of the brittle-star *Ophiothrix fragilis*, *J. Mar. Biol. Ass. UK*, *51*, 267-82

Warner, G.F. (1982) Food and feeding mechanisms: Ophiuroidea, in M. Jangoux and J.M. Lawrence (eds), *Echinoderm nutrition*, A.A. Balkema, Rotterdam, pp. 161-81

Warner, G.F. and J.D. Woodley (1975) Suspension-feeding in the brittle-star

Ophiothrix fragilis, *J. Mar. Biol. Ass. UK*, *55*, 199-210
Watts, S.A., R.E. Scheibling, A.G. Marsh and J.B. McClintock. (1983) Induction of aberrant ray numbers in *Echinaster* sp. (Echinodermata: Asteroidea) by high salinity, *Florida Sci.*, *46*, 120-5
Weber, J., R. Greer, B. Voigt, E. White and R. Roy. (1969) Unusual strength properties of echinoderm calcite related to structure, *J. Ultrastructure Res.*, *26*, 355-366
Webster, S.K. (1975) Oxygen consumption in echinoderms from several geographical locations, with particular reference to the Echinoidea, *Biol, Bull.*, *148*, 165-80
Webster, S.K. and A.C. Giese. (1975) Oxygen consumption of the purple sea urchin with special reference to the reproductive cycle, *Biol. Bull.*, *148*, 165-80
Weihe, S.C. and I.E. Gray. (1968) Observations on the biology of the sand dollar *Mellita quinquiesperforata* (Leske), *J. Elisha Mitchell Sci. Soc.*, *84*, 315-27
Wilkie, I.C. (1980) The systematic position of *Ophiocomina* Koehler and a reconsideration of certain interfamilial relationships within the Ophiuroidea, in M. Jangoux (ed.), *Echinoderms*: *present and past*, A.A. Balkema, Rotterdam, pp. 151-7
Wilkie, I.C. (1984) Variable tensility in echinoderm collagenous tissues: a review, *Mar. Behav. Physiol.*, *11*, 1-34
Wilkie, I.C., R.H. Emson and P.V. Mladenov. (1984) Morphological and mechanical aspects of fission in *Ophiocomella ophiactoides* (Echinodermata, Ophiuroidea), *Zoomorphology*, *104*, 310-22
Wolcott, T.G. (1981) Inhaling without ribs: the problem of suction in soft-bodied invertebrates, *Biol. Bull.*, *160*, 189-97
Wolfe, T.J. (1978) *Aspects of the biology of* Astrophyton muricatum (*Lamarck, 1816*) (*Ophiuroidea*: *Gorgonocephalidae*). Unpublished MS thesis, University of Puerto Rico, Mayaguez
Wolfe, T.J. (1982) Habitats, feeding, and growth of the basketstar *Astrophyton muricatum*, in J.M. Lawrence (ed.), *Echinoderms*: *Proceedings of the International Conference*, *Tampa Bay*, A.A. Balkema, Rotterdam, pp. 299-304
Woodley, J.D. (1967) Problems in the ophiuroid water-vascular system, *Symp. Zool. Soc. Lond.* (20), 75-104
Woodley, J.D. (1975) The behaviour of some amphiurid brittle-stars, *J. Exp. Mar. Biol. Ecol.*, *18*, 29-46
Woodley, J.D. (1980) The biomechanics of ophiuroid tube-feet, in M. Jangoux (ed.), *Echinoderms: present and past*, A.A. Balkema, Rotterdam, pp. 293-9
Worley, E.K., D.R. Franz and G. Hendler. (1977) Seasonal patterns of gametogenesis in a north Atlantic brooding asteroid, *Leptasterias tenera*, *Biol. Bull.*, *153*, 237-53
Yakolev, S.N., V.L. Kasyanov and V.V. Stepanov. (1976) Ecology of spawning of the black sea urchin in Peter the Great Bay (Sea of Japan), *Ecologia* (5), 85-91
Yamaguchi, M. (1975) Coral-reef asteroids of Guam, *Biotropica*, *7*, 12-23
Yamaguchi, M. and J.S. Lucas. (1984) Natural parthenogenesis, larval and juvenile development, and geographical distribution of the coral reef asteroid *Ophidiaster granifer*, *Mar. Biol.*, *83*, 33-42
Yamanouchi, T. (1926) Some preliminary notes on the behavior of the holothurian, *Caudina chilensis* (J. Müller), *Sci. Rep. Tôhoku Imp. Univ.*, *ser. 4* (*Biol.*), *2*, 85-91
Yamazi, I. (1950) Anatomy and regeneration in Japanese sea-stars and ophiurans. I. Observation on a sea-star, *Coscinasterias acutispina* Stimpson and four species of ophiurans, *Annot. Zool. Japon.*, *23*, 175-86
Yazaki, M. (1930) On the circulation of the perivisceral fluid in *Caudina chilensis*

(J. Müller), *Sci. Rep. Tôhoku Imp. Univ.*, *5*, 403-14

Yingst, J.Y. (1982) Factors influencing rates of sediment ingestion by *Parastichopus parvimensis* (Clark), an epibenthic deposit-feeding holothurian, *Estuarine, Coastal and Shelf Science*, *14*, 119-34

Index